ALCOHOL AND PUBLIC POLICY: Beyond the Shadow of Prohibition

PANEL ON ALTERNATIVE POLICIES AFFECTING
THE PREVENTION OF ALCOHOL ABUSE AND ALCOHOLISM

Mark H. Moore *and* Dean R. Gerstein,
Editors

Committee on Substance Abuse and Habitual Behavior
Assembly of Behavioral and Social Sciences
National Research Council

NATIONAL ACADEMY PRESS
Washington, D.C. 1981

NOTICE: The project that is the subject of this report was approved by the Governing Board of the National Research Council, whose members are drawn from the councils of the National Academy of Sciences, the National Academy of Engineering, and the Institute of Medicine. The members of the committee responsible for the report were chosen for their special competences and with regard for appropriate balance.

This report has been reviewed by a group other than the authors according to procedures approved by a Report Review Committee consisting of members of the National Academy of Sciences, the National Academy of Engineering, and the Institute of Medicine.

The National Research Council was established by the National Academy of Sciences in 1916 to associate the broad community of science and technology with the Academy's purposes of furthering knowledge and of advising the federal government. The Council operates in accordance with general policies determined by the Academy under the authority of its congressional charter of 1863, which establishes the Academy as a private, nonprofit, self-governing membership corporation. The Council has become the principal operating agency of both the National Academy of Sciences and the National Academy of Engineering in the conduct of their services to the government, the public, and the scientific and engineering communities. It is administered jointly by both Academies and the Institute of Medicine. The National Academy of Engineering and the Institute of Medicine were established in 1964 and 1970, respectively, under the charter of the National Academy of Sciences.

Library of Congress Cataloging in Publication Data

Main entry under title:
Alcohol and public policy.
Includes 7 studies commissioned by the panel.
1. Alcoholism—Government policy—United States—Addresses, essays, lectures. I. Moore, Mark Harrison. II. Gerstein, Dean R. III. National Research Council (U.S.). Panel on Alternative Policies Affecting the Prevention of Alcohol Abuse and Alcoholism. [DNLM: 1. Alcoholism—Prevention and control. WM 274 P191b]
HV5292.B49 362.2′9256′0973 81-11217
ISBN 0-309-03149-4 AACR2

Available from

NATIONAL ACADEMY PRESS
2101 Constitution Avenue, N.W.
Washington, D.C. 20418

Printed in the United States of America

PANEL ON ALTERNATIVE POLICIES AFFECTING THE PREVENTION OF ALCOHOL ABUSE AND ALCOHOLISM

MARK H. MOORE (*Chair*), John F. Kennedy School of Government, Harvard University

GAIL B. ALLEN, Department of Psychiatry, St. Luke's-Roosevelt Hospital Center, New York

DAN E. BEAUCHAMP, Department of Health Administration, School of Public Health, University of North Carolina

PHILIP J. COOK, Institute of Policy Sciences, Duke University

JOHN KAPLAN, School of Law, Stanford University

NATHAN MACCOBY, Institute for Communication Research, Stanford University

DAVID MUSTO, Child Study Center and Department of History, Yale University

ROBIN ROOM, Social Research Group, School of Public Health, University of California, Berkeley

THOMAS C. SCHELLING, John F. Kennedy School of Government, Harvard University

WOLFGANG SCHMIDT, Social Sciences Department, Alcoholism and Drug Addiction Research Foundation, Toronto

NORMAN SCOTCH, Department of Socio-Medical Sciences and Community Medicine, School of Medicine, Boston University

DONALD J. TREIMAN, Department of Sociology, University of California, Los Angeles

JACQUELINE P. WISEMAN, Department of Sociology, University of California, San Diego

DEAN R. GERSTEIN, *Study Director*

Preface

The Panel on Alternative Policies Affecting the Prevention of Alcohol Abuse and Alcoholism was charged by its sponsor, the National Institute on Alcohol Abuse and Alcoholism (NIAAA), to produce a systematic analysis of alternative policies affecting the prevention of alcohol abuse and alcoholism. Prevention is not, of course, a new idea in the alcohol field. It has turned up with increasing frequency in recent years as an attractive way to think about an intractable issue. There are a number of recent reports from committees, both domestic (Plaut 1967, Wilkinson 1970, Joint Committee of the States 1973, Medicine in the Public Interest 1979) and international (Bruun et al. 1975, World Health Organization 1980), in which prevention ideas have had a prominent place. A chapter on prevention was included in each of the last three official reports to Congress on alcohol and health (National Institute on Alcohol Abuse and Alcoholism 1974, 1978, 1981).

Despite all this attention, it seemed to us—and to those at the NIAAA and on the Committee on Substance Abuse and Habitual Behavior who chartered our project—that the concept of prevention policies had not been systematically developed. The particular kinds of interventions, the nature of the arguments that could be made, and the assessment of existing evidence on the effectiveness of various instruments could all profit from sustained consideration. The panel was assembled, comprising experts in the study of alcohol problems, prevention methods used in other fields, and other relevant academic disciplines, to lend structure and content to this inquiry. Our concerns were not entirely abstract or academic, however. In formulating ideas about prevention policies, we sought to ground our thinking in the existing historical and institutional context.

In view of the expansive territory of policies affecting prevention, we could hardly expect to cover intensively every possible track within it. We have therefore taken the role of scientific survey party: systematically noting the lay of the land; describing the prominent features; examining closely—and marking for further attention—conformations that seemed especially interesting and accessible. We have had to invent some of our survey equipment along the way. In particular, our typology of policies for modifying alcohol's effects did not come to us ready-made but was crafted over the course of writing, intensive discussion, and rewriting. The result, we hope, strikes a useful balance of the mandates to be systematic, analytical, disciplined—and crisp.

We have had considerable help in writing this report from seven studies commissioned by the panel, discussed at length during our deliberations, revised by the authors, and published in this volume. Two of these studies were written to help get us under way by concisely summarizing the relevant historical and scientific background. The first of these, by Paul Aaron and David Musto, is an overview of drinking in U.S. political and cultural history, beginning with colonial times and concluding with some speculations about the possible impact of current social movements on drinking policies and practices. This paper was an important element in building the perspective of the first, historical chapter of this report and especially in tempering our understanding of the Prohibition experience. In the second paper, Dean Gerstein describes the principal methodological and conceptual controversies that surround the measurement and evaluation of alcohol use and its consequences, including the interaction of drinking with a variety of environments: biological, physical, and social. This paper was a starting point in the development of our second chapter, on the structure of the alcohol problem.

Dan Beauchamp's analysis of the closely studied 1969 Alcohol Act in Finland provides a general perspective on the strategic limits to alcohol-related policy choices available to democratic governments. This analysis points toward the concerns involved in the transition from definition of the problem to specific policy proposals. In this sense, Beauchamp's work speaks to our third chapter, regarding perspectives on current policies.

The papers written by Philip Cook, John Hochheimer, and David Reed address the efficacy of several of the better-known policy instruments that might be (and have been) used to prevent alcohol problems. Cook focuses on the effects of controlling alcoholic beverage market prices by taxation. Hochheimer analyzes the use of educational media and personnel (and the educational use of media and personnel) to reduce alcohol abuse. Reed discusses the use of deterrence, education, risk reduction, and mixed strategies for handling drinking-driving problems. Finally, James Mosher and Joseph Mottl examine the surprisingly extensive roles of government agencies not specifically concerned with alcohol in affecting drinking behavior and consequences. These four papers were quite instrumental in developing the analyses of specific policy alternatives in chapters 4 through 6.

In short, the commissioned papers, with their more extensive consideration of many of the points covered in our report, have enabled the report to be written more efficiently and with a broader understanding than might otherwise have been possible. While the panel does not necessarily concur in every detail of what the supporting papers have to say, we are pleased to include them, to indicate points of concurrence in the text of the report, and to encourage readers to turn to these studies for further insight into the specific areas covered.

We are grateful to the many other people who have contributed to this venture. The participants in the panel's Workshop on Alcohol Policies held May 15-17, 1980—Kettil Bruun, Daniel Horn, H. Laurence Ross, Gerald Wilde, Joseph Gusfield, Klaus Mäkelä, Phil Davies, and Irmgard Vogt—helped to inform and shape the panel's report. Individual reviewers read and constructively criticized drafts of the report. The library and administrative offices of the National Academy of Sciences provided needed assistance. In the Assembly of Behavioral and Social Sciences, David Goslin, executive director, and Eugenia Grohman, associate director for reports, gave expert guidance; Christine McShane, editor, and Elaine McGarraugh, editorial assistant, polished the report and papers and prepared the volume for publication. And at NIAAA, the project management staff, especially David Promisel, provided support and encouragement.

This project has benefited from start to finish from the good judgment and dedication of the staff of the Committee on Substance Abuse and Habitual Behavior: Marie Clark, administrative secretary; Beverly Blakey and Charlotte Simpson, secretaries; Deborah Maloff, research associate; and the committee's study director, Peter Levison. It only remains for me to acknowledge the hard, essential work of the panel itself, and of the study director and coeditor of this volume, Dean Gerstein.

MARK H. MOORE, *Chair*
Panel on Alternative Policies Affecting the
Prevention of Alcohol Abuse and Alcoholism

Contents

REPORT OF THE PANEL

INTRODUCTION 3

1 SIMPLIFYING CONCEPTIONS OF ALCOHOL PROBLEMS AND POLICIES 6
Governing Ideas, 8
Minority Conceptions, 12
Conclusion, 15

2 THE NATURE OF ALCOHOL PROBLEMS 16
Effects of Drinking, 20
Underlying Patterns of Drinking Practices, 24
Definition of Drinking Practices, 24; Consumption: Quantity and Frequency, 27; Stability of Individual Drinking Practices, 39; Causes of Drinking Practices, 40
The Problem of Attributing Effects, 42
Conclusion: Avenues for Affecting Alcohol Problems, 44

3 PERSPECTIVES ON CURRENT ALCOHOL POLICIES 48
Current Alcohol Policies and Institutions, 48
The *Prima Facie* Case for Prevention, 50
The Definition of Prevention Policies, 52
Objections to Prevention Policies, 55

Categories of Prevention Policies, 58
Summary and Conclusion, 59

4 REGULATING THE SUPPLY OF ALCOHOLIC BEVERAGES 61
The Lessons of Prohibition, 62
Current Institutions and Policies, 64
The Single Distribution Theory, 66
Taxation and Prices, 68
Controlling Off-Premise Retail Sales, 73
Controlling On-Premise Sales, 75
Minimum Age Restrictions, 76
Conclusion, 78

5 SHAPING DRINKING PRACTICES DIRECTLY 79
Concepts of Safer and More Appropriate Drinking, 81
Drinking Practices and the Law, 83
Drunken Driving and the Law, 83; Public Drunkenness, 87
Education, Information, and Training Programs, 89
School Education Programs, 90; Mass Media Information Campaigns, 92; Community-Based Health Information and Training, 93; Conclusions About Education, Information, and Training Programs, 96
Setting a Good Example: A Note on Symbolic Effects of Government Actions, 96
Conclusion, 98

6 REDUCING ENVIRONMENTAL RISK 100
Damage in the Physical Environment, 102
Physical Safety Measures, 102; Social Safety Measures, 105
Damage in the Social Environment, 106
Victimization, 106; Public Drunkenness and the Urban Environment, 106; Deemphasizing Hostility, 107
Conclusion, 109

SUMMARY 112

REFERENCES 117

COMMISSIONED PAPERS

TEMPERANCE AND PROHIBITION IN AMERICA: A HISTORICAL OVERVIEW 127
Paul Aaron and *David Musto*

ALCOHOL USE AND CONSEQUENCES 182
Dean R. Gerstein

THE PARADOX OF ALCOHOL POLICY: THE CASE OF THE 1969 ALCOHOL ACT IN FINLAND 225
Dan E. Beauchamp

THE EFFECT OF LIQUOR TAXES ON DRINKING, CIRRHOSIS, AND AUTO ACCIDENTS 255
Philip J. Cook

REDUCING ALCOHOL ABUSE: A CRITICAL REVIEW OF EDUCATIONAL STRATEGIES 286
John L. Hochheimer

REDUCING THE COSTS OF DRINKING AND DRIVING 336
David S. Reed

THE ROLE OF NONALCOHOL AGENCIES IN FEDERAL REGULATION OF DRINKING BEHAVIOR AND CONSEQUENCES 388
James F. Mosher and *Joseph R. Mottl*

BIOGRAPHICAL SKETCHES OF PANEL MEMBERS AND STAFF 459

Tables

1. Distribution of Alcohol Consumption by Adults in the United States 28
2. Average Daily Consumption and Number of "Drunk Days" Per Year in a Sample of Air Force Men 30
3. "Drunk Days" Within Different Levels of Average Consumption in a Sample of Air Force Men 33
4. Changes in Apparent Consumption of Absolute Alcohol in 20 Countries by the Population Aged 15 Years and Over 36
5. Where, When, and With Whom People Drink (Drinking Contexts), by Age of Drinker 38
6. Alcohol Incidents by Total Consumption Level in a Sample of Air Force Men 44
7. Percentage of U.S. Population Experiencing Different Problems Associated With Drinking, by Level of Consumption 45
8. Contribution of Different Consuming Groups to Reported Experience of Problems 45
9. Ranking of 39 "Test Cases" of Alcohol Tax Increases by Net Percentage of Change in Rates of Liquor Consumption, Highway Deaths, and Cirrhosis Deaths 71

Figures

1. Schematic View of the Relationship Between Drinking and Consequences of Alcohol 18
2. Important Effects of Drinking and Policies for Dealing With Drinking 22
3. Relationship Between Average Daily Consumption and "Drunk Days" in a Sample of Air Force Men 31
4. Time Trends of U.S. Consumption of Absolute Alcohol, Cirrhosis Death Rate, and Alcoholism Death Rate in the Drinking Age Population, 1830–1977 35
5. Reported Drinking Status of Men and Women in College and 25 Years Later 39
6. Schematic View of the Causal System Underlying Alcohol Problems 46

REPORT OF THE PANEL

Introduction

As representatives of the scientific community examining and considering the implications of available evidence about alcohol policies, we recognize three peculiar aspects of our topic. First, alcohol is hardly an obscure, socially neutral topic. Two amendments to the Constitution concern alcohol. Intense controversy about drinking has surfaced several times in our history. The controversies have pitted cherished but divergent values about the proper role of the state in a free society against concern for vivid human suffering. The resulting strong sentiments make objective evaluation of policies difficult.

Second, drinking and its consequences have been the object of careful scientific study for only a short time, and the coverage is spotty. About a few things scientists know quite a lot, because intense public, political, or scientific concern has produced bursts of research. About many other things knowledge is rather superficial, and for some important things only speculation is possible.

Third, personal and institutional responses to drinking have been varied. Medical intervention, regulatory action, law enforcement, commercial activity, and moral mobilization have all played a role in shaping the contemporary context in which alcohol policy is managed. Given the divergent circumstances of these different responses, there is much honest intellectual disagreement about the nature of the problem, about what information is relevant, and about how to interpret relevant information.

In planning and framing our inquiry, we have first had to escape—or at least loosen—constraints that follow from the peculiarities noted above. The first type of constraint comes from prior history, which conditions how one conceives the problem and subtly biases one's judgment about it. We have found that social policies on alcohol—and other matters—tend to form and be formed by "governing ideas," in which specific ends and means are fused together into distinct agendas for action, agendas that often crystallize in specific historical programs or movements. When this happens, it is easy to overlook gaps in the seams that bind policy targets, instruments, and outcomes together in a specific historical instance. The seemingly inevitable connection of the temperance movement with Prohibition and the disease model of alcoholism with Alcoholics Anonymous furnish convenient, familiar examples. If only one or a small number of potent governing ideas exists in a policy area, any search for alternatives may be reduced to a very narrow range of all-or-none (or all-of-one versus all-of-the-other) choices. To canvass an interesting array of policy options, one must loosen the hold of these governing ideas.

The second type of constraint derives from a common confusion about the ends and means of alcohol policy. For many people, drinking as such is the thing to be evaluated as good or bad, apart from its specific consequences. Some see it as fundamentally immoral or morally weak; others see it as a positive sign of modernity, liberal social values, or traditional conviviality. Thus, they evaluate policies as good or bad depending on whether they discourage or encourage drinking. These moral values are important and relevant, since they determine much of the political and social context of drinking and drinking policies. But in guiding an objective analysis of alcohol use and its consequences, they cannot be decisive. It becomes necessary to look beyond these simple conceptions, to evaluate the specific effects of drinking on a more complex set of criteria, including the health, satisfaction, and well-being of drinkers; the economic status of families of drinkers—and of those whose livelihood involves the selling of alcohol; the magnitude of public expenditures dealing with alcohol problems and of public receipts from taxes on alcoholic beverages; and the like. Whether drinking increases or decreases is then considered important only as these attributes are affected. Distinguishing drinking per se from the socially relevant consequences of drinking and considering drinking itself and attitudes toward drinking as intermediate variables in the analysis are difficult but essential steps in studying alternative alcohol policies.

The third type of constraint involves the breadth of evidence that might be considered. Once one has held current governing ideas at arm's length, framing them within a history of such ideas and a careful explanation of the complex attributional structure of the problem, it becomes possible to conceive and seriously consider some alternative or additional approaches. At a conceptual level, one is tempted by a variety of appealing, hypothetical policy ideas. At an empirical level, one is strongly motivated to search the experience of the world for relevant policies and their effects, making both historical and cross-cultural evidence interesting. These searches, both logical and empirical, must be disciplined by the limitations of time and resources. This discipline requires a conceptual ordering of different kinds of approaches, so that the search for information becomes a systematic *sampling* of policy approaches within broad categories that seem interesting and appealing.

These considerations have guided our inquiry and given shape to this report. Chapters 1, 2, and 3 of the report seek to create some critical distance from current conceptions of and approaches to the problem. Chapter 1 examines the ideas that have guided alcohol policy in the past as well as those now making claims on our credibility. Chapter 2 develops a complex conception of alcohol use and its effects, which seeks to hold the moral qualities of drinking to one side while it explores the underlying causal systems. Chapter 3 critically reviews current policy approaches in light of the historical experience and the complex conception of the problem, and in light of institutional and normative factors that need to be taken into account. It identifies the broad policy approaches in the domain of prevention that seem worth investigation.

Chapters 4, 5, and 6 of the report pick up this agenda, presenting arguments and evidence useful in evaluating the potential of three broad classes of alcohol-related prevention policies. Chapter 4 examines strategies focused on the supply of alcoholic beverages and the regulation of drinking premises. Chapter 5 addresses policies that seek to shape drinking practices as such without operating through or on supply channels. And chapter 6 considers policies designed to change the consequences of drinking by altering features of the physical and social environment that now create hazards for drinkers.

The report concludes with a summary of our judgments about the evidence that is available and the research that is needed; about the practical, normative, and historical considerations that condition our current options; and about the actions that we think are possible, rational, and legitimate in light of our investigations.

Readers who are interested above all in our conclusions may proceed directly to the summary. Those who are principally interested in the detailed analysis of particular policy instruments may turn to chapters 4, 5, and 6. Those whose interests in prevention policies are general should begin at the beginning.

1 Simplifying Conceptions of Alcohol Problems and Policies

In a democracy, government policy is inevitably guided by commonly shared simplifications. This is true because the political dialogue that authorizes and animates government policy can rarely support ideas that are very complex or entirely novel. There are too many people with diverse perceptions and interests and too little time and inclination to create a shared perception of a complex structure. Consequently, influential policy ideas are typically formulated at a quite general level and borrow heavily from commonly shared understandings and conventional opinions.

Alcohol policy is no exception to this rule. Current policy is profoundly shaped by a body of conventional wisdom, including the belief that alcohol problems are created largely by a small group of alcoholics who require intensive, prolonged treatment and that any effort to restrict drinking practices in the general population is doomed to failure. The power of these ideas is apparent in that they are widely treated as the most obvious and incontrovertible facts—the foundation of an informed and intelligent discussion of alcohol policy.

Much can be said for the wisdom of governing through commonly shared perceptions. If many people understand and agree with an idea, a *prima facie* legitimacy much valued in democratic society is established. Moreover, widespread understanding and acceptance establishes a necessary condition for effective implementation in a society in which governmental power is broadly dispersed. Finally, it is often the case that although the shared simplifications fail to reflect or capture all the important aspects of a problem, they often do focus attention on some of the most important. Thus, the simplifications help to concert social attention and action—something that a more complicated idea could not achieve.

There is also a price to be paid for simple ideas. Simplification inevitably distorts our perception of a problem. While some important features of the problem are elevated, others that could plausibly claim equal importance are subordinated. While some avenues for social intervention are brightly illuminated, others plausibly as effective are obscured or condemned to darkness. Moreover, precisely because the simplifications are so powerful and seem so much a part of our current society, it is hard to be skeptical about them and to ask effective and probing questions about their limitations.

Such limitations can be of two sorts. One simplifying strategy is to focus on a limited set of effects. Thus, one can focus on adverse health effects and promote policies to reduce cirrhosis and accidental injuries associated with alcohol, taking everything else as secondary in importance. Alternatively, one might be primarily concerned with the social degradation accompanying chronic alcohol dependency and concentrate on policies to locate and rehabilitate (or at least provide humane care to) alcoholics. Or one might consider the appearance of public sobriety to be of overriding importance and choose policy instruments that will simply but effectively keep drunkenness out of public view. In short, by choosing a limited range of effects to be the dominant objective of alcohol policy—effects that are the largest, or most important, or the only ones that are conceived to be an appropriate concern of government—the problem can be simplified sufficiently to gain confidence in designing and recommending alcohol policy.

A different simplifying approach is to decide which causal variables are most important in generating the effects of drinking, then to choose policy instruments that operate most directly on these causes. Thus, one might judge (on the basis of available evidence) that the quantity of alcohol drunk is itself the major causal variable determining the observed pattern of effects and try to develop policies that will ration the amount of individual access to it. Alternatively, one might conclude that the aggressiveness with which alcohol is marketed and the kinds of settings in which drinking occurs are capable of making otherwise moderate and sensible people into dangerous, risky drinkers. This judgment might lead to efforts to "take the profit out of selling alcohol" or to carefully shape the times, places, and settings in which drinking takes place. Then again, one might determine that whenever there are alcohol problems, they are due to a relatively small number of unusually reckless or vulnerable drinkers and consequently tailor alcohol policy to prevent such people from drinking or to "treat" them so that they are more resistant to it. In short, by limiting one's attention to a small set of causal variables, one can find a comfortable basis for supporting a given policy.

The most successful simplifications of the alcohol problem have involved both kinds of limitations simultaneously: the major objective of the policy and the judgment about what causes this particular effect are sewn together into a neat conceptual bundle. A few such bundles have had widespread and durable appeal in U.S. society, because they have proven compatible with common social views, individual experiences, and the interests and purposes of organized groups concerned with alcohol. We refer to these cognitive bundles as *governing ideas*. A few other such bundles have considerable intellectual appeal and have at times claimed the interest and loyalty of "experts" who influence alcohol policy, but they have not succeeded in capturing the imagination of the broad population or in shaping their actions. We will refer to these as *minority conceptions*. Before looking at the kinds of policy choices that our analysis suggests are available, it is well worth understanding the basic structure of the three ideas that have succeeded in profoundly shaping alcohol policy as well as two others that are interesting and have appeared historically but have made lesser claims on credibility.

GOVERNING IDEAS

A review of the history of alcohol in the United States reveals three dominant conceptions associated with its use (see Aaron and Musto in this volume; Beauchamp 1973; N. Clark 1976; Levine 1978, 1980; Room 1974; Rorabaugh 1979; Wiener 1980). Each of these governing ideas was initiated and became most prominent in a distinct historical period, but none has disappeared from American consciousness, politics, or scientific discourse.

The first governing idea, dominating the 150-year colonial era of American history, focused on customary attachment to drinking and the moral qualities of drunkenness. During this period, drinking and drunkenness aroused as much interest as eating and obesity—which is to say, there was plenty of both and they were widely accepted as normal. Drinking was done largely at the local inn, and the local innkeeper was a community notable. Drinking was a social affair—largely public and under responsible community control.

In these times, those few who raised alarms about drinking did so on religious grounds: that habitual drunkenness was sinful, a dissipation of the moral energies that the colonial ministry expected to be devoted to

God's work on earth. The fault was not in the rum or cider, however, but in the defective moral character of those who besotted themselves when they should have been soberly at work or prayer. The solution was not in controlling alcohol, but in disciplining—visibly in the stocks, audibly from the pulpit—the moral character of such drinkers.

During the central expansionary period of American history (approximately the century-and-a-half from the Revolutionary War to the first decades of the 20th century), a quite different notion took hold: that alcoholic beverages, formerly held to be benign and healthful, were in fact toxic and addicting. In this concept, alcohol itself (rather than the character of the drinker) became the focus of concern. As befits this concentration on the substance, extensive discussion occurred about which forms of alcohol were dangerous. Liquor (distilled spirits) was held to be the central evil, but there was considerable dispute whether the milder drinks—cider, beer, and wine—could be considered temperate and acceptable. The preoccupation with alcohol itself altered the views of the consequences of drinking. Instead of viewing drunkenness as an annoying personal habit, the excessive drinker came to be seen as someone who was ravaged and transformed by an alien substance. Otherwise decent people could be transformed by drink to become dissolute, violent, or degenerate. Moreover, since alcohol was an addicting substance, even the most moderate drinker flirted with danger at the rim of every cup.

The institutional carriers of this conception were the temperance pledge societies—voluntary associations whose core tenet was the mutual renunciation of liquor by their memberships. Most of these societies were evangelical in tone and middle class in origin. Membership was open to all, and these clubs appealed to many with their optimistic and communal creed. The growth of pledge societies in the 1820s and 1830s constituted a remarkable social movement, the archetype for de Tocqueville's observation of America as a nation of joiners.

Eventually these groups developed political aims and sought to embody their views in legislation at federal, state, and local levels. The purveyor of drink was viewed with increasing suspicion, rather than as a respected (indeed a leading) citizen. The grogshop or tavern, contractually tied to a manufacturer, pushing drink to its profit-making limit, and attracting to itself prostitution, vote-buying, thievery, and murder, became a stock image. In the early 1850s, 13 states passed prohibitions on the sale of hard liquor. Although these bans were in most cases soon rescinded, they were an early efflorescence of the main aim of temperance activity between 1875 and 1920: the reform of the corrupt political and moral culture associated with the urban saloon.

The political aims of the second governing image, which viewed alcohol and its sale as a public menace, were carried into fullest bloom by the skilled single-issue politicking of the Anti-Saloon League. By 1916, prohibitionist laws of various sorts had been established (mainly by referendum) in 23 states and were finally extended to the nation as a whole by the 18th Amendment and the Volstead Act. The consequent unwillingness of most jurisdictions to adopt Draconian enforcement measures, or (in the days of Harding, Coolidge, and Hoover) to commit more than a bare minimum of public funds to such activities, ensured that illegal marketeers, buoyed by the willingness of affluent drinkers to pay three to four times the prewar going rates, developed a strong black market in booze.

Federal prohibition was swept off the books in the first months of the Roosevelt Administration's New Deal. With it went the future of radical legal controls over the supply of alcohol as devices to control alcohol problems. In the half-century since repeal, a third governing idea has steadily gained adherents: the modern "disease" view of alcoholism. In this conception, both the drinker and the supplier of alcohol were stripped of their moral imputations. The problems associated with alcohol that attracted attention were those involving the social collapse of the chronic, heavy drinker. These problems were seen to result neither from moral weakness in the drinker nor from the universally addicting power of alcohol itself, but from a little-understood chemistry that occurred between the substance and certain drinkers. In contrast to the colonial view that although alcohol is physically and morally innocuous, some morally defective individuals take to perpetual drunkenness as a sign of their dissipation, this modern view holds that although alcohol is innocuous for most people, a minority—fine people in all other respects—cannot use it without succumbing to alcoholism, a disease process for which there is no known cure except total abstinence. This view makes it the responsibility of the alcoholic, and those who care personally or professionally about him or her, to see that a treatment or recovery process is initiated and that abstinence is maintained.

These three governing ideas can be summarized as follows:

- *The Colonial View*: drinking is a valued social custom; overindulgence is a weakness in moral character; public discipline is the appropriate response.
- *The Temperance View*: alcohol (at least, strong liquor) is an addicting poison; its sale is a public hazard; use of the law to restrict its sale is the appropriate response.
- *The Alcoholism View*: alcoholism is a disease; its causes are as yet

unknown; treatment of those who are vulnerable to it is the appropriate response.

Each of these ideas, aged by historical experience and colored by current affairs, has its residue in contemporary public policy.

The first, the colonial view, survives most visibly in laws regarding public drunkenness. Many jurisdictions have recently reduced the sanctions for drunken demeanor from criminal penalties (fines, jail terms) to civil ones (compulsory education or treatment). These results of the societal drift toward greater tolerance of unconventional life-styles on one hand, and therapeutic justice on the other, serve only to soften, not to efface, the lines of moral judgment on the sidewalk drunk.

The second notion, the temperance view, has been shaped by two key forces: the widespread post-repeal conviction that Volsteadism was a failure and the growth of government regulation of commerce as a central feature of national life (itself the legacy of Progressivism, the New Deal, and war mobilization). The alcoholic beverage industry, from manufacture through retail sale, is thoroughly regulated, principally by the U.S. Bureau of Alcohol, Tobacco, and Firearms (BATF), the Food and Drug Adminstration (FDA), and the many state and county Alcoholic Beverage Control (ABC) boards. Part food, part drug, part hazardous substance, alcohol is subject to a system of regulatory control that includes strict purity, packaging, and labeling rules; special (and unusual) principles of taxation; and a differentiated system of licensing restrictions, including (in 18 states) government liquor monopolies.

The third idea, the alcoholism view, has only recently moved from the realm of voluntary organization and private clinical practice to establishment in public policy. Its main institutional basis has been the National Institute on Alcohol Abuse and Alcoholism (NIAAA) and the current system of formula grants to state agencies that supports and encourages local treatment efforts. The disease view has successfully allied modern medical science with the organizational form of the mutual pledge societies, as in Alcoholics Anonymous (AA). The emphasis of NIAAA, AA, and a host of related institutions has been on the refinement, financing, and legitimation of the treatment of alcoholism. At the same time, there have been efforts to institutionalize alcoholism treatment in occupational programs and in health care financing by favorable federal regulation as well as private-sector action. It is also notable that this governing idea, projecting the conception of vulnerability onto a small part of the population, has been able to establish and maintain support from the alcohol industry itself.

The alcoholism treatment idea is still relatively new in the context of our historical perspective. It has become clear in the last decade, however, that this governing idea and its constituency are capable of building a powerful institutional base. Whether the current base can be defended or enlarged in the face of current government fiscal retrenchment cannot be vouchsafed. Nevertheless, there is little question about the hardiness and sustained growth of the concept of alcoholism in public attitudes, interested organizations, and formal government policy.

MINORITY CONCEPTIONS

In addition to these three dominant conceptions of alcohol and alcohol policy, two other conceptions are worth noting. Although they have not achieved the prominence and influence of the governing ideas, they have the same synthetic qualities. Moreover, they have exerted considerable force in the communities of "experts" who are influential with respect to conceptions of the problem and implied solutions. One idea—the concept of alcohol control—was strong at the time of repeal of Prohibition, played an important role in establishing the system of regulation of drinking at the state level, then quietly disappeared. The other idea—the public health perspective—has made its appearance more recently.

Following the repeal of Prohibition, the direction of alcohol policy was strongly influenced by an organization called the Association Against the Prohibition Amendment, a business-based alliance that drew heavily on the intellectual resources of Columbia University. In the eyes of this group, the principal problem was neither the alcohol itself (as most of the temperance movement believed) nor certain drinkers (which the colonists and later the alcoholism movement believed), but the aggressive, ruthless, and at times criminal way in which commerce in alcohol had been developed. As a result, the contexts in which drinking occurred became the natural targets of policy. Since many of these contexts are created by small, informal, private arrangements, there were limitations to what could be accomplished. But the public contexts of drinking (bars, taverns, saloons, restaurants, etc.) could be brought under governmental scrutiny. And it was possible that private informal settings could be influenced by regulating the hours, accessibility, and prices charged by stores selling alcohol. Thus, in this conception, the alcohol problem could be controlled by governing the terms and conditions on which alcohol was available for both on-site and off-site consumption and, using this control, temperate drinking practices could be promoted.

This idea never achieved the sustained prominence that the idea of alcoholism has gained. But the concepts (and the group that pressed them) were powerful enough to provide a rationale for replacing the Volstead Act with the institutional apparatus of state ABC systems and to give early shape to these institutions. Originally, the state boards were authorized to regulate the marketing of alcohol to inhibit vicious or excessive drinking as well as to rid the business of criminal influence. The first part of this idea was vague and lacked a potent constituency, and the state ABC systems have become concerned mainly with the protection, promotion, and orderly development of the legitimate industry per se. The institutional legacy of this perspective thus remains hardy but is generally unconcerned with a major component of its original justification.

The concern that proponents of repeal expressed in seeking to inhibit abusive drinking through the management of social contingencies around drinking has been revived—with a different institutional base—in the public health perspective. As with the other ideas, the public health perspective focuses attention on selected aspects of alcohol problems and carries with it a variety of normative and empirical assertions. Two basic principles seem fundamental to this idea, which is still developing.

First, the public health perspective focuses on the health consequences of alcohol use (particularly cirrhosis and traumatic deaths) as the most visible aspect of alcohol problems. In doing so, it shifts away slightly from the historical preoccupation with the social collapse of individual heavy drinkers. In this view the health consequences of drinking seem compelling because: (1) the objective conditions (mortality and morbidity rates associated with alcohol-induced disease and trauma) are large enough and serious enough to warrant attention; (2) the evidence on alcohol's contribution to such problems is fairly convincing; (3) it is generally preferred that people die later rather than earlier; and (4) social intervention in health problems draws on a strong (although not uncontested) tradition of public concern.

Second, in this view alcohol problems arise, not from a small group of chronic dependents, but from the drinking habits of the general population. This is true in part because certain health consequences of concern can occur in a large segment of the drinking population (not just those who are chronically dependent) and in part because some empirical evidence suggests that the absolute number of chronically dependent drinkers is not fixed simply by an underlying distribution of "vulnerability to alcoholism," but can be importantly affected by factors that govern the consumption patterns of the general population. This evidence argues that the more drinking there is in the general population, the greater the number of people chronically dependent on alcohol and the greater the extent, therefore, of serious alcohol-related health effects.

In important respects this public health conception of alcohol problems fits well into some current American ideologies and moods. To a great degree, the approach escapes those moralistic qualities of other approaches that have made them currently suspect in the public mind. It does not focus attention on the work ethic, the importance of the family, or individual life-styles. Of course, there is no small amount of moralism involved in asserting the importance of remaining healthy. But it appears to be characteristic of our epoch that health is a cause that government has been allowed and even encouraged to pursue to considerable lengths. By focusing on alcohol-induced mortality and morbidity, this perspective points to a problem that clearly costs society a great deal of personal pain and loss, to say nothing of economic resources. Given the current mood, the public health conception of the problem is a tempting one.

On the other hand, this conception has some significant liabilities as a governing idea of the problem. Perhaps its greatest liability is the fact that it includes in the problem many people who do not see themselves as being at risk and do not desire protective measures. It is possible that some of those who had regarded themselves as not at risk are surprised by the outcome; they may end up regretting the absence of barriers between them and their unhappy fate. Before the fact, however, they may feel that their prerogatives are being infringed without any benefit to them, since they are the kinds of individuals who can live healthily, even in a world that is not designed to protect them from risk; or that the political or economic costs of protection are not justified by the benefits. In addition, the public health perspective faces the problem of competing with other governing ideas. In going beyond the alcoholism view and its preoccupation with chronically dependent drinkers, it departs from another conceptually satisfying and institutionally entrenched idea. In addition, since it focuses on drinking in the general population and tinkers with the idea of restricting alcohol availability, it cannot quite shed the taint of "the great experiment." Since Prohibition has been so widely repudiated as a clumsy, ineffective, if not malicious social intervention, anything that looks like a move toward Prohibition will now be stubbornly resisted.

CONCLUSION

These five different simplifying conceptions of the nature, source, and solution of alcohol problems have been presented as distinct, explicit ideas embraced by particular groups at particular times and exercising measurable degrees of influence over governmental (and, more broadly, public) responses. To this extent the recounting of these ideas may be viewed as an intellectual history of alcohol policies, recognizing that ideas are a limited part of the full history of alcohol in U.S. society. Our purpose here is to use these ideas to orient our investigation of the problem. At a minimum, the ideas reveal the diverse perspectives one can adopt as one circumnavigates alcohol problems. The diversity, in turn, suggests the fundamental complexity of the underlying problem.

More particularly, however, these simplifying conceptions point to the crucial dimensions in which simplifying choices must be made. The most vivid choice concerns the question of what exactly is causing the concrete problems created by drinkers: is it the alcohol, the drinker, or the environment around the drinker? Much seems to depend on which of these things one chooses to emphasize. Only slightly less vivid is the choice about which effects of alcohol are the most important: is it the capacity of the drinker to meet economic and social responsibilities, the health of the drinker, or the risks that the drinker creates for others? Again, both the image of the problems and the implied response change as one becomes more or less preoccupied with a given idea. Finally, the question of what forms of government intervention are tolerable and efficient is always in the background. The roles of individual freedom, collective responsibility, and the social contract are subtly changed as one shifts the emphasis from economic problems to breaches of public order to acts (or accidents) of nature.

Thus, the underlying complexity of the problem gives relatively free rein to simplifications; many different ones can be supported. Which one(s) should be encouraged is the issue. Presumably, this depends on a set of empirical and normative judgments about which causal mechanisms are powerful and which effects of drinking are both large and socially significant. Only against the backdrop of a fairly detailed investigation of the complex structure of the underlying problem can a judgment be made about the virtues and liabilities of any given simplification. Consequently, it is to the complex structure of the problem, as best we can describe it, that we next turn.

2 The Nature of Alcohol Problems

For most people most of the time, even the immediate effects of drinking, to say nothing of the long-term effects, are hardly noticeable. When effects are noted, they vary across a broad range of kinds, degrees, and levels of desirability. As the rich vocabulary of descriptive terms, some of them centuries old, suggests, one can be mellow, tipsy, or pleasantly plastered; stoned, skunked, or sloshed; or dead drunk, wrecked, or under the table. Whatever the degree, however, intoxication alters the mental, behavioral, and physiological capacities of drinkers. An intoxicated person is generally less mentally alert in scanning the environment for hazards, less reliable in interpreting what is observed, and inclined to relax or forget about some things that might be the focus of intense concern in a sober state (Cappell and Herman 1972). On the behavioral dimension, the drinker is likely to become distractible and clumsy.

Of course, these effects of alcohol intoxication are all matters of degree, and the degree depends on various factors. Among these, the amount of alcohol consumption itself can be very important. Given enough alcohol, one can reliably produce a stupor—even death—no matter what other factors are operating on the drinker (Poikolainen 1977). In the more usual case, however, the effects depend on such factors as the spacing of drinks, the drinker's size and weight, how recently he or she has eaten, his or her own hopes and expectations about the effects of drinking, and even the expectations and demands of the people present in addition to the amount consumed. An excited, skinny teenager anticipating a big night with pals can become quite exhilaratingly "drunk" on a quantity of alcohol that would produce no effect or perhaps a mild relaxation in a heavy, middle-aged man who had just finished a large dinner and had no greater aspiration than to pass the evening quietly.

Of course, some people regard even the most temporary departure from sobriety as a significant moral and social problem—a willful denial of individual responsibility in a society that both values and depends fundamentally on a sturdy, universal commitment to this standard. But for most people, short-lived infrequent periods of intoxication, to say nothing of drinking events that stop short of noticeable intoxication, do not create or indicate substantial harms (Cahalan et al. 1969). Many people regard drinking and intoxication as beneficial—a harmless, pleasant indulgence when they feel beset or entitled or a way of turning an ordinary event into a festive occasion. To create problems for drinkers and others, some special characteristics beyond drinking and drunkenness must come into play.

The idea of an alcohol problem brings some paradigmatic situations to mind: the drunk driver who causes a serious accident by ignoring a road sign or losing control of the car; the domestic fight that, fueled by alcohol and the ready availability of a weapon, flares into a bloody assault; the previously responsible husband, father, and employee whose ability to meet the needs and expectations of his family and employer deteriorates as a result of increasingly frequent and ill-timed drunkenness and hangovers; the aging heavy drinker whose damaged liver becomes a chronic health problem and ultimately a cause of death; and perhaps even the rowdy teenagers or the skid row resident whose behavior offends others' sense of propriety and order.

These paradigmatic situations may capture only a portion of an accurate accounting of alcohol problems. But, if we can take these as typical, we can make an important observation: few of the bad consequences result from drinking only. Instead, they emerge from relatively complex causal systems in which drinking is a major—but only a single—contributing factor. Some of the consequences depend on unfortunate combinations of episodic drunkenness with dangerous or demanding environments. Others depend on being drunk frequently enough in situations in which sobriety is expected and demanded that the drinker comes to be regarded as irresponsible and unreliable. In many cases, however, the environments in which the drinking is done as well as the frequency and degree of intoxication play major roles in determining whether bad effects occur. Figure 1 presents a schematic view of the hypothesized relationships.

This observation should not be used to minimize the role that drinking

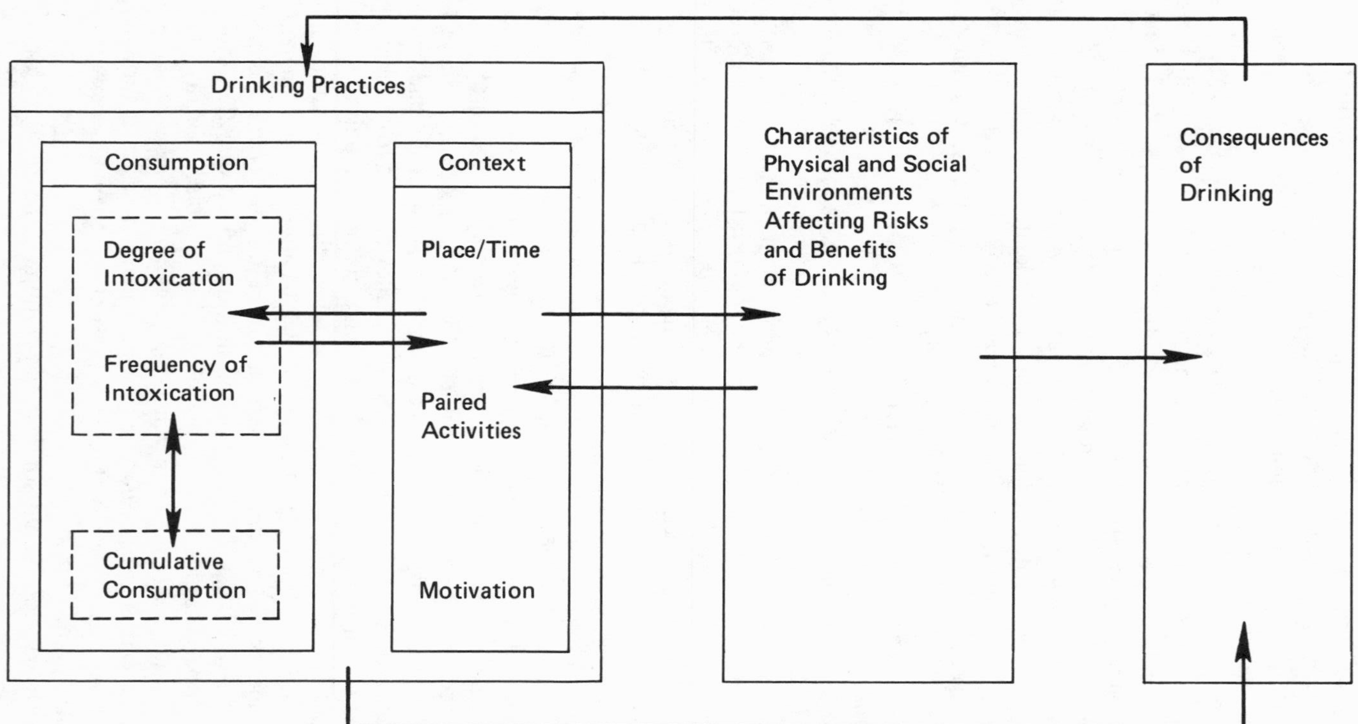

FIGURE 1 Schematic view of the relationship between drinking and consequences of alcohol.

itself can play in creating problems for drinkers and others. Sustained frequent drinking can be directly responsible for some of the more important physiological consequences of drinking (e.g., fatty liver, physical dependence) without regard to circumstance and even without notable drunkenness. Moreover, for many important consequences (such as drunken driving and drunken falls), some contributing features of the environment (nighttime roads, steep stairways) may be so common and resistant to change that they might just as well be treated as inalterable features of nature, with all the variation in accident experience attributed to drinking.

What the observation does suggest is that a simple equivalence of drinking with problems, and more drinking with more serious problems, while often true, can be somewhat misleading. Much seems to depend on how and where alcohol is absorbed in a drinker's life: how much drunkenness is generated and how that drunkenness is fitted into more or less demanding physical and social environments. To be sure, one would expect heavy drinkers to cause and suffer many of the bad consequences of drinking. But moderate and light drinkers could also cause and suffer problems if they consumed their smaller quantities of alcohol in unusually dangerous patterns. Moreover, some kinds and degrees of risk confronting drinkers of all kinds depend on certain characteristics of the social and physical environment.

These considerations have two crucial implications for policy. One is that alcohol problems may be distributed over a large segment of the drinking population. They may not be only, or even mainly, problems of alcoholics. To the extent that this is true, focusing exclusively or even primarily on the treatment of alcoholics is not sufficient. A second implication is that there may be other avenues open to controlling alcohol problems besides reducing alcohol consumption. For example, one can think of influencing the drinking practices of individuals so that a given quantity of alcohol could be distributed more safely and appropriately within a given individual's activities, or one can think of altering the structural characteristics of the environment to make it safer for drunken people.

We have now sketched the main outline of a complex conception of alcohol problems. In this view alcohol produces a variety of significant consequences—some good and some bad. Moreover, the quantity of alcohol consumed is not the only cause producing the bad consequences; also significant are: (1) "drinking practices" that generate different degrees and frequencies of drunkenness from given quantities of alcohol and place these episodes of drunkenness in different environments and (2) characteristics of social and physical environments that can make drunkenness more or less risky. Since drunkenness, drinking practices, and environments all play a role in creating and shaping alcohol problems, some of the problems may occur among drinkers who drink, not heavily and continuously, but simply unwisely or unluckily.

The remainder of this chapter develops this conception with greater rigor and more detail, then tests the conception against observed facts. We first build an account of the important consequences commonly associated with alcohol and present some information about their magnitudes. We then analyze drinking practices: how much Americans drink; how drinking is distributed across the population; where current patterns stand in historical and cross-cultural perspective; and in what contexts drinking (and intoxication) typically occur. Finally, we analyze the relationship between the observed effects and the underlying pattern of drinking practices: specifically, the role that things other than alcohol itself play in producing the varied effects and how alcohol problems are distributed across the population of drinkers. We conclude with the major policy implications of this conception.

EFFECTS OF DRINKING

A complete, evenhanded accounting of the important consequences of current alcohol use is beyond current intellectual resources. We cannot report the simple frequencies of all the important events and conditions in which alcohol is implicated, to say nothing of gauging the infinitely more subtle issue of how much alcohol alone contributes to the character of the events and conditions we observe. But to facilitate the design of an effective policy, it is important that we develop an accounting scheme that strains to be comprehensive and orderly in arraying relevant effects, even if we know at the outset that the data will be incomplete. In particular, the scheme should be constructed in accordance with two principles.

The first principle is that in identifying relevant effects of drinking, it is better to err on the side of inclusiveness rather than exclusiveness. Otherwise the scheme loses its value as a device for alerting us to (potentially) important aspects of the problem. The accounting scheme should include social as well as economic effects, collective as well as individual ones. It should include effects that are easily quantified as well as those that are not. It should include effects that have already been well measured, those that could in principle be studied but still await careful investigation, and those that cannot in principle be precisely or reliably measured.

This principle suggests that we include not only the effects of drinking in the accounting scheme, but also the effects created by the current policies and programs designed to manage drinking. Drinking never occurs in a vacuum; it occurs in a social setting in which it is influenced by private attitudes and practices and by governmental policies and programs. Thus, when we observe the effects of current alcohol consumption, we are inevitably looking simultaneously at the effects of drinking and of the policies and programs we maintain to shape drinking practices and to cope with whatever problems emerge.

This first principle also requires us to recognize good effects of drinking as well as bad. That benefit results from drinking is usually conceded even by those who are most appalled by the damages. Over $35 billion was spent in 1980 on alcohol by people who could have chosen to spend the money on better housing, new clothes, roast beef, or vacations. (For that matter, they could have chosen to spend the money on cocaine, marijuana, casino bets, or sexual services.) In describing the nature of alcohol problems the beneficial effects usually get short shrift: analysts are far more industrious in seeking out, enumerating, and marshalling impressive statistical evidence revealing the damages (e.g., Berry and Boland 1977; National Institute on Alcohol Abuse and Alcoholism 1978; Institute of Medicine 1980). We are not major exceptions to this rule, first, because our concern is primarily with damage reduction and, second, because our discussion of beneficial effects is constrained by the dearth of available evidence. Still, we insist on the principle that protection of benefits must be weighed if we are to know whether a policy's net effect is favorable (see Walsh and Walsh 1970, Mäkelä and Österberg 1979).

The second principle is that the accounting scheme should keep track of who is experiencing the varied effects as well as what the various effects are. For our purposes, it is useful to distinguish among the effects on (1) drinkers themselves, (2) people who are intimately connected to (or dependent on) the drinker, and (3) people who are relatively remote from the drinker (i.e., the general population). Such distinctions are important because they provide clues about who might be interested in controlling the drinking. For example, people who take losses from drinking problems might well be motivated to help solve at least the aspect of the problems that affects them.

The distinctions are also important, however, because this society attaches different significance to effects that seem consciously chosen by individuals and those that individuals receive involuntarily from others. To be sure, the boundaries between state intervention and individual choices are far from clearly established, and we do not propose to adjudicate them here. Our purpose is to array information so that one can see how such concerns are reflected in alcohol problems.

Figure 2 presents the important effects of alcohol use and of relevant

EFFECTS ON DRINKERS

HEALTH

- Medical (e.g., liver, heart, brain; nutrition)
- Traumatic (accidents, assaults, suicide)
- Psychological (mood, self-esteem)

ECONOMIC INDEPENDENCE/SECURITY

- Expenditure (on drinking, damage repair)
- Income (earnings; wealth)
- Dependency on others

SOCIAL FUNCTIONING OF INDIVIDUALS

- Discharge of roles and responsibilities
- Ability to develop, sustain, enjoy social relationships
- Vulnerability to state supervision

REPUTATION WITH RESPECT TO DRINKING

EFFECTS ON INTIMATES

HEALTH

- Medical (nutrition, development, including fetal)
- Traumatic (accidents, assaults)

ECONOMIC INDEPENDENCE/SECURITY

- Economic expenditures due to drinking
- Economic contributions of drinker
- Family working patterns
- Workgroup productivity

SOCIAL FUNCTIONING OF FAMILY

- Family roles and responsibilities
- Family morale

DRINKING PRACTICES OF INTIMATES

EFFECTS ON OTHERS IN SOCIETY

HEALTH

- Traumatic (accidents, assaults)

ECONOMIC POSITION

- Expenditures on law enforcement, medical care, education, welfare
- Changes in productivity (human and material capital)
- Tax revenues, consumer surplus, return on investment due to alcohol use

SOCIAL RELATIONS

- Between government and individual
- Societal morale

GENERAL DRINKING PRACTICES

FIGURE 2 Important effects of drinking and policies for dealing with drinking.

policies for dealing with drinking. In principle, one can track any significant changes in the size and character of the alcohol problem by measuring these effects over time. Thus, if drinkers started to do appreciably less driving while drunk, one would see decreases in the number of fatalities, injuries and, economic losses associated with traffic accidents: these categories (and possibly others) would all move in favorable directions. However, if this change in drivers' habits had resulted from more aggressive enforcement of traffic laws and a major effort to mobilize public opinion, increases in public spending and losses in the social independence of certain drinkers would also be recorded. In practice, it is difficult to find or develop reliable empirical estimates of all the phenomena of interest; the figure provides a basic orientation. (More detailed discussion and review of many of these effects are included in the paper by Gerstein in this volume.)

It is important to understand that the detailed entries in each of the categories may respond negatively or positively—or both—with respect to drinking. For example, some evidence now suggests that low or moderate levels of alcohol intake may be associated with reduced levels of ischemic heart disease and fatal coronary events; other evidence suggests that heavy drinking may be associated with elevated incidence of other cardiovascular diseases such as cardiomyopathy and hypertension (deLint and Schmidt 1976, Hennekins et al. 1978, Klatsky et al. 1978, Regan and Ettinger 1979). Similarly, the effects of drinking on the ability to develop, sustain, and enjoy personal relationships include some positive contributions of drinking, even though for many people drinking interferes with developing and enjoying relationships, inducing losses in this category. The positive gains may be judged to be small or large relative to the risks, yet the effects listed in Figure 2 do seem to include beneficial effects as well as harms and costs associated with drinking.

Which effects of drinking should be the primary focus of our concern is an important but ultimately ambiguous issue. To a degree, this judgment can be usefully informed by empirical evidence on the frequencies and magnitudes of the various effects (e.g., the number of people who die of alcohol-induced cirrhosis, the frequency of alcohol-induced criminal assaults among strangers in public locations, the number of people who drink so much that their spouses seek divorces, etc.). Ultimately, however, reliable information of this sort exists for only a few of the effects. In his paper in this volume, Gerstein finds a high degree of confidence in a figure between 20,000 and 25,000 for annual deaths from alcohol-induced cirrhosis of the liver. Reed (in this volume) provides a stable Bayesian estimate of about 12,000 annual traffic fatalities causally attributable to (not just "associated with") drunken driving and somewhat less than $.5 billion in property damage from the same cause. Estimated medical expenditures for sequelae of alcohol abuse and alcoholism have been consistently placed in the neighborhood of $10 billion annually (Berry and Boland 1977), a figure that very roughly equals the annual receipts from state and federal excise taxes on alcoholic beverages (Hyman et al. 1980). For other categories, the intervals of estimation are too open for numerical summary or do not now exist as operationally defined concepts. Even if this kind of information were available for all the effects of drinking, however, the question of which effects are most important would remain ambiguous. The reason is simply that "social importance" is ultimately a value question about which reasonable people can disagree.

Thus, while Figure 2 does not pretend to give a precise, quantitative account of alcohol problems, it prevents one from thinking too narrowly about it. Drinking is not just a health problem or a social problem or an economic problem; it is not even just a problem. Moreover, it involves more than just drinkers and their intimates. Figure 2 alerts the reader to the variety of effects of choosing or having a given policy toward alcohol use.

UNDERLYING PATTERNS OF DRINKING PRACTICES

Beneath the effects outlined in Figure 2 lie drinking practices. Drinking practices produce these effects, not necessarily directly, but probabilistically in combination with some important characteristics of the physical and social environments. In order to see how the effects of drinking are associated with (or are the result of) drinking, one must begin with the drinking practices of the general population. To the extent that we expect to affect the current shape of the problem by altering drinking practices, it is important to understand what aspects of drinking behavior should be included in useful descriptions of drinking practices, what the current practices are, and what kinds of factors seem to be shaping them.

DEFINITION OF DRINKING PRACTICES

At the outset, it is important to be clear about what we mean by drinking practices. This may appear to be largely a semantic issue, but the definition is of crucial importance in structuring our conception of the problem and how it might be controlled. Specifically, it emphasizes the complexity of the link between the concepts of "drinking" and "alcohol problems." Moreover, the problem is not simply that characteristics of the physical and social environment must operate to shape and create the consequences; choices about how much drinking is done, how much intoxication is generated, and how the periods of intoxication are woven into settings and activities that bring drinkers into contact with dangerous parts of the environment seem to be at least as important in creating risks for drinkers as the general level of hazards in their environment. We include these additional aspects of drinking behavior in the idea of drinking practices.

If one could describe how the effects in the various dimensions of Figure 2 were linked to specific characteristics of drinking practices, our ability to design effective alcohol prevention policies would be significantly enhanced. For one thing, we would then be able to recommend or prescribe certain drinking practices and discourage others. In addition, we would be able to see more precisely how risks and benefits of drinking are distributed across the population of drinkers. Unfortunately, at this stage the analytic or empirical work is still insufficient to fully define drinking practices as we want to use the concept. What can be done is to outline the dimensions that are likely to be linked importantly to risks and benefits and report what is known about how the population is distributed across these dimensions.

Drinking practices as we define them involve two distinct dimensions: one concerns the description of alcohol consumption and the other concerns settings and activities (contexts) commonly associated with drinking. With respect to consumption, as we have said, it is useful to distinguish three characteristics. First, many effects depend on the *degree* of intoxication. At very high levels of blood alcohol content (BAC: above 0.30, absent any other drug effects), a drinker can die of acute alcohol poisoning. At somewhat lower levels of BAC, in the 0.15 to 0.30 range, the drinker does not risk death by poisoning but is likely to be so clumsy and inattentive that even benign physical environments become hazardous. At even lower levels, the drinker may still be sufficiently clumsy and inattentive to be unable to cope with moderately taxing physical or social demands; these lower levels generally are those at which psychological mood is most favorably affected (Mello 1972). Thus, other things being equal, the degree of intoxication itself will probably increase the risks and decrease the benefits a drinker creates for others and himself or herself, beyond the threshold at which effects are noted.

Second, certain effects of drinking depend on how often one is drunk (or how much time one spends above a given level of BAC). The chance of being drunk at the wrong place at the wrong time increases as a person is drunk more often. But it also seems plausible that some important consequences of drinking (specifically, bad social consequences in the areas defined in Figure 2) begin to occur only when a person is drunk sufficiently often (or at sufficiently awkward moments) to become "undependable." After all, in social terms, a person can be drunk infrequently (or even fairly frequently if drunkenness is confined to "appropriate" times and places) without being seen as irresponsible and undependable. But if a person is drunk often and allows drunkenness to intrude on times when and in situations in which others need and expect him or her to be sober, he or she will become "undependable" and cause harm to those who count on him or her. The eventual result will be a loss of self-esteem for the drinker as well.

Third, it seems fairly certain that the total quantity of alcohol consumed has important independent effects on certain attributes. It is fairly well established that the best-known medical risks of alcohol use (e.g., liver damage) are primarily linked to total quantity consumed and not the particular way in which it is consumed. The extent to which alcohol becomes a financial drain on the drinker or his or her dependents also depends in large part on the total quantity consumed.

The specification of contexts involves greater complexity. The times and places one chooses for drinking are important in determining the effects of the practice. Drinking while at work is generally much riskier (in its health, social, and economic dimensions) than drinking at home during leisure time (Aarens et al. 1977). Being drunk in a neighborhood bar where one is known well is probably safer and more satisfying than being drunk in a bar in which strangers fight (Bruun 1969). Similarly, the activities one combines with drinking can be more or less risky or rewarding. Even a safe home can be a dangerous place if one combines late-night smoking or cleaning rain gutters with drinking. And it is especially dangerous to operate moving equipment when one is drunk. In general, to describe drinking contexts in ways that reveal riskiness, it is necessary to talk about how well a given degree of intoxication is welcomed or accommodated in a given context.

It is important for our scheme to distinguish drinking episodes from drinking practices. A drinking episode describes a discrete period of time in which a given degree of intoxication is combined with a given context and associated activities. A drinking practice refers to a characteristic clustering of drinking episodes. The distinction is important for the simple reason that many of the effects defined in Figure 2 depend on accumulated experience. Therefore, a drinking practice not only generates the sum of effects associated with each episode, but also has effects that depend on the cumulative result of repeated episodes.

To summarize, then, our definition of drinking practices is built on the following concepts. A drinking episode combines a degree of alcohol use or intoxication with a given context (a physical and social setting and set of activities and attitudes). Cumulative consumption of alcohol emerges across time as one views the succession of drinking episodes. A drinking practice is a characteristic, more or less durable propensity to combine drinking episodes in certain frequencies and combinations over time. Drinking practices carry with them different probabilities of effects in the various dimensions of Figure 2.

There is not a great deal of scientific data about drinking practices in the United States—which is to say, the research territory on alcohol use is full of virgin or barely turned soil (Institute of Medicine 1980). Nonetheless, some indicators are firm enough to elicit conclusions, particularly, conclusions of relevance to our inquiry about policy alternatives. This information and summary views on what additional data resources would be most important to develop further are presented below.

CONSUMPTION: QUANTITY AND FREQUENCY

Current knowledge of drinking practices rests heavily on general sample surveys of households in the general population. Various limitations adhere to the use of such survey data. Since large-scale survey research is a post-World War II phenomenon, detailed quantitative data about drinking practices of the general population are confined to this period. In addition, the precision of the surveys depends fundamentally on how accurate respondents are about their drinking, how willing and able they are to report what they know about their own drinking to an interviewer, how skillfully the survey instrument is designed, and how adequately the sample represents all drinkers (see Midanik 1980). Since drinking rates reported in such surveys do not account for all of the alcohol sold, it is likely that heavy drinkers are underrepresented and that some respondents tend to underestimate their consumption. Some information exists on how the tendency to underreport consumption is distributed across drinkers, but the issue is not settled. Despite these difficulties, surveys remain the best way of projecting detailed quantitative information about how individuals are distributed across varying patterns of alcohol consumption.

Quantity Consumed

Table 1 summarizes the results of several recent surveys of the general household-based population. Alcohol consumption is expressed in terms

TABLE 1 Distribution of Alcohol Consumption by Adults in the United States

Estimated Average Quantity of Pure Alcohol Consumed Daily in Ounces		Mean Percentage (Standard Deviation) 1971–1976	1979
0	("Abstainers")	35 (1.4)	33
0.01–0.21	("Light Drinkers"—up to three drinks weekly)	32 (3.4)	34
0.22–0.99	("Moderate Drinkers"—up to two drinks daily)	22 (3.0)	24
1.0+	("Heavy Drinkers")	11 (1.5)	9
1.0–2.0	(2–4 drinks daily)	6.5 (1.1)	
2.01–5.0	(4–10 drinks daily)	3.4 (0.7)	**
5.01+	(10+ drinks daily)	1.2 (0.5)	
Number		*	1,758

* 7 national surveys, total number = 12,139, mean number = 1,734, range 1,071–2,510.
** Breakdown not available.
Sources: Adapted from Johnson et al. (1977), Clark and Midanik (1980).

of the average daily consumption of pure alcohol. A typical drink of alcohol (e.g., a 12-ounce can of beer, a 4-ounce glass of wine, or a 1-ounce shot of distilled liquor) amounts to about half an ounce of pure alcohol, so one can conveniently relate the numbers presented in Table 1 to our natural experience with drinking: an ounce of ethanol a day amounts to about 2 drinks a day; 5 ounces implies about 10 drinks, etc. However, reported consumption from national survey populations amounts to at best two-thirds of all alcohol sold (Room 1971), indicating a certain degree of underrepresentation and underreporting. However, this does not make much difference with respect to the table, since only small changes in the percentage distribution would be sufficient to reflect this "missing" consumption.[1]

We can draw three conclusions from these data. First, the aggregate

[1] There are two alternative patterns that seem most likely to account for the missing third. One alternative is that a large fraction—in the neighborhood of three-fourths—of the heaviest category of drinkers (5+ ounces daily) are virtually denying any alcohol use; this pattern of denial is supported by extensive clinical observation and a study in Ontario (Popham and Schmidt, 1981). The other alternative is that a great many people in all consumption ranges are underreporting, but in a sufficiently uniform or predictable way such that the rank ordering of drinkers by quantity is reasonably well preserved (Boland and Roizen 1973, Fitzgerald and Mulford 1978). For example, if 5 drinkers who consume 25, 16, 9, 4, and 1 gallon(s) annually were each underreporting by one-third (a simple linear transformation: 17, 11, 6, 2.7, and 0.7), or were even scaling their drinking claims in a more complicated nonlinear way (e.g., to read 14, 11, 7.5, 3.5, and 1), their rank

pattern of alcohol consumption has been fairly stable over the first half of the 1970s. Second, it is remarkable how much of the population either is completely abstinent or drinks very little. We can calculate that close to two-thirds of the adult population drinks three or fewer drinks per week. Third, the skewness of the distribution implies that much of the consumption of alcohol is concentrated in a small fraction of the drinkers. The heaviest-drinking 5 percent of the population accounts for roughly 50 percent of total consumption, and the heaviest third accounts for over 95 percent of the total alcohol consumed. It is worth noting that among clinical alcoholic populations, daily consumption of 5 ounces or more is reported by virtually all patients (Schmidt and Popham 1975). Hence, alcoholics are drawn from the upper 1-4 percent of the consumption range in the United States (Armor et al. 1976)—but they are not equivalent to this fraction, since quantity of drinking per se is not sufficient qualification for a diagnosis of alcoholism.

Frequency of Drunkenness

These data on quantity provide important insights about the distribution of cumulative consumption. Moreover, at the extremes, the data allow us to make some informed guesses about the frequency of drunkenness as well as cumulative consumption. The 35 percent of the population that is abstinent (in a given year) is never drunk. The small percentage of the population whose reported average daily consumption is in excess of 4 ounces a day is probably drunk for some period of time most days of the year. Between these extremes, however, it is difficult to guess how often drinkers get drunk, since it depends crucially on how the

---

order could still be preserved, even though the reported consumption would fall short by a third.

If one recalculates the consumption aggregates in Table 1 using these three possible patterns (denial, linear, nonlinear), the result is a modest readjustment except in the very highest consumption category, in which a few percentage points represent a large relative error; in every other category the adjustment falls within the range of sampling error. The range of values for the 1971-1976 column of Table 1 and the three transformations are as follows:

Daily Ounces of Alcohol	Percentage (Range)
0	33-35
0.01-0.21	30-32
0.22-0.99	20-22
1.0-2.0	5.9-8.5
2.01-5.0	3.4-4.7
5.0+	1.2-4.3
	(100)

TABLE 2 Average Daily Consumption and Number of "Drunk Days" Per Year in a Sample of Air Force Men

Annual "Drunk Days" (More Than 8 Drinks/Day)		Average Daily Consumption (Ounces of Ethanol)					
		0	0.1–1.0	1.1–2.0	2.1–3.0	3.1–5.0	5.0
Interval	Midpoint						
0	(0)	100%	40.0%	7.4%	1.7%	2.4%	0%
1–11	(5)	0%	46.0%	26.0%	14.0%	2.4%	2.7%
12–23	(18)	0%	6.1%	8.1%	5.2%	7.1%	2.7%
24–47	(36)	0%	6.6%	26.0%	21.0%	9.5%	5.4%
48–77	(63)	0%	0.5%	11.0%	8.6%	2.4%	7.7%
78–181	(130)	0%	0.2%	21.0%	38.0%	33.0%	16.0%
182	(273)	0%	0%	0%	12.0%	43.0%	70.0%

Note: The term "drunk day" is defined here as more than 8 drinks/day. If one spaced these drinks evenly over a 16-hour waking period (e.g., a drink every two hours from 6 a.m. to 10 p.m.), one might never be visibly drunk, since the usual healthy adult's liver can detoxify alcohol fast enough to completely neutralize each drink before the next arrives. A more conservative term than "drunk day" might be "intensive drinking day." However, the evidence offered by Polich and Orvis indicates that little drinking occurred during the day (especially on weekends). The description "drunk for at least a few hours out of a day" is a fairly accurate interpretation of the 8-drink-daily category: hence, "drunk day."
Source: Adapted from Polich and Orvis (1979).

light, moderate, and heavy drinkers distribute their consumption over time. Since many important consequences of drinking are related to frequency of drunkenness rather than cumulative quantity consumed, it is useful to know how the population is distributed with respect to drunkenness as well as total consumption.

A recent survey of drinking practices among U.S. Air Force personnel (Polich and Orvis 1979) provides some very useful data on this issue. While the Air Force population cannot be taken as representative of the U.S. population as a whole, it is quite representative of the young employed male population. For this group, these survey results provide a good estimate of the relationship between the overall quantity of consumption (measured in terms of average daily ounces) and the frequency of drunkenness (measured in terms of the number of days in which more than eight drinks were consumed). It is notable that this survey, taken in a very "wet" drinking climate and based on a complete enumeration of personnel rather than a household census base, appeared to account for about 85 percent of alcohol consumed. Table 2 presents the relevant data.

As one would expect, the frequency of drunkenness is highly correlated with total consumption: the more one drinks overall, the more

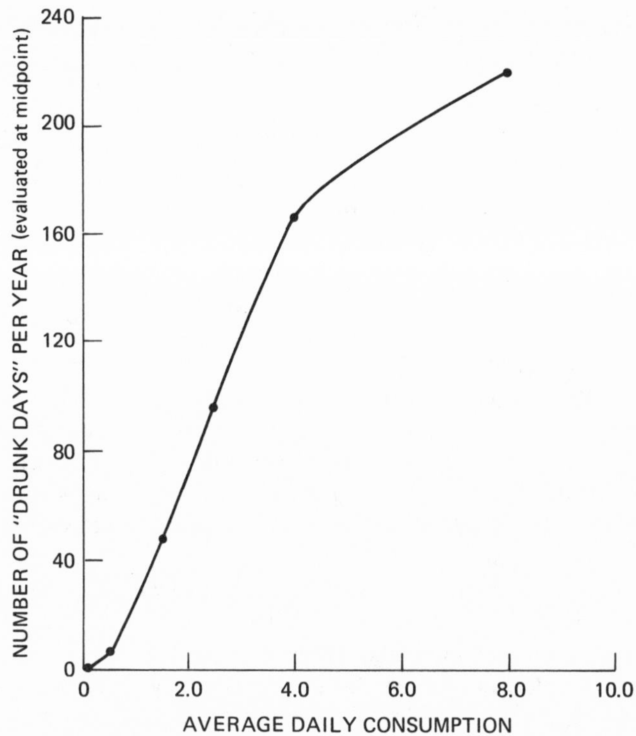

FIGURE 3 Relationship between average daily consumption and "drunk days" in a sample of Air Force men. For an explanation of the term "drunk day," see the note to Table 2.

Source: Based on data from Polich and Orvis (1979).

days one is likely to spend being drunk for at least a part of the day (see note to Table 2). A close inspection of these data reveals a more interesting and less obvious result. It appears that those who drink an average of 1.0-3.0 ounces per day have proportionately more "drunk days" than do very light drinkers or very heavy drinkers. Figure 3 illustrates this phenomenon by plotting the average number of drunk days per year against average daily consumption. The slope of this curve represents the "efficiency" with which cumulative consumption is converted to drunk days: the steeper the slope, the more drunk days are wrung out of a given level of consumption. Inspection suggests that this slope is at a maximum between 1.0 and 3.0 ounces of pure ethanol per day. The data suggest that moderate to heavy drinkers in this population tend to mass their drinking in binges rather than spacing their drinking evenly over time.

The degree to which drinking is massed rather than spaced by these drinkers is more typical of young men (not only those in the military) than characteristic of the rest of the population (Cahalan and Cisin 1968). Still, this pattern reminds us that drunkenness is different from total consumption and is not necessarily proportional to total consumption. To the extent that social consequences of drinking are associated with drunkenness rather than with total consumption and to the extent that the many moderate to heavy (but hardly alcoholic) drinkers get drunk out of proportion to their consumption, these drinkers will figure more prominently in causing and experiencing the social consequences of alcohol than one might expect.

Table 3 presents another useful calculation: the fraction of the total number of drunk days contributed by the different consuming groups. The very heavy drinkers (4 percent of the sample) account for about one-quarter of the total number of drunk days (26.1 percent). This proportion of drunk days is very large relative to their proportion in the community, of course. But if drunk days expose *anyone* to various risks, then three-fourths of the risks are borne by relatively unremarkable drinkers. In fact, even those consuming at the low rate of between 0.1 and 1.0 ounces per day generate about one-tenth of the total drunk days—more than a negligible contribution.

There are two reasons why the lighter-drinking categories have an appreciable proportion of drunk days. First, there are so many more people in the lighter-drinking categories than there are in the very heavy-drinking categories that even small individual contributions to total drunkenness will cause the lighter-drinking categories to rival the overall contributions of the smaller, very heavy-drinking group. Second, as we discussed above, the moderate to heavy drinkers are more "efficient" producers of drunk days than the very heavy drinkers. Both the number of these drinkers and their relative tendency to binge result in their contributing an unexpectedly large share of the total drunkenness experienced by this population.

In summary, the aggregate pattern of consumption in the United States has been relatively stable in the 1970s (Clark and Midanik 1980). Approximately one-third of the adult population is abstinent. Another third averages no more (most average much less) than 3 drinks per week and seldom gets drunk. The next fifth or so of the population averages about 2 drinks a day, and of this fifth about 1 in 12 gets drunk (8 drinks or more in a day) more than 6 times a year. The next tenth or so of the population averages 3 drinks or more a day and gets drunk more than once a week. The remaining 1-5 percent of the population (a more exact percentage is difficult to ascertain) averages 10 drinks or more per day

TABLE 3 "Drunk Days" Within Different Levels of Average Consumption in a Sample of Air Force Men

Annual Number of "Drunk Days"[a] Evaluated at Midpoint	Average Daily Consumption of Ethanol (Midpoint)					
	0.0	0.5	1.5	2.5	4.0	8.0
0	0	0	0	0	0	0
(N)	(477)	(710)	(34)	(3)	(3)	(1)
5	0	4,075	550	130	15	15
(N)	(0)	(815)	(110)	(26)	(3)	(3)
18	0	1,944	612	162	162	36
(N)	(0)	(108)	(34)	(9)	(9)	(2)
36	0	4,212	3,888	1,332	432	180
(N)	(0)	(117)	(108)	(37)	(12)	(5)
63	0	567	2,898	945	189	126
(N)	(0)	(9)	(46)	(15)	(3)	(2)
130	0	390	11,570	8,840	5,590	2,340
(N)	(0)	(3)	(89)	(68)	(43)	(18)
273	0	0	0	6,006	15,015	21,840
(N)	(0)	(0)	(0)	(22)	(55)	(80)
Total People (3,078)	477	1,761	421	180	128	111
Total Drunk Days (94,061)	0	11,188	19,518	17,415	21,403	24,537
Percentage		11.9	20.7	18.5	22.7	26.1

[a] For an explanation of the term "drunk day," see the note to Table 2.
Source: Adapted from Polich and Orvis (1979).

and spends most days of the year above the level of 8 drinks. If the bad consequences of alcohol consumption began when one reached a yearly average of more than 8 drinks per day, we could calculate that virtually all of these bad effects would be concentrated in the worst 5 percent (or less) of the drinkers. On the other hand, if a large share of these effects are linked to days of actual drunkenness rather than the average quantity of consumption, the adverse consequences would be distributed more evenly across the population of drinkers. The 5 percent of the heaviest drinkers would account for only about 25 percent of the drunk days. The people averaging 2 drinks or more a day would account for 60 percent of the drunk days.

Historical and Cross-Cultural Comparisons

While comparable information about the aggregate pattern of consumption for other times and other countries is scarce, we can say something about how typical our current position is by comparing our current average (per-capita) consumption with that of other times and other countries. While per-capita drinking statistics cannot fully substitute for precise information about the distribution of consumption practices, they have been shown to capture more information about the distribution of drinking than one might expect. In particular, the proportion of very heavy drinkers (consuming 5 ounces or more of pure alcohol daily) varies roughly as the square of average consumption in the adult population as a whole (Bruun et al. 1975). On this basis, the current position of the United States with respect to quantity of drinking is unremarkable in both historic and cross-cultural terms.

Figure 4 presents a record of the 150-year trend of average annual per-capita consumption in the United States. Generalizing broadly, there are four periods. First, through 1850, the trend was downward from the high volume (6-7 gallons per capita) of drinking characteristic of revolutionary America. Second, from 1850 through 1915, the level varied between about 1.75 and 2.75 gallons, increasing to the high point in the decade just prior to World War I. Third, between the world wars, the dislocations associated with the prohibition laws and then the Great Depression served to lower the quantity to well below the level in the Victorian era. Finally, after 1945 the prewar increase was repeated, to about 2.75 gallons.

For comparative purposes, Table 4 supplies a two-point time series for the postwar era for a number of industrial (mainly European) countries. While some caution must be used in comparing consumption statistics among countries, we may safely make two general observations. First, the United States is relatively unremarkable in its rate of mean alcohol use in this period: it is neither especially "dry" nor especially "wet." Second, the recent rise in U.S. consumption is also unremarkable: with little exception, increases have been common throughout the industrialized world and, for that matter, virtually every other country for which statistics have been compiled (Sulkunen 1976, Mosher 1979). In both the static and comparative perspectives, the United States is in the middle of the pack formed by the array of mean consumption figures.

Aggregate Patterns of Drinking Contexts

In addition to facts about quantities and rates of alcohol consumption, we would like to know something about the contexts in which Americans

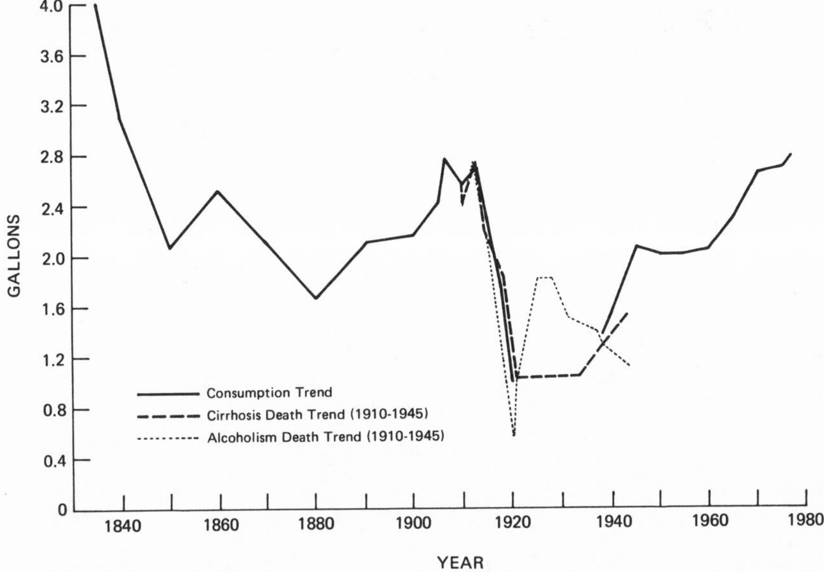

FIGURE 4 Time trends of U.S. consumption of absolute alcohol, cirrhosis death rate, and alcoholism death rate in the drinking age population, 1830-1977.

Note: Data points are less dense and less reliable prior to 1900. All estimates adjusted to reflect the population 15 years and older.

Sources: Consumption statistics from Hyman et al. (1980); alcoholism death trend from Jellinek (1947-1948); cirrhosis death trend from Bureau of the Census (1975). In estimating Prohibition-era consumption Jellinek favored use of cirrhosis-based estimators; Warburton (1932) used principally alcoholism death rates and production estimates.

do their drinking and how they have been changing over time. Do Americans do most of their drinking at home in the company of their families, with friends at private parties, or in bars and restaurants as part of a weekly (or daily) celebration? Do many people drink during working hours and has this changed over time? Similarly, how often do people combine drinking with driving, and has this been affected by the campaigns against drinking and driving?

We have already stressed one reason for our interest in these issues: drinking contexts have a very important effect on the kinds of risks and benefits drinkers experience as a result of their drinking. It is worth noting that contexts have this effect on risks through two different mechanisms. The context chosen by the drinker (or thrust on him or her by peers) defines the social and physical environment he or she will en-

TABLE 4 Changes in Apparent Consumption of Absolute Alcohol in 20 Countries by the Population Aged 15 Years and Over[a]

Country	Year	Total Absolute Alcohol (Gallons Per Capita)	Year	Total Absolute Alcohol	Percentage of Change
France	1955	6.8	1972	5.9	− 13
West Germany	1959	2.2	1974	3.9	+ 80
Belgium	1958	2.2	1973	3.8	+ 72
Switzerland	1956–1960	3.0	1971–1973	3.7	+ 23
Italy	1958	2.8	1973	3.6	+ 28
Australia	1957–1958	2.4	1972–1973	3.5	+ 44
New Zealand	1957	2.5	1972	3.3	+ 31
Denmark	1958	1.4	1973	2.9	+112
Canada	1958	1.9	1974	2.8	+ 51
Netherlands	1958	1.8	1974	2.8	+ 53
USA	1960	2.1	1975	2.8	+ 34
United Kingdom	1959	1.6	1974	2.7	+ 71
USSR	1957	1.3	1972	2.3	+ 83
Ireland	1959	1.1	1973	2.2	+100
Poland	1959	1.7	1974	2.1	+ 22
Finland	1957	0.8	1973	2.0	+157
Sweden	1958	1.3	1973	1.8	+ 43
Norway	1959	0.9	1974	1.5	+ 68
Iceland	1956	0.7	1973	1.3	+ 84
Israel	1959	0.1	1974	0.9	− 12

[a] Arranged in order of consumption in most recent year available. Compared time is the available date nearest to 15 years prior in each country. See Gerstein (in this volume) on problems of comparisons between countries. Here even a 1-year difference in date could substantially alter any percentage change. The country in which consumption was highest in 1974, Portugal, is absent due to missing data for the earlier period.
Source: Adapted from Keller and Gurioli (1976).

counter while intoxicated and thus the constellation of environmental risks. This is the mechanism we have so far emphasized. But another mechanism is at work as well. Substantial empirical evidence exists showing that the context exerts a separate, independent effect on the quantity and rate of alcohol consumption. Drinkers take cues from the people around them about how much and how fast they should drink. Sometimes (as is often the case when one drinks at home in the company of relatives) these influences tend to moderate drinking. Other times (as, for example, when one is out on the town with friends) the influences work to increase drinking. To the extent that the context shapes patterns of consumption and to the extent that the risks associated with given drinking episodes and practices depend on the *combination* of degrees of intoxication with given features of the environment, the context in which drinkers do their drinking can shape the risks confronting drinkers by affecting amounts of consumption as well as by differentially exposing them to environmental hazards. We are interested in drinking contexts because they shape some of the effects associated with drinking.

Despite the potential interest of the subject and despite frequent surveys and observational studies that capture some qualitative notions about U.S. drinking practices, we have surprisingly little quantitative information about how periods of drinking are distributed among given contexts and virtually nothing systematic on how the contexts are combined within a given individual's drinking history. Investigations of drinking contexts have been of two types. One type has focused on the micro question of whether the local contingencies of drinking contexts can shape short-term quantities and rates of alcohol consumption; the answer to this question has been a definite yes. Rates of drinking can be influenced in given contexts by such factors as price, the example of other drinkers in the same context, the relationship of the drinker to others in the same context, and the setting of the drinking (Babor et al. 1978, Harford 1979, Mass Observation 1970). The other type has sought to establish differences in the contexts of drinkers who drank different amounts and suffered different kinds and degrees of problems (Cahalan and Room 1974). Again, a fairly clear answer has emerged: heavier drinkers (and those drinkers with problems) use a much broader array of drinking contexts than light drinkers and are, in particular, more frequent users of bars and restaurants as drinking places (Clark 1977).

Such studies have their uses, but they fail to provide any clear answer to the questions of how different kinds of risks and benefits attach to different kinds of contexts and how the contexts of drinking episodes have been changing in quantitive density over time. All we have is a tantalizing suggestion that the context seems to shape alcohol consumption and that heavy drinkers seem to frequent on-premise establishments.

From our perspective, two important facts about drinking contexts are now available. The first is that sales for off-premise consumption now account for approximately three-quarters of total consumption—up from less than half at the end of World War II. This predominance of off-premise consumption shows up in survey results indicating that home is the most preferred drinking site and that drinking episodes at home or at friends' homes are much more common than drinking in bars and restaurants (Harford 1979). It is hard to know whether this is good or bad news, however, from the point of view of reducing the bad consequences of drinking. To the extent that this trend takes drinking contexts out of the direct control of state-regulated enterprises, we may

TABLE 5 Where, When, and With Whom People Drink (Drinking Contexts), by Age of Drinker

	Weekdays				Weekends					
	Private		Public		Private		Public			(Total
Age	Relatives	Friends	Relatives	Friends	Relatives	Friends	Relatives	Friends	Total*	Events)
18–25	11%	14%	2%	16%	11%	20%	4%	23%	100%	2,107
26–35	17%	12%	4%	13%	18%	16%	7%	13%	100%	2,172
36–48	28%	6%	3%	10%	25%	10%	5%	13%	100%	2,513
49–62	33%	5%	6%	5%	29%	9%	6%	6%	100%	1,414
>63	43%	7%	2%	3%	33%	8%	2%	4%	100%	1,013

*Addition to less or more than 100 percent is due to rounding error. Solitary drinking episodes were quite rare and were coded under "friends" (none present).
Source: Adapted from Harford (1979, p. 171).

judge that public capacity to shape drinking contexts is being reduced. On the other hand, privately created drinking contexts may tend to moderate drinking and insulate the drinker from certain dangerous consequences of drunkenness more effectively than publicly sanctioned drinking opportunities. To the extent that this is true, the shift toward off-premise sales and consumption is a good sign, but not enough is now known about this change (Partanen 1975).

The second fact is that preferred drinking contexts change dramatically with age (Table 5). Drinking for all ages is heavily concentrated on weekends—about half of all drinking episodes are crammed into the relatively short weekend period—but this concentration tends to diminish with age. In addition, young people tend to prefer public locations with friends, while older people do more of their drinking in private with relatives. Again, because we do not know exactly how these different contexts are linked to different risks, these findings are difficult to interpret. One might hypothesize that this aggregate pattern makes young people more vulnerable to risks of auto accidents, fights, and other disturbances linked to public intoxication. This is speculation, however; the data presented in Table 5 are useful primarily to indicate the kind of information and data that must be developed to discover current aggregate patterns of drinking contexts and to explore the relationship among them and the risks and benefits of drinking.

We think a major opportunity exists to do additional research on drinking contexts. In fact, we think much data could be reanalyzed to discover important facts about aggregate patterns of drinking contexts. The crucial questions involve how the contexts are linked to certain kinds of risks and whether the contexts are changing in more dangerous or safer directions.

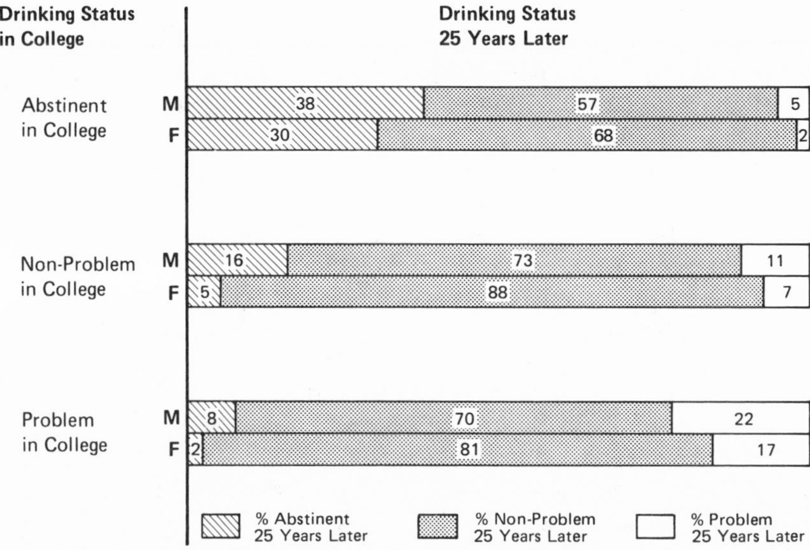

FIGURE 5 Reported drinking status of men and women in college and 25 years later. *Source*: Adapted from Fillmore et al. (1977).

STABILITY OF INDIVIDUAL DRINKING PRACTICES

A crucial question affecting the feasibility of controlling the alcohol problem by shaping drinking practices is simply how stable these practices are. It is common to think of drinking practices, particularly those that cause problems, as relatively deeply rooted and difficult to alter. There is evidence, however, that drinking practices and associated problems are not so constant. A major source of evidence on changing drinking practices comes from a study in which individuals surveyed about their drinking practices in college during 1949-1952 were resurveyed 20 and 25 years later (Fillmore 1974, 1975; Fillmore et al. 1977). Figure 5 presents one set of the results of the follow-up. The figure indicates a clear but modest relationship between drinking problems in college and drinking practices 25 years later. Certainly, the probability of later problems is considerably higher, in terms of relative risk, for problem drinkers in college than for others. But the vast majority of problem drinkers in college were not problem drinkers 25 years later, just as the majority of those who were abstinent in college were not problem drinkers 25 years later. Thus, while the drinking patterns of early adults do influence subsequent patterns, the influence is far from decisive. Such differences across time are apparent during shorter periods of a few years as well (Clark and Cahalan 1976). Cahalan et al. (1969) found that retrospective reports of changes in consumption were common in an adult national sample, especially among men. Two-thirds of all male drinkers recalled broad variations in their quantity of consumption at different periods. The least lifetime variation was reported by abstinent women, 80 percent of whom were consistently so.

Evidence about the volatility of drinking practices of clinical problem drinkers comes from follow-up studies of drinkers who have been in treatment. In a 4-year follow-up study, Polich et al. (1980, p.168) found that "although there is frequent improvement [in terms of drinking practices], there is also frequent relapse and much instability." Apparently, the drinking practices of even alcoholic drinkers are not etched permanently into their lives: there are good periods and bad periods lasting for several months or years. Moreover, the changes do not bear a necessary relationship to treatment.

The fact of impermanence in drinking practices has several important policy implications. It suggests that effective influence might be brought to bear on current drinking practices: they are not so deeply embedded in personal characteristics that it is fruitless to think of trying to change them. On the other hand, it suggests that there are few permanent victories in the efforts to shape drinking practices. Nor can people with serious problems be treated then discharged with the expectation that the drinking (or nondrinking) practices of more than a few will retain for long or without further ado the pristine quality they may have had at the end of treatment. Apparently, the factors that shape drinking practices are varied, transient to some degree, and under the control of many different parts of society. They are neither permanent features of individual character nor monopolized by government agencies.

CAUSES OF DRINKING PRACTICES

To this point we have treated drinking practices descriptively, asking what they are rather than posing the issue in explanatory terms: i.e., why do people drink and why do they drink in these ways? The only certain answer to these latter questions is that no single explanation is adequate. The complexity of causes of drinking is indicated by Sulkunen's (1976, 1978) discussions of the three "use values" of alcohol: as a food and beverage, an intoxicant, and a symbol of sociality.

As a food, alcohol is itself a carbohydrate; most alcoholic beverages contain additional nutrients; these beverages are generally quite resistant to spoilage or infestation; and they are free of most water-borne diseases. Thus, fermented products have a long history as useful, durable basic commodities. People make and use them for reasons as diverse as biological thirst and hunger, personal taste, and cuisinary custom (Gastineau et al. 1979).

In addition, however, alcohol is an intoxicant, and its multiple effects on psychological mood and alertness are well known. Alcohol is variously a stimulant and depressant, euphorigen and soporific, irritant and anxiety reducer—depending on a myriad of factors such as dose and schedule of use, individual metabolism, personality factors, and situation. Alcohol, like other intoxicants, can produce such dependency phenomena as persistent search behavior, withdrawal, relapse, and loss of control, although these terms are controversial (see the forthcoming report on Commonalities in Substance Abuse and Habitual Behavior from the Committee on Substance Abuse and Habitual Behavior, National Research Council, National Academy of Sciences).

Finally, alcohol carries symbolic values that are independent of its nutritional value or its intoxicating effects—for example, the religious symbolism of communion and sabbath wines; the celebration that calls for champagne; the slave-trade heritage of "demon rum"; and the laughter in numberless jokes about drunkenness.

The drinking practices we observe in society are shaped by a variety of factors—those noted above and others—acting simultaneously. Some factors are individually based (e.g., genetic proclivities, psychological and developmental needs); some are rooted in intimate, informal social processes (e.g., family, ethnic, or religious traditions; the practices of one's spouse; the rituals of one's working companions); some are based on the marketing efforts of alcohol producers; and some are managed explicitly by government (e.g., taxes on alcohol, restrictions on availability, laws regulating drinking conduct, and a variety of educational messages about drinking). We assume that these factors operate with varying degrees of force and varying degrees of persistence. Some factors (such as general personality traits and restrictions on availability) may operate continuously, with diffuse rather than precise impact on drinking practices. Other factors (events such as an illness, a divorce, or a firing) may operate with great force in the short run, with relatively transient effect. Still others (such as strong biological aversions or family drinking traditions) may be both powerful and persistent. Unfortunately, little is known about the relative power of the different kinds of factors shaping drinking practices, and discussion of such factors draws heavily on speculative assumptions (Institute of Medicine 1980). Moreover, expert views on the most important factors seem to depend more on disciplinary training than on persuasive facts currently in hand. Alcohol policy would benefit from the identification of those factors that are powerful but can be efficiently influenced by government—a difficult combination to locate.

Although we have stressed observations of the impermanence of drinking practices over individual drinkers' lifetimes, it is clear that in the short run, at least, drinking practices do have some inertia. Past practices exercise important influence over future practices, particularly in the short run. This is true in part because some of the factors shaping practices are powerful, persistent, and not easy to manipulate or escape, and in part because people generally tend to maintain in their lives habits and routines that work. In general, then, we hypothesize that the longer a given practice has continued, the more meaning the individual attaches to that practice; and the more stable the social environment surrounding the individual is, the more stable the drinking practice will be. On the other hand, we assume that drinking practices are to some degree modifiable by changes in the experiences of the drinker. Often the experiences that lead to modification of drinking practices are the result of small-scale, private initiatives: e.g., a husband begins encouraging his wife to drink more often with him in the evening; an employer complains about tardiness and threatens firing; a favorite neighborhood tavern closes. At other times, the experiences that cause individuals to change can be the result of broader efforts of social management (e.g., a persistent and persuasive mass media campaign designed to make people more self-conscious about drunken driving, a tax increase on alcohol, or a reduction in availability). In any event, we assume that drinking practices can change—first as immediate circumstances change, then as persisting new conditions force continued adjustments and the development of new practices supported in part by the surrounding circumstances and in part by recent experience.

THE PROBLEM OF ATTRIBUTING EFFECTS

It is one thing to determine how much, how often, and how people drink. It is quite another to pinpoint how much the different dimensions of drinking practices and different characteristics of the environment contribute to the important social effects of drinking. For the purpose of informing policy judgments, the question of attribution seems fundamental: we need to know what interventions targeted at different aspects of drinking practices or different features of the physical and social environment can be expected to yield in terms of positive and negative effects.

Unfortunately for this purpose, a main conclusion of this chapter is that a variety of factors shape the effects associated with drinking. Drinking is a major factor, but the degrees and frequency of intoxication as well as cumulative consumption must be considered to develop any clear sense of the kinds of risks and benefits that a drinker may encounter.

In addition, the contexts in which drinkers do their drinking play an important role in creating effects, in part by differentially exposing drinkers to the hazards of the environment and in part by exerting an independent influence on quantities and rates of alcohol consumption. Finally, general features of the envirnments into which drinking may be fitted by the choice of context also play a role in producing different kinds of effects. The sheer complexity of this conception defeats any simple effort to attribute quantitative risks or benefits to specific factors.

What our analysis suggests that is useful in guiding policy choices is a more general idea: namely, that various avenues for reducing the bad consequences of drinking exist, and that pursuit of each avenue results in slightly different gains (and losses) in terms of the dimensions of Figure 2. This suggests that policies toward alcohol use require complex choices about the relative importance of different consequences as well as choices about which instruments to select from a large and diverse array—some targeted at cumulative consumption, some at drunkenness, some at contexts, and some at characteristics of the social and physical environment.

A separate question about the attribution of risks and consequences of drinking can be usefully addressed: how the risks and consequences of Figure 2 are distributed across the population of drinkers. The adverse consequences are doubtlessly at highest concentration in the small fraction of the drinking population who drink a lot, get drunk frequently, and do so with little respect for time and place. This expectation has justified the current focus of policy on alcoholics and chronic, intensive users of alcohol. What is less obvious, however, is how much of the total social problem described in Figure 2 lies beyond this small fraction of the heaviest drinkers.

Current answers to this question are preliminary but revealing. Table 6 presents data from the Air Force survey showing the fraction of people in given drinking patterns who experienced two or more serious incidents associated with alcohol in the past year. As we would expect, the fraction of people having these problems increases noticeably with given levels of consumption. But if we calculate the fraction of all the people experiencing two or more serious incidents related to alcohol who were in the highest consumption groups, we find that the heaviest drinkers account for only 24 percent of the population having problems. The reason is that there are so many more people in the lower-consuming groups that even small proportions in trouble can swamp the larger fraction of people who are in trouble from the higher-consuming groups.

These qualitative results also stand up when we look at data from national surveys conducted in 1967 and 1977. Table 7 shows the fraction of people at different levels of consumption who experienced certain

TABLE 6 Alcohol Incidents by Total Alcohol Consumption Level in a Sample of Air Force Men

Average Daily Consumption Level in Ounces of Alcohol	Total Number	Number of People with Two or More Serious Incidents	Percentage of All People with Two or More Serious Incidents
0.1–1.0	1,761	37	26.8
1.1–2.0	418	33	23.9
2.1–3.0	181	13	9.4
3.1–4.0	79	14	10.1
4.1–5.0	49	8	5.8
5.0+	111	33	23.9
TOTAL	2,599	138	99.9

Source: Adapted from Polich and Orvis (1979).

kinds of problems. Again, it is clear that the chance of experiencing difficulties associated with drinking increases as one drinks more. But Table 8 shows how much each portion of the drinking population contributed to people who have problems associated with drinking. Again, it turns out that the heavy drinkers typically account for less than half of the people with problems.

Thus, in attributing the consequences of alcohol consumption to various causal factors and types of drinkers, we can make two important observations. First, alcohol consumption itself is important as a *sufficient* cause for some effects. Other consequences of drinking can be attacked by reducing drunkenness, motivating people to change drinking contexts, or changing the external environment as well as by reducing cumulative individual consumption. Second, while chronic drinkers with high consumption both cause and suffer far more than their numerical share of the adverse consequences of drinking, their share of alcohol problems is still only a fraction—typically less than half—of the total. Alcohol problems occur *throughout* the drinking population. They occur at lower rates but among much greater numbers as one moves from the heaviest drinkers to more moderate drinkers.

CONCLUSION: AVENUES FOR AFFECTING ALCOHOL PROBLEMS

The main conclusions of this chapter are that controlling alcohol and treating relatively small numbers (even as many as 1 or 2 million) of the most troubled drinkers are not the only, and may not be very effective, ways of coping with alcohol problems.

We have developed these conclusions largely by analyzing the complex

TABLE 7 Percentage of U.S. Population Experiencing Different Problems Associated with Drinking, by Level of Consumption

	Health Prob-lems	Bellig-erence	Prob-lems with Friends	Prob-lems with Spouse	Job Prob-lems	Accidents/ Legal Problems	Total Number of Respon-dents
1967 National Survey							
Light drinkers	6	1	*	*	*	*	394
Moderate drinkers	8	5	2	1	4	1	499
Heavy drinkers	14	20	8	5	11	0	133
1979 National Survey							
Light drinkers	1	1	1	*	1	*	444
Moderate drinkers	5	10	4	2	6	2	597
Heavy drinkers	10	31	6	6	24	7	124

* Less than 0.5 percent.
Source: Clark and Midanik (1980).

TABLE 8 Contribution of Different Consuming Groups to Reported Experience of Problems

	Health Problems	Bellig-erence	Problems with Friends	Problems with Spouse	Job Prob-lems	Accidents/ Legal Problems
1967 National Survey						
Light drinkers	26%	9%	*	*	*	*
Moderate drinkers	51	45	42%	30%	54%	100%
Heavy drinkers	23	46	58	70	45	0
Total percentage	100	100	100	100	99	100
(Total number)	(83)	(56)	(19)	(10)	(33)	(6)
1977 National Survey						
Light drinkers	13%	4%	12%	5%	9%	0%
Moderate drinkers	59	59	66	62	49	55
Heavy drinkers	28	37	22	33	43	45
Total percentage	100	100	100	100	101**	100
(Total number)	(46)	(103)	(32)	(21)	(70)	(20)

* Less than 0.5 percent.
** Totals do not add to 100 percent due to rounding error.
Source: Clark and Midanik (1980).

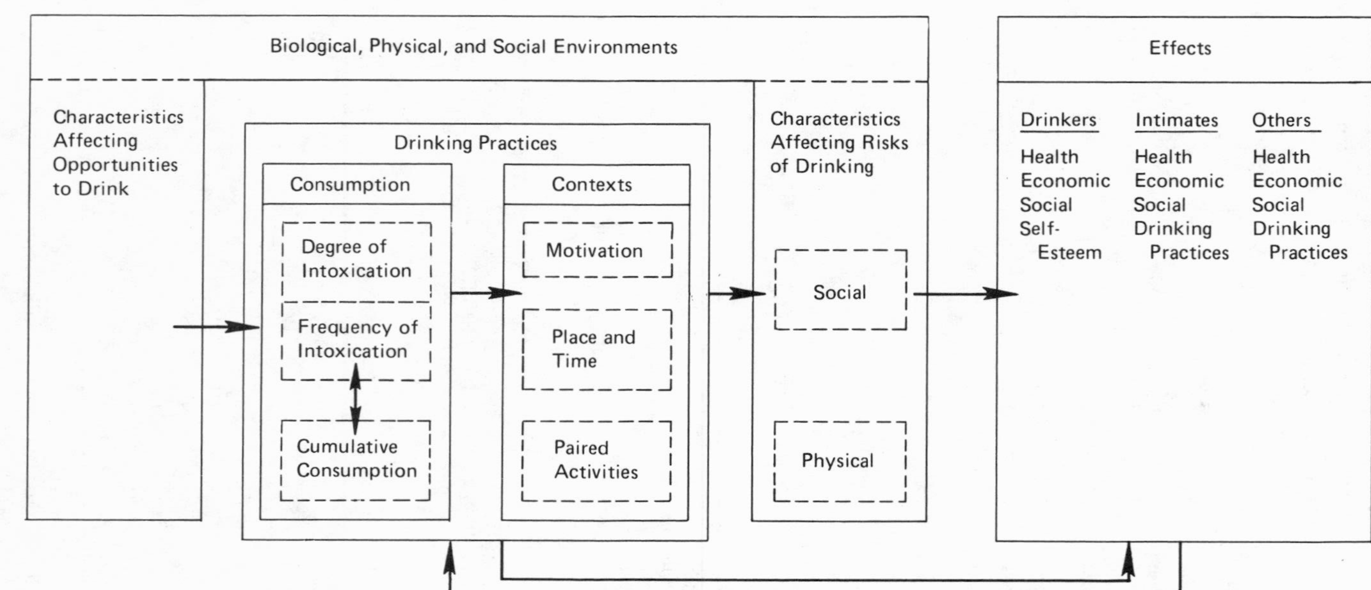

FIGURE 6 Schematic view of the causal system underlying alcohol problems. For a fuller explanation of the "Effects" column, see Figure 2.

Note: This figure is an expansion of Figure 1.

causal machinery that linked drinking with the important consequences of drinking. Figure 6 presents a schematic view of our conception of the broad causal system. In this scheme, drinking is incorporated into the concept of a drinking practice, which specifies drinking in several dimensions (degree of intoxication, frequency of intoxication, and cumulative consumption) and includes commonly chosen contexts for drinking as well. Drinking practices mediate the effects of drinking but, for many effects, do so only in combination with some features of the physical and social environment. Note that the environment appears in this scheme twice: as something that combines with a given drinking practice to produce consequences (see Figure 2) and as something that shapes drinking practices by making alcohol available, by defining attitudes toward drinking practices, and so on. Often, the same features of the environment play a role in both shaping practices and producing consequences; still, one can analytically distinguish among the different causal roles. Contexts appear in this scheme as a description of how drinkers incorporate pieces of the environment into their drinking activities. We assume that the contexts exert an independent influence on consumption and that the environment influences the frequency with which certain contexts appear in drinking practices.

The sheer complexity of this system argues for the existence of alternative routes for controlling the dimensions and shape of the alcohol problem, apart from reducing total consumption. At a minimum, one can conceive of reducing consumption levels, shaping drinking practices, and changing features of the physical and social environment. Although the theoretical existence of these alternative approaches can hardly be taken alone as a strong argument supporting their use, their existence does stir our curiosity about whether they might be exploited.

The conclusion that treating alcoholics is not sufficient response to the problem is based on an empirical result: namely, that the problems associated with drinking are distributed rather broadly through the population of drinkers. It is true that the small proportion of the most troubled drinkers cause and suffer far more than their numerical share of alcohol problems; it is also true that if all the clinically diagnosable alcoholics were to stop drinking tomorrow, a substantial fraction of what we understand as alcohol problems would still remain. The general alcohol problem, then, is more than what the standard view of "alcoholism" suggests, because it affects many more of us than we are accustomed to believing. The problem may have less to do with alcohol than the alcoholism concept suggests and have much more to do with how alcohol is fitted into daily life and the external environment.

3 Perspectives on Current Alcohol Policies

The simplifying conceptions of alcohol policy presented in chapter 1 and the diverse avenues for affecting the problems outlined in chapter 2 provide a broad perspective on conceivable alcohol policies. It is useful to place current policies toward alcohol in this context to see how many of the conceivable alternatives are now being exploited and what pieces of the problem lie outside current efforts.

CURRENT ALCOHOL POLICIES AND INSTITUTIONS

Characterizing current alcohol policies with precision is no simple task; the very term "policy" is ambiguous. It may refer to publicly stated goals, to specialized institutions predominantly concerned with alcohol, or to the net effect of all institutions. Another difficulty is that in the complex institutional structure of the United States, many different policies are being pursued independently by many separate institutions—each with independent purposes, resources, clientele, and authority. Still, a survey of the current institutional setting reveals three strategically important features.

First is the heavy regulation of the production and distribution of alcohol. The federal government, operating through the Bureau of Alcohol, Tobacco, and Firearms (BATF) of the Department of the Treasury, collects approximately $5 billion per year from taxes on alcohol. The 51 state ABC boards exercise substantial influence over the price and availability of alcohol in off-site premises and, through regulation, shape some aspects of the contexts of drinking in on-site premises. These institutions exert influence on alcohol problems despite the fact that they are not now explicitly being managed for this purpose. The objective of BATF's operation is largely to collect revenue and to ensure the legal and orderly operation of the commercial system for distributing alcoholic beverages. Thus, although substantial authority over the supply of alcohol is vested in these institutions and they are both active and influential in shaping drinking practices (for good or ill), they are not now being managed for this purpose.

The second strategically important characteristic of the current institutional setting is the continued existence of powerful cultural beliefs and attitudes that tend to moderate drinking practices in the United States. It is easy to forget that one-third of the adult population of the United States is abstinent and another third drinks very little over the course of a year. These drinking practices, no less than the heavier drinking practices of the remaining third of adults, are created and sustained by powerful, deeply rooted cultural and family traditions. The majority disapprobation of immoderate drinking is still embodied in state and local statutes prohibiting drinking by minors, driving while drunk, and public drunkenness. These laws have recently been subject to liberalizing trends. Age restrictions on drinking have been lowered, convictions for drunken driving have been difficult to secure, and, in perhaps the most significant of the changes, public drunkenness has been widely "decriminalized." But the continued existence of these laws and the very moderate drinking practices of most of our population testify to the continued vitality of public intolerance of immoderate or reckless drinking.

The third feature of the institutional setting pertaining to alcohol control policies is the National Institute for Alcohol Abuse and Alcoholism and the network of state agencies and treatment programs partially spawned and supported by NIAAA's most influential client groups. This institutional nexus is important in part because it absorbs a substantial volume of economic resources. Even more important, however, NIAAA is the most prominent government agency concerned primarily with alcohol problems. It is therefore the primary institutional focus for conceiving, articulating, and, to a degree, managing the government's explicit efforts to cope with alcohol problems. For broad and innovative thinking to be done in this area and for managerial efforts to be launched to exploit underutilized institutional capacities, NIAAA must be the lead agency. Since NIAAA is the primary custodian of alcohol policies at the level of goals and current understandings of the nature of the problem, it becomes important to know how NIAAA understands the problem and the appropriate responses.

The answer to this question is difficult to pin down. In the articulated aspirations of representatives of NIAAA and in its organizational structure, there is a variety of conceptions and approaches. At the level of agency operations, however, it seems clear that the prevailing underlying assumption is that the general alcohol problem is largely one of alcoholism and that the most effective way of dealing with it is to locate and successfully treat alcoholics. Of course research on alcoholism may well throw useful light on drinking problems of a less intensive or less chronic nature, but most of NIAAA's funds, institutional apparatus, and managerial attention have been devoted to treatment delivery, not research programs. NIAAA searches for more effective modes of treatment and broader methods for identifying and recruiting "problem drinkers" or those likely to become alcoholics. Thus, it has tended both to reflect and sustain the governing idea that alcoholism is a disease for which treatment is the most appealing policy.

The overall institutional picture, then, is one in which a few major institutions play important but largely unintentional roles in shaping drinking practices; and in which the government institution that is most responsible for shaping our social response to alcohol problems reflects the current conception of the problem as largely one of alcoholism. The juxtaposition of this institutional setting with the analytic conception of the problem developed in chapter 2 prompts an important question: Is full advantage being taken of the current institutional capacity to cope with the diverse aspects of alcohol problems, or is the current concept of alcoholism narrowing our conception of the appropriate objectives of alcohol policy and diverting society from some important policy alternatives? More specifically, the question is whether the regulatory apparatus and cultural commitments to temperate drinking could be utilized more effectively to ameliorate alcohol problems, and if so, whether current policies (and conceptions of the problems) should be diversified to take better advantages of these institutional opportunities.

THE *PRIMA FACIE* CASE FOR PREVENTION

We believe that a significant opportunity for diversifying current policies does exist. Specifically, there is both a need and an opportunity for alcohol policies to be diversified to prevent alcohol abuse and problems in the general population of drinkers. This argument rests on three observations.

First, as chapter 2 indicates, there are clear limits to what treatment policies can accomplish by themselves. Treatment policies can be justified by a humanitarian commitment to care for the casualties of drinking and by the contribution (whatever its size) that treatment makes to prospects for recovery. But it is also clear that bad consequences of drinking are strewn so widely among the general population of drinkers—including many for whom treatment would be entirely inappropriate—that treatment alone could at best deal with only a portion of the important social consequences of drinking.

Second, the analysis of existing institutions suggests that while there is little institutional focus for prevention efforts, there is a surprising amount of institutional capability. The regulatory apparatus governing commercial availability remains intact but underutilized for the prevention of alcohol problems. Social sentiment in favor of moderate drinking remains strong and ensures a supportive climate for broadly targeted efforts to shift drinking practices in desirable directions. Thus, much of the institutional basis for prevention programs already exists.

Third, it seems clear that treatment and prevention can be seen as complements to one another rather than competitors. There is ample evidence indicating that the number of chronically dependent drinkers is not a fixed number and is importantly influenced by general drinking practices and opportunities (Bruun et al. 1975). Moreover, efforts to protect lighter drinkers from the worst consequences of occasional episodes of drunkenness could also protect heavier drinkers in their more frequent drinking episodes. In effect, prevention efforts might usefully supplement treatment efforts by decreasing the size of the population needing treatment and by shielding the treatment population in ways that treatment could not accomplish. Furthermore, treatment and prevention programs do not seem to compete for resources. Most prevention programs do not cost the government a great deal of money. They involve taxation, regulation, and the design of persuasive communications that could be used over and over again. While not all of these things are inexpensive to the society as a whole, most of them are relatively inexpensive in federal budget terms. Besides, the possibility of important positive interactions between treatment and prevention would make it wise to invest in both, even if they were equally expensive.

Thus, it seems sensible to consider diversifying alcohol policies by exploiting a variety of opportunities to prevent drinking problems in the general population. Note that in considering diversification, we are thinking, not of an abrupt shift, but of an incremental addition of new capacities that can be enlarged if they turn out to be successful in handling some aspect of the problem. As discussed below, existing evidence about the efficacy of prevention policies is not so strong and compelling as to justify "great experiments," yet, opportunities for dealing with alcohol problems through prevention programs seem sufficiently attractive to merit close investigation. In the second part of the report, we look at the available empirical evidence about the efficacy of prevention programs. It is first necessary, however, to become more precise about what we mean by prevention policies.

THE DEFINITION OF PREVENTION POLICIES

The general idea of a prevention policy toward alcoholism and alcohol abuse is at once too narrow and too broad for our purposes. It is too narrow because it focuses attention on the harms to be limited and obscures the fact that benefits of alcohol use exist and should be protected. It is too broad because all instruments of alcohol policy can be understood as devices to prevent some bad effects of drinking. Treatment programs, for example, can be understood as devices for preventing cirrhosis from reaching terminal proportions, forestalling suicides, and preventing accidents by motivating chronic alcoholics to reduce their alcohol use and change other aspects of their lives. To serve our purposes, then, we need a more precise definition of how we mean prevention policies to be understood.

Perhaps the most convenient way to develop the concept of prevention policies is to distinguish the policies we have in mind from clinical alcoholism treatment programs. From our point of view, such programs (as a class) have two important characteristics. First, they operate through a continuing face-to-face relationship with discrete, identified individuals. Individualized attention begins at intake, when a detailed client history is taken; continues with a treatment plan calling for continuing, intensive involvement; and becomes the basic mode of operation as the progress of a client is continuously monitored on a variety of dimensions. The services and supervision that operate within this individualized relationship are the heart of treatment programs. Second, treatment programs tend to be directed at the relatively small proportion of drinkers who have experienced the most severe problems with their drinking and have either asked for or been referred to treatment. This concentration occurs in part because treatment is sufficiently expensive that its use is reserved for those who are in the most serious trouble. The distinctive features of treatment, then, are individualized services and a focus on drinkers who are already in trouble.

One can logically imagine extending such individualized services to people who are on some grounds thought destined for trouble as well as those who are already in it. Examples of such at-risk or high-risk populations are those beginning to experience problems on the job or in school related to excessive or ill-timed drinking; young populations that are just starting to drink but have sufficient problems in other areas of their lives (for example, an alcoholic parent) to induce concern about their futures; and demographically defined groups (for example, native Americans) that have previously shown especially high rates of clinical or other problems with alcohol. Such programs do in fact exist, and are generally called early detection and intervention, occupational programming, or preventive services. Moreover, they do have a preventive aspect in that they are directed at populations that are not yet deeply in trouble with alcohol. On the whole, however, these are really treatment programs with different kinds of outreach mechanisms. They still depend on personalized services. Although not confined to people who are already in severe trouble related to or dependent on drinking, the ability of such efforts to reach people who may be potentially at risk will be strongly limited by reliance on personalized services that are expensive, vaguely defined, and demanding of the client and on detection capabilities that are still at the basic research stage (Institute of Medicine 1980).

The prevention policies on which this report focuses are different from either treatment programs or preventive services. They include: taxes on alcohol, regulation of the availability of alcohol, liability rules that would make bartenders or hosts more responsible for the safety of their guests, improved enforcement efforts against drunken driving, education programs that present a view of unsafe drinking practices and encourage bystanders to comment on unsafe practices, and the design of workplaces and homes to make them safe for people who are intoxicated. This is a very heterogeneous set of policies, but they are all alike in the following respect: *They are all policies that operate in a nonpersonalized way to alter the set of contingencies affecting individuals as they drink or engage in activities that (when combined with intoxication) are considered risky*. In effect, they are designed to manipulate conditions that will influence either patterns of drinking or the consequences one can expect from any given drinking practice. The differences between these programs and treatment are three.

First, the prevention policies operate on a fundamentally nonpersonal basis. None of these policies depends on a continuing personalized relationship between a program and a drinker. Instead, they operate through the remote manipulation of a relevant set of incentives and contingencies: the terms and circumstances under which alcohol is available, the attitudes of people surrounding the drinkers, and the benignity of the physical and social environment toward drunkenness.

Second, the programs operate generally throughout the society. The incentives and contingencies are established for everyone. They become activated when one begins drinking or engaging in behavior that would be risky if one had been drinking. Thus, drinkers in many patterns of consumption are affected, in addition to people who may become clumsy and inattentive as a result of drunkenness, fatigue, senility, or anger.

Third, with the one exception of education programs directed at drinkers themselves, the programs that we focus on do not operate by seeking major changes in the personalities and orientations of drinkers. Instead they seek to alter the set of opportunities, risks, and expectations that surround drinkers in society. To be sure, the orientations of individuals may change as they adjust their attitudes to the altered conditions surrounding drinking, but the attitudes and drives of drinkers are not the immediate target. Prevention programs seek to avoid the intrusions associated with therapy and concentrate on managing the set of external contingencies that operate on drinkers as they drink, pair their drinking with other activities, and accept a variety of risks associated with their drinking from the external environment. In fact, even education programs avoid the most penetrating kinds of intrusions simply because, although they carry personal significance for drinkers, they do not seek out specially identified individuals and track their behavior. Thus, the prevention programs that are the focus of our study avoid sustained personalized efforts to alter the individual attitudes and drives of individual drinkers.

These characteristics seem to give the prevention policies described above some important advantages. First, because they operate impersonally without a continuing relationship between an agency and an identified individual, they do not create the problems that may arise from "labeling" individuals as deviants. To be sure, a drunken-driving arrest is an adverse personal effect, and encouraging third parties to help regulate the drinking of others may exacerbate social conflict over drinking (as similar efforts have done in the area of smoking). Yet prevention programs involve many fewer of society's explicit labeling activities than treatment programs do.

Second, because these programs broadcast their effects generally throughout society, they may be a low-cost way of reaching the general population of drinkers. A tax increase affects all drinkers in proportion to their consumption. An effective media program can help educate large numbers of people at low per-capita costs. Safer highways protect everybody who drives whenever they are driving—not just a few problem drinkers. Of course, the fact that large numbers of people can be reached at a relatively low per-capita cost does not necessarily make such programs valuable. If they produce no important changes in the pattern of drinking or in the set of consequences that result from any given pattern of drinking, the lower unit cost is hardly a virtue. If, however, as we think the arguments and evidence of chapter 2 warrant, it is important to reach a large population of drinkers, these kinds of programs represent an inexpensive way to start. The question of whether any of them are likely to produce any important, beneficial results is the dominant focus of the second part of this report.

Before looking at this question, however, it is important to consider some basic normative and pragmatic arguments against such broad efforts to shape drinking practices. Unless these basic objections can be overcome or at least placed within limits pending the receipt of evidence, the attractiveness of these general prevention policies is in jeopardy.

OBJECTIONS TO PREVENTION POLICIES

The most fundamental objection to the idea that government should seek to shape drinking practices is that such action may violate common understandings about the proper role of government in a free society. In this view, government may place constraints on individual liberty only when one person's actions materially affect the welfare of another who is unable to protect himself or herself from the unwanted intrusion; voluntarily assumed personal risks are beyond government bounds. Moreover, our legal system is designed largely to restrain government from acting until after the fact of intrusion has been established, not on mere presumption or potential. Since drinking produces adverse external effects only occasionally, general drinking practices seem for the most part inappropriate matters for official concern. Thus, any governmental effort to influence drinking practices may be seen as an inappropriate paternalistic restriction on freedom—irrespective of how gentle or heavy-handed the intrusion.

One can disagree with this conclusion at several points in the argument. First, individual freedom is not the only premise defining the proper role of government. There are other (equally venerable) notions of politics in which the government is called on to enhance the general welfare, promote the spread of knowledge, and encourage civil behavior among its citizens as well as guarantee various liberties. Alternatively, one could cite the extent of harmful, irreversible, external effects of drinking practices that clearly do warrant government interference (e.g., reckless drunken driving or piloting a plane while under the influence of alcohol). Thus, the basic principle that government should constrain private conduct only when that conduct affects others in important ways need not be, and has not been, an absolute bar to governmental efforts to shape drinking practices.

Beyond this fundamental normative issue, several more pragmatic issues arise. The most important rest on the judgment that a representative government is unable to exercise any independent influence on drinking practices. This argument appears in varied forms. Sometimes it is argued that government influence on drinking can do nothing more than reflect prevailing practices. When we observe a change in government policy, it is not because someone has decided on rational grounds that a change is desirable, but because large cultural forces are at work. Government actions are corks bobbing in great cultural tides. If by some chance a government policy were established at odds with prevailing practices, it would work substantial mischief—making deviants or criminals of the large groups in the population who stubbornly refused to alter their habits, fueling the development of illegal operations to meet the needs of those who are now outlawed, and generally sacrificing the legitimacy and effectiveness of government. But then, the policy would shortly collapse in the face of continued private intransigence. The implication of these judgments is that since government policy must mirror prevailing practice, it is fruitless to rely on the government to shape those practices. Such efforts will be at best redundant and at worst self-defeating.

A slightly less extreme view holds that government may indeed influence current drinking practices, but in unexpected and potentially dangerous ways. Since current drinking practices are supported by informal but nonetheless strong and long-standing networks of beliefs, relationships, and other social practices, and since drinkers will seek to maintain their current patterns against both new inconveniences and the risks of deviance, the best the government can hope to accomplish is to set in motion many modest but unpredictable changes in actual drinking practices. It is certain that at least some people will try to maintain centuries-old practices against new prescriptions and will suffer inconvenient and painful losses in some dimensions of their lives as a result of the new policies. With such an uncertain (and potentially bad) result, it is foolish for the government to act.

The central insight in these views is that informal social controls are much more powerful in shaping collective drinking practices than the government could ever be and that it is dangerous to have government conceptions of drinking be wildly at odds with the drinking practices that emerge from the informal system. Clearly, there is wisdom in these views. But the general observation nevertheless implies that properly chosen governmental actions could be valuable. It leaves open the possibility, for example, that government prescriptions that were *slightly* more constraining than current practices, that put a gentle brake on dangerous shifts in current practices or accelerated favorable trends, might operate to shape drinking practices in a positive way without doing much harm. Moreover, to point out that some drinkers will probably be made worse off by a particular policy does not eliminate the value of the proposed policy: it depends on how many people move in what directions in response to the policy. It is by no means obvious that the largest effect will be for drinkers to remain intransigent and therefore suffer. It is also possible that over time the informal controls will change under consistently applied pressure, and drinkers will adapt as though things had "always been this way."

A final pragmatic objection is that prevention efforts cannot effectively alter drinking practices at low cost. Instead, to be effective such policies would have to "over-deter": to penalize and restrain drinking behavior that is not or would not be troublesome, interfering with the long-standing use of alcohol as a pleasant accompaniment to occasions of relaxation and festivity, far out of proportion to whatever bad behavior and undesirable consequences are averted. Certainly, in sizing up the effects of policies, it is essential to look closely at the untroublesome, beneficial conduct that might be discouraged and disrupted, to see how much good behavior might be suppressed along with the bad. But this is an empirical judgment, and it is best to let the available evidence speak for itself.

In summary, the strongest objections to governmental efforts to shape drinking practices are not insurmountable barriers. Instead, they suggest useful principles to recognize in considering whether and how government policies should be formed. The pragmatic objections point to the subtlety with which specific policy instruments must be chosen and employed: they must be sensitive not only to the consequences of given practices, but also to the kinds and degrees of support for prevailing practices. Interestingly enough, because the vast majority of the population drinks safely most of the time, the government, if it did nothing more than reflect prevailing practices back to the population, would be exercising an important moderating influence.

The normative objections suggest a second important principle: We should be aware of who and how many people are the intended beneficiaries of certain kinds of prescriptions and select the particular policy instruments accordingly. It is of great importance to recognize commonly understood lines that divide ideal or appropriate areas for government interference from less appropriate or inappropriate areas. Proportionality must be maintained between the coerciveness of the measures used and the extent to which policy goals are consistent with prevailing practices and common ideologies about state intervention. A line defended by criminal statutes must command wider compliance (i.e., admit more of current drinking practices) and be focused on behavior that produces more adverse external effects than a line defended by weaker measures such as economic incentives, civil sanctions, advisory educational programs, or exemplary actions by government. This suggests that one might want to establish a variety of lines with varying degrees of force. Criminal statutes should be sparsely used to discourage only the rarest and most dangerous conduct. Other programs could be used more liberally and establish somewhat more controversial goals.

CATEGORIES OF PREVENTION POLICIES

The objections to prevention that we have just reviewed and the principles of selection to which they lead indicate that prevention policies must be fashioned from a variety of materials. We need therefore an orderly way to identify the different approaches that are available, in order to judge their suitability for the normative climates and practical occasions for which they might be used.

The conception of alcohol problems presented in chapter 2 invites a sorting of prevention policies according to how they seek to ameliorate the problems (cf. Bruun 1971).

First are those policies that, by regulating the terms of commercial availability, operate primarily on the dimensions of alcohol consumption. These include taxes, restrictions on the number of outlets and hours of sale, limits on the kinds and quantities of alcohol sold, etc. Of course, many of these policies react to the contexts of drinking as well as the quantities of alcohol consumed. Relative proportions of on-premise and off-premise drinking might be importantly influenced by commercial regulation as well as many details of on-premise drinking, such as location, activities paired with drinking, and so on. But the dominant and most important thrust of policies regulating commercial availability is to shape broad patterns of consumption over time with the management of contexts as an important additional factor.

Second are policies that seek to influence drinking practices directly by suggesting (with varying degrees of force) what are unsafe or inappropriate practices. Such policies can be written into law (such as laws against public intoxication or drunken driving). They can be conveyed through explicit educational programs. Or they can emerge implicitly from the accumulated actions of government. The particular vehicle used is important, of course. Sensitive issues about proper relationships between the state and private drinking conduct are raised in choosing one vehicle over another. It is equally true that the force and precision of a control is influenced by the forms through which it is disseminated. Perhaps the most interesting thing at stake in the choice of vehicle is the enthusiasm with which the public can be mobilized to change their customary drinking practices. The clearer the message, the stronger the government's commitment to it, and the closer the conception is to widely shared private attitudes toward drinking—the greater will be the effort made by millions of private individuals to promote compliance. If images of unsafe drinking patterns can be persuasively communicated, they may begin to influence observable drinking practices and their consequences.

Third are policies that are designed to make the external environment less hostile to drinkers—to make the world a safer one in which to be drunk or similarly impaired. This set of policies involves changes in both the physical and the social environments. Changes in the physical environment could include safer structures, consumer products, and machines. Changes in the social environment could include more effective emergency medical services or more sensitive police reponses to domestic disturbances. Note that there is an important interaction between policies designed to reprove a specific drinking practice and policies designed to influence the social environment of drinkers. It is perhaps impossible to motivate individuals to refrain from a given way of drinking but to be perfectly tolerant of others who continue to drink in that way. Thus, extensive reliance on policies promoting certain conceptions of drinking will guarantee greater problems for those whose behavior is not consistent with the conception.

SUMMARY AND CONCLUSION

When current alcohol control policies are viewed in the context of the current institutional setting and against the backdrop of the analytic conception of the problem developed in chapter 2, an interesting opportunity appears. It may be desirable to diversify our policies in the direction of prevention programs that operate on the general population, through the management of contingencies that affect drinking, the contexts commonly paired with drinking, and the general features of the environment that make drinking in given contexts more or less risky. At a minimum, these policies may succeed in reaching a population that is having trouble with drinking, but would be unlikely to be involved in (and for the most part would be unsuitable subjects for) treatment. Beyond this aim, however, these prevention policies may be effective measures for reducing the number of clinical alcoholics.

In the remaining chapters of this report, we take a closer look at the available evidence about the efficacy of these prevention policies, following the lines of inquiry suggested in chapter 2. In chapter 4, we examine the potential of policies influencing price and availability. In chapter 5, we look at the potential for trying to shape drinking practices directly through laws, education, and symbolic action. In chapter 6, we consider what could usefully be done to make the environment safer for drinkers (and others). At the conclusion of the second part of the report, we shall have a more sharply defined view of the potential of what are here described as prevention policies.

4 Regulating the Supply of Alcoholic Beverages

For much of U.S. history, commerce in alcoholic beverages has been regulated and subject to relatively high taxes. A disparate set of objectives has motivated this historical regulatory effort, the relative importance of these objectives varying with shifts in public attitudes. The current regulatory structure reflects a paramount concern with maintaining an "orderly" commercial trade in alcoholic beverages and maintaining tax revenues. Also reflected in current regulations, though less visibly, is a concern for promoting temperance and protecting the public from adverse consequences of drinking. Minimum age restrictions on sales, limits on the number and nature of sales outlets, and (to some extent) high taxes are intended to limit the availability of alcohol and thereby reduce the harm engendered by high-volume or inappropriate drinking.

Public enthusiasm for restrictive regulation of alcohol commerce peaked in the early decades of this century, when many states and then the nation adopted a prohibition on the supply of "intoxicating liquors," the term actually used in the 18th Amendment. While nearly all parties agreed that this meant potable distilled spirits, there were continual arguments during Prohibition about whether lower-proof alcoholic beverages should be covered. In addition, only "manufacture, sale or transportation," but not possession, consumption, or home production were covered (Aaron and Musto, in this volume). Thus, even during the period of greatest legal constraint on supply, the national laws governing alcohol use were by no means "bone-dry."

Since repeal in 1933 the regulatory emphasis has been on maintaining orderly markets and collecting the legal tax revenues. The national experiment with Prohibition has been deemed an overwhelming failure. But very recently there has been renewed interest in using commercial regulation of the alcohol beverage industry to advance prevention goals.

This chapter reviews the regulations governing trade in alcoholic beverages. In each case, the main concern is with the potential efficacy of these regulations in preventing alcohol-related problems. We begin with a brief review of the effects of Prohibition, both as an influence on subsequent policy and as a source of evidence on the effectiveness of strategies to reduce supply in preventing alcohol-related problems. We then provide an overview of the current federal, state, and local regulatory structure. Subsequent sections discuss taxation, the myriad of regulations governing retail outlets, and minimum drinking age restrictions.

THE LESSONS OF PROHIBITION

It is widely believed that Prohibition was a failure and that it demonstrated once and for all the futility of attempts to legislate morality. The legacy of Prohibition has been to strongly discourage the use of alcoholic beverage control (ABC) laws as preventive instruments.

There is no question that the Volstead Act was widely violated and that smuggling, moonshining, and speakeasies all thrived during the Prohibition era. The crime fostered by Prohibition was of paramount importance in the public mind at the time of repeal. Fosdick and Scott, after interviewing a number of opinion leaders in the early 1930s, found substantial support for their first "principle" (1933, p. 15):

> At all costs—even if it means a temporary increase in consumption of alcohol—bootlegging, racketeering, and the whole wretched nexus of crime that developed while the Eighteenth Amendment was in force must be wiped out. The defiance of law that has grown up in the last fourteen years, the hypocrisy, the break-down of government machinery, the demoralization in public and private life, is a stain on America that can no longer be tolerated.

But while Prohibition was a failure in the sense that government authorities failed to suppress the emergence of a vigorous illegal supply network, there is nevertheless strong evidence to suggest that the "noble experiment" was an instructive failure. Law enforcement during this period was effective enough to raise alcohol prices and to reduce the ease of availability. Clark Warburton (1932), in his seminal study *The Economic Results of Prohibition*, demonstrated that consumption declined considerably, especially among the "working class"; various indications of the prevalence of heavy drinking, such as arrests for public drunkenness and admissions to mental hospitals for alcoholic psychosis, declined to low levels during the first few years of Prohibition. The most reliable indicators of this sort, acute alcohol overdose deaths and the rate of mortality due to liver cirrhosis, dropped well below their pre-Prohibition levels, reaching minima lower than any observed before or since in the 20th century.

This partial result of Prohibition in preventing alcohol-related problems has been overshadowed by its dramatic failures. As Room and Mosher note (1979-1980, p. 11):

> In American political and intellectual life since Repeal, Prohibition, an attempt at a structural and societal solution to alcohol problems, has been seen as an entirely negative experience, and those interested in alcohol problems and in helping alcoholics have often been concerned to dissociate themselves from the taint of temperance.

In particular, it appears that ABC administrators have little faith in the possibility of promoting temperate drinking habits through supply regulations. In a recent study, a number of these ABC administrators were interviewed by researchers for Medicine in the Public Interest (1979), to determine their perceptions of the goals of existing ABC laws (p. 27). The overwhelming majority of those interviewed feel that their primary, if not exclusive, purpose is regulatory and not in any way related to the public health aspects of consumption. Interviewees stressed the importance of revenue collection, maintaining orderly markets, and excluding criminal elements from the business. The report also concluded that state legislators are "generally skeptical about the effect of regulations, including taxation, on the incidence, patterns, or circumstances of use," (p. 31) and focus instead on issues related to tax revenues and economic regulation. The repeal advocates of the 1930s did not take this position. Fosdick and Scott report that the opinion leaders they talked to were very concerned with promoting temperate drinking practices through supply regulation. They advocated a state monopoly on retail trade and/or moderate taxation to promote this end.

To sum up, the real lessons of Prohibition are threefold:

(1) Drinking customs in the United States are strongly held and resistant to frontal assault. It is well beyond the will or capacity of government ever to eradicate the customary demand for alcoholic beverages.

(2) A criminal supply network emerges—if not instantly, then within

a few years—if production and sale of alcoholic beverages are outlawed. The prices and extent of this criminal supply depend on the degree of public support for the law and the resources devoted to law enforcement.

(3) The quantity of alcohol consumption and the rates of problems varying with consumption can, however, be markedly reduced by substantial increases in real price and reductions in the ease of availability.

In contrast, the lesson that has apparently become ingrained in conventional wisdom is something like the following: It is futile and mischievous to legislate drinking morals. Prohibition and, by extension, even moderate supply restrictions create a criminal industry and are not effective in reducing the consumption or the problems of alcohol.

Neither reading favors a return to prohibition. But there is much to be gained by unburdening discussion of other prevention strategies from the distorted image of the Prohibition experience that currently prevails.

CURRENT INSTITUTIONS AND POLICIES

Since the repeal of Prohibition the task of regulating commerce in alcoholic beverages has been left primarily to the states, but the federal government does play an important role. The U.S. Bureau of Alcohol, Tobacco, and Firearms (BATF) licenses importers, manufacturers, and wholesalers and regulates the advertising, size of containers, and labeling of alcoholic beverages. Although states have authority in this area of the market, they have largely left it to the federal government to manage. BATF is also responsible for collecting federal taxes on alcoholic beverages (which are ordinarily higher than state taxes) and suppressing illegal production. The U.S. Food and Drug Administration is responsible for overseeing the purity and cleanliness of alcoholic beverages. In addition to this general regulatory role, the federal government has direct control over the supply of alcoholic beverages on military reservations, a jurisdiction that is bigger than most states (8 million people are currently eligible to buy at outlets on military reservations) as well as direct control over national parks and waterways, rail, and air carriers. Other federal agencies have an indirect influence on the availability of alcohol: the Small Business Administration has been active in lending money to tavern and liquor store owners, and the Internal Revenue Service's rulings on business-related expense deductions have encouraged the notorious three-martini lunch (Mosher and Mottl, in this volume).

State systems for controlling the alcoholic beverage trade regulate almost every aspect of retail sale. All states set a minimum age for legal consumption, ranging from 18 to 21, and prescribe penalties for retailers who knowingly sell to underage customers. All states have enacted special excise taxes on alcoholic beverages and most have placed restrictions on advertising, hours of legal sale, selling on credit, and so forth. The 21st Amendment left the "dry" option to the individual states. While all states now permit alcoholic beverage sales, most have passed the option on to local jurisdictions: in 1976, 3.5 percent of the population still resided in dry counties (DISCUS 1977).

During the 1930s, 18 states chose to separate private profit from one part of the trade by creating state or county monopolies to control both wholesale distribution and (except in Wyoming) retail sales of (at least) spirits. The remaining 32 states plus the District of Columbia have adopted a licensing system, whereby a state regulatory agency is empowered to decide which wholesalers and package retailers will be permitted to operate in the state. This agency, called the Alcoholic Beverage Control Board in most states, promulgates detailed rules to implement the state's ABC laws and sets policy on the density, location, and nature of outlets through its licensing activities.[1] These agencies also license outlets that sell alcoholic beverages for on-premise consumption, and the counterpart agencies in monopoly states have this responsibility as well. States differ in the extent to which local governments are part of the licensing process. Fair trade laws govern pricing policies in some license states, and these are enforced by ABC boards.

State and local governments have promulgated a wide variety of ordinances governing the nature and operation of establishments that sell drinks. Of greatest interest here are the "dramshop" laws, which establish civil liability for employees and owners of establishments in which a drunken or underage patron is served and then causes an accident. Civil liability has been established either by statute, by court decisions, or both in 27 states (Mosher 1979).

In this era of extensive economic regulation by federal and state government, the alcoholic beverage industry remains one of the most heavily regulated. But the trend in regulatory activity during the last two decades has been toward permitting increased supply; taxes and prices have declined (relative to the overall price level), and drinks are being sold at an increasing number and range of outlets. During the early 1970s a number of states lowered their minimum drinking age (although several of these have since reverted to a higher age). The

[1] Some license states require that liquor be sold in stores that specialize in liquor or alcoholic beverages. One state (Missouri) limits liquor sales to stores that sell some other product line. The majority of license states do not place severe restrictions on the nature of package store outlets.

central question of this chapter is whether the regulatory apparatus that has gradually been weakening might be used to improve alcohol problems. In the next section, we state the general case for using regulatory powers in this way. In subsequent sections, we review the available evidence on specific kinds of control mechanisms.

THE SINGLE DISTRIBUTION THEORY

A fundamental postulate of economics states that, other things being equal, an increase in the cost of a commodity to consumers will reduce the quantity they demand. For a given cost increase, the extent of this reduction in quantity demanded may vary across a broad range of "elasticities" for different individuals. Nevertheless, an increase in taxes or a general reduction in the availability of alcoholic beverages should reduce overall alcohol consumption for both moderate and heavy drinkers. Such a reduction, particularly to the extent that it involves heavy drinkers, would in turn yield public health benefits.

This line of argument is rejected by those who equate alcohol problems with alcoholism and believe that the alcoholic is insensitive to economic (market) factors such as price and availability. In this view, alcoholic beverage control measures simply make alcoholics' lives more difficult without helping to cure them of their disease. In response, we make two general observations: (1) As we have argued above, alcoholism is only one of several consequences of alcohol consumption deserving of public concern, and (2) there is considerable evidence, both direct and indirect, to suggest that the volume of heavy drinking is sensitive to economic factors. The indirect evidence is reviewed here and the direct evidence in subsequent sections.

The indirect evidence in support of the influence of market forces on the prevalence of heavy drinking rests on an analysis of the distribution of alcohol consumption. If a group of alcoholics exists whose drinking habits are not sensitive to alcohol-related features of their socioeconomic environment, then we would expect the amount of alcohol consumed by the top 5 or 10 percent of drinkers to have little relationship to the drinking habits of the remainder of the population. This hypothesis has been evaluated and rejected by Ledermann (1956) and his followers.

Ledermann's studies of the distribution of consumption persuaded him that there exists a precise relationship between per-capita consumption and the proportion of drinkers consuming in excess of a specified level (e.g., three ounces of pure alcohol per day). Ledermann's evidence for the existence of such a "distribution law" was survey and other evidence on the drinking habits of several groups of people; he found that these groups differed widely with respect to average consumption (and in other respects), but in each case the distribution of consumption among the group's members was approximated by a "log normal" curve, with approximately the same variance for each group. Ledermann concluded that a sufficient and virtually necessary condition for reducing the volume of heavy drinking in a population was to reduce per-capita consumption.

Since Ledermann's pioneering work, additional data on the distribution of drinking have been analyzed for a number of population groups in North America and Europe (Bruun et al. 1975). These distributions, like those of Ledermann, for the most part are reasonably approximated by a log normal curve. In all cases they exhibit a single mode and are highly skewed to the right, with the top 10 percent of drinkers consuming in the neighborhood of 40-50 percent of the total. However, Skog (1971) and others (Popham et al. 1976) who have followed Ledermann's path express their conclusions in terms of tendencies rather than rigid laws of nature. The definitive statement on this subject by an international panel (including both Skog and Popham) was expressed as follows (Bruun et al. 1975, p. 45):

> 1. A substantial increase in mean consumption is very likely to be accompanied by an increased prevalence of heavy users. . . .
>
> 2. If a government aims at reducing the number of heavy consumers, this goal is likely to be attained if the government succeeds in lowering the total consumption of alcohol. . . . As the distribution of consumption is highly skewed, a substantial proportion of the total amount drunk by a population is consumed by heavy drinkers, so that a sizeable reduction in total consumption will not occur unless some of the heavy drinkers reduce their consumption.

This conclusion is stated conservatively, since it claims simply that per-capita consumption is a good indicator of (covaries substantially with) the prevalence of heavy drinking; this claim is entirely responsible, given the arithmetic importance of heavy drinkers in determining the mean consumption level. But the results suggest a stronger conclusion, i.e., a close link between the median consumer's drinking level and the consumption level of the consumer at the 90th or 95th percentile of the distribution. That is, to some degree the same factors that determine the intake of the relatively moderate median drinker also appear to influence the intake of the relatively heavy consumer. The drinking level of the typical, relatively extreme consumer is sensitive to many of the same cultural, socioeconomic, and legal forces as influence other consumers. This relationship between median and 95th-percentile consumption has not been directly tested, but its probability is strong enough to force consideration in designing policy.

This conclusion does not preclude the possibility of designing policies that are specifically focused on heavy consumers. For example, a tax policy designed to raise the price of the least expensive source of alcohol (cheap fortified wine) would probably have a greater effect on destitute alcoholics than on other drinkers. Our assessment of the distribution literature does suggest that overall trends in price and availability are likely to affect the entire distribution of consumption; there is no evidence that a large fraction of drinkers is entirely immune to such forces. It is worth noting that the evidence is comprised of econometric studies of aggregate indicators. Firmer guides to policy may be provided by additional, less aggregated studies, such as the effects of "naturally occurring" price changes on longitudinal panels of individual consumers. Direct evidence on these matters must be derived from evaluations of specific changes governing price and availability within a jurisdiction. Some evidence of this sort is available and is included in the review of alternative control strategies presented below.

TAXATION AND PRICES

Alcoholic beverages were first subject to federal taxation in the United States in 1791 (Hu 1950). Indeed, a liquor excise was the first internal revenue law enacted by Congress under the Constitution. Alcohol tax revenues were a major source of income for the federal government until Prohibition; these revenues constituted 80 percent of all federal internal tax collections in 1907 and about 10 percent at the beginning of World War II. Currently the federal tax on alcoholic beverages has a large effect on alcohol beverage prices but figures very lightly (less than 1 percent) in the federal budget.

Federal tax rates on alcoholic beverages were last changed in 1951, when they were set at \$.29 per gallon of beer, between \$.17 and \$3.40 per gallon of wine depending on alcohol content and type, and \$10.50 per proof gallon for distilled liquor. (A "proof gallon" is defined as 1 gallon of 100 proof [50 percent] liquor, or 1.2 gallons of 80 proof liquor, or, in general, the volume of liquor of any proof that contains 2 quarts of ethanol.) The states also levy taxes on alcoholic beverages; in license states, the liquor tax rates range from \$1.50 to more than \$4.00 per proof gallon. Total tax collections from liquor sales constituted about 0.5 percent of total expenditures by consumers in 1977; much less went for beer and wine taxes. Of all consumer expenditures for alcoholic beverages, just over one-third went to government as tax receipts (Hyman et al. 1980).

States influence alcohol prices through fair trade laws (in all but a few license states) and set them by administrative fiat in monopoly states. In recent years, tax and price decisions by state governments, combined with federal inaction with respect to federal tax rates, have contributed to a steady downward trend in the prices of beer, wine, and spirits, when compared with the consumer price index. Between 1960 and 1980, the "real" cost to the consumer of a bottle of liquor declined by 48 percent, of beer by 27 percent, and of wine by 19 percent (Cook, in this volume). If drinking is responsive to price, then declines of this magnitude should have provided considerable stimulus for increased consumption.

Obtaining reliable measures of the responsiveness of alcohol consumption to changes in prices is difficult for a number of reasons, as discussed by Cook (in this volume) and Ornstein and Levy (no date). A number of econometric studies provide estimates of price and income elasticities for each of the three beverage types, for both the United States and other countries. A comprehensive review of these studies is presented by Ornstein (1980). Most of these studies conclude that the demand for alcoholic beverages, like other commodities, is responsive to price. The demand for beer is probably moderately inelastic in the United States today, meaning that an increase in price will result in people budgeting more money for beer but buying less of it in total, all other things remaining equal. The demand for spirits appears to be somewhat more responsive to price than the demand for beer. There has only been one study of the demand for wine in the United States (Niskanen 1962), but evidence from other countries suggests that it tends to be fairly sensitive to price (Ornstein 1980.) Sulkunen (1976) has speculated that the less popular beverage types in a given country will generally prove to be more price-elastic than the national favorite. The results of econometric studies differ widely, due to differences in the data and methods of analysis as well as national, regional, and local variations in preferences; therefore, more precise quantitative conclusions cannot now be drawn.

The effect of changes in alcohol prices on consumption is of interest only if there are corresponding effects on at least some of the adverse health and social consequences of drinking. For example, it is logically possible that alcoholic or problem drinkers could be relatively insensitive to alcohol prices in their consumption decisions, so that price-induced changes in aggregate consumption could result exclusively from a subset of drinkers who do not in any case cause or experience drinking problems. This logical possibility was examined and rejected for the Canadian province of Ontario (Seeley 1960), and similar findings, consistent with Seeley's result, have been derived from other data series (Popham et al. 1978).

The most direct test of the connection between tax-increased alcohol prices and consequences of drinking is the analysis by Cook (in this volume), using an elaboration of the quasi-experimental approach introduced by Simon (1966). Cook examined changes in liquor taxes among 30 license states between 1961 and 1975, in order to ascertain whether liquor tax increases led to statistically discernible changes not only in the indicator of liquor consumption, but also in two alcohol problem indicators: highway crash fatalities and deaths due to cirrhosis of the liver.

There were 39 cases of tax increases, ranging from $.25 to $1.75 per proof gallon, during this 15-year period. One can view each tax increase as a "test" case and use these 39 tests (a few happening each year) to see whether such small tax increases are effective, as hypothesized. To do this, Cook first calculated the percentage changes in annual rates of liquor consumption, highway fatalities, and cirrhosis deaths in each of the 30 states in each of the 15 years, then rank-ordered these changes from largest decrease (rank 1) to largest increase (rank 30) for each year. If the tax increases had had no independent effect on the drinking or death rates, then one would expect the test cases to be evenly distributed above and below the middle; that is, on each indicator, about half of the 39 test cases would have been above the median rank, half below it.

The result is summarized in Table 9 (see Cook, in this volume, for additional tests and details). On each indicator, an excess of cases falls below the median. The excess is very pronounced for liquor consumption (77 percent ranked below), as one would expect. But the differences extended to highway fatality and cirrhosis death rates; the results hover about the conventional border of statistical significance (p 0.05). Even a marginally significant influence from a relatively small pressure (1 percent to 10 percent) on retail price is a suggestive result, since: (a) liquor represents less than half the alcohol consumed in these states, generally; (b) it is estimated by Reed (in this volume) that only about one-fourth of highway fatalities are causally dependent on alcohol use; and (c) cirrhosis death rates are a complicated precipitate of a long drinking history, in addition to the effect of the most recent consumption.

These results indicate that even relatively small changes in prices may influence not only the quantity of consumption but also the most serious health effects as well. The few historical instances in which there has been a large sustained price change lend credence to this conclusion. A tenfold increase in the price of aquavit in Denmark (the national beverage at the time) during World War I, resulting from tax increases, transformed Denmark into a predominantly beer-drinking country, and in the near term greatly reduced total per-capita alcohol consumption and the prevalence of heavy drinking (Popham et al. 1976). The sub-

TABLE 9 Ranking of 39 "Test Cases" of Alcohol Tax Increases by Net Percentage of Change in Rates of Liquor Consumption, Highway Deaths, and Cirrhosis Deaths[a]

Rank Among 30 States During the Test Year	Liquor Consumption	Highway Deaths	Cirrhosis Deaths
Below median	30	25	24
Above median	9	13	14
Number of tests	39	38[b]	38[b]

[a] States in which a liquor tax increase occurred.
[b] One case, in 1975, was deleted for lack of sufficient data.
Source: Adapted from Cook (in this volume).

stantial reductions in the prevalence of heavy drinking, cirrhosis mortality, and alcohol overdose deaths in the United States during the 1920s could reasonably be viewed as, in large part, a response to sustained Prohibition-induced market price increases, roughly triple to quadruple prices in 1915 (Warburton 1932).

We conclude that alcohol consumption and the problems caused by it respond to the price of alcoholic beverages, and we infer that the large reductions in the real cost of alcohol to consumers in recent years are likely to have exacerbated drinking problems. The downward trend in alcohol prices could be reversed by indexing federal excise taxes to inflation or by making the tax proportional to wholesale price rather than volume. A more extreme action would be to restore the federal tax in real terms to, for example, the 1951 level, which would mean an increase from $10.50 to roughly $30.00 per proof gallon.[2]

If alcohol taxes are to be viewed as principally a preventive rather than a revenue measure, the appropriate structure of tax rates across different types of alcoholic beverages needs review. Currently, liquor (distilled spirits) is taxed much more heavily than beer or table wine, not only per gallon of beverage, but also per gallon of alcohol content. State and federal ABC laws reflect the widely held view that beer and wine are more temperate or innocuous than liquor; that beer, in particular, is the "drink of moderation," while liquor is more likely to cause problems. However, epidemiological research evidence raises questions about the appropriateness of this distinction. These studies suggest that

[2] A question of relevance to tax strategies is whether increased tax rates will lead to massive criminal evasion and fraud on the part of buyers, manufacturers, or distributors. There is no single answer to the question. Insofar as the taxes are considered legitimate by the public, as the additional revenues are used to support proportionately intensified enforcement efforts, and as increases are applied uniformly, it seems doubtful that new major criminal activity would be incurred by the relatively modest increases that could be practically considered during an inflationary peacetime.

the key factor in the incidence of most major diseases associated with lifetime consumption is simply the daily intake of ethanol, irrespective of the type of beverage (Popham et al. 1976). Research on the physiology of drunkenness has failed to find important differences among beverage types, and studies of the beverage choice of drivers with high blood alcohol content and of alcoholics demonstrate that all three beverage types are well represented (Borkenstein et al. 1974, Popham et al. 1976). For poor alcoholics, at least, relative prices are the major determinant of beverage choice (deLint 1962, Mäkelä 1971).

These results cast doubt on such grounds for tax discrimination among beverage types as the assertion that liquor is intrinsically more dangerous than beer. A more appropriate tax strategy might be to tax alcoholic beverages not on the basis of beverage type, but according to ethanol content or whatever other characteristics might be shown to create problems.[3]

A different type of concern generally taken into account in evaluating tax strategies is the relative distribution (incidence) of costs and benefits across the citizenry. While we cannot here present a comprehensive analysis of this complex issue, the following observations seem basic and germane.

The bulk of alcohol taxes are paid by a small fraction of the population. Half or somewhat more than half of the alcohol is bought by one-tenth of the U.S. consumer population. Members of this fraction suffer proportionately greater health damage, trauma, and personal financial loss associated with the adverse effects of alcohol consumption, and hence would also be major beneficiaries of reduced consumption that resulted from an increase in taxes. Thus, the incidence of direct benefits and costs would both be correlated and concentrated. These statistical associations would be modified to the extent that Medicaid, Medicare, and health insurance collections from all consumers might be reduced as a result of health cost savings. Measuring the incidence of cost and

[3] Each of the following three tax strategies could be viewed as neutral with respect to beverage type:

1. A tax simply proportional to ethanol content.
2. A tax structure designed to equalize the average wholesale or retail price of an ounce of ethanol from each beverage.
3. A tax structure designed to equalize the price of an ounce of ethanol from the cheapest brand of each beverage type.

Even more complex alternatives (from the point of view of computation) emerge if the tax structure is designed to reflect that demands for beer, wine, and liquor may differ with respect to their sensitivity to price.

benefit resulting from taxes on alcohol is a difficult research problem for which current data are thin and methodological issues require much more attention. However, the current imprecision does not seem to us a major stumbling block for policy evaluation.

Poor households that are burdened by the heavy drinking habit of a member may be further deprived if tax-led price increases induce an increase in alcohol expenditures. By some estimates (see Cook, in this volume, for a review), the price elasticity of demand for alcoholic beverages is high enough that expenditures would not increase much, if at all, in the average family. To the extent that the heavy-drinking member reduces his (in most cases) or her alcohol consumption in response to higher prices, earnings may increase and health care expenditures may be reduced, thus tending to improve the net economic position of the household.

However, alcohol taxes are probably regressive, in the sense that they constitute a higher fraction of poor households' budgets than other households. The net incidence of alcohol taxes depends on how the revenues are used. If they were used, for example, to create an income deduction in the social security tax or to underwrite additions to minimum income maintenance programs, the net incidence would be more progressive. It is probably already the case that the external costs of medical care incident to alcohol use, supported by public expenditure, are balanced by current tax receipts.

Finally, a comprehensive pricing policy for alcoholic beverages would take notice of mechanisms other than the excise tax by which government regulation influences the cost of alcoholic beverages to consumers. For example, the income tax code subsidizes consumption by allowing tax deductions for alcoholic beverages purchased in connection with business-related meals (Mosher 1980). A second example are the armed services, which sell alcoholic beverages at greatly discounted prices at post exchange stores and at clubs on base. This policy encourages drinking, not only by uniformed and civilian employees and their families, but also by reservists and others eligible to shop at post exchange stores, about 8 million people in all (Mosher and Mottl, in this volume). The armed forces have long been concerned with the high incidence of alcohol problems among career personnel, the rate of which is clearly exacerbated by these pricing policies.

CONTROLLING OFF-PREMISE RETAIL SALES

Since World War II, there has been a secular increase in the fraction of alcohol purchased for consumption at home (off-premise), rather than at taverns and restaurants (on-premise). State, municipal, and federal jurisdictions can regulate retail sales for home consumption in two major respects. First, they control the number and location of retail outlets, either directly, as in the monopoly states, or through licensing or other regulation. Second, the operation of retail outlets is regulated by specifying legal hours of sale, minimum and/or maximum purchases per customer in any one visit, prohibition of sale to inebriated customers, and so forth. Retail outlets also differ among jurisdictions with respect to allowable merchandising practices.

With respect to the number and location of retail outlets, it is doubtful that variation in the density of outlets, within the ranges observed in North America, has much effect on beverage sales. A number of studies (Harford et al. 1979, Smart 1977b) have attempted to measure the effect of outlet density on drinking by correlation or regression analysis, comparing, for example, consumption and outlet density across states. As Ornstein and Hanssens (no date) point out, it is difficult to interpret the results of these studies due to the confusion of cause and effect. We would expect outlet density to be primarily a result of the demand for alcoholic beverages.

A recent quasi-experimental study of retail sales in a rural area of Ontario compared sales to residents of two cities located some miles apart, both of which were serviced by a package store located in one of them. Per-capita sales were about equal for these cities, despite the differences in accessibility (Popham et al. 1976). More extreme distances may make a difference; rural alcohol consumption in Finland increased appreciably when state monopoly stores were first opened in rural areas in 1968 (Beauchamp, in this volume). The same phenomenon would be expected to occur when outlets first open in formerly "dry" counties in the United States. For most consumers living in "wet" counties, the opening of a new package store in an especially frequented shopping zone does not appear to change their consumption levels measurably, although it may cause them to keep a smaller home inventory while making more frequent purchases. Still, there is not a great deal of detailed knowledge about variations in availability. The effects of permitting sales of alcoholic beverages in food stores and other high-density outlets as well as more specialized locations deserve more study.

Very little is known about the effect of merchandising practices on consumption. A recent study of Ontario found that customers made larger purchases on the average in a self-service outlet than in a nearby outlet in which orders were filled by clerks; the customers in the self-service store were also more likely to report to interviewers that they made unplanned purchases. However, it is possible that the larger inventories thus acquired by the self-service customers did not influence their actual consumption. Other merchandising practices have not been studied with respect to alcoholic beverage sales.

CONTROLLING ON-PREMISE SALES

Public drinking places have two distinct images. On one hand are the neighborhood bar or tavern, the sophisticated cocktail lounge, the British pub, the German beer hall—places where neighbors, friends, and workmates gather for relaxation and conviviality. On the other hand is the residual temperance-era image of the saloon, where criminal activities are hatched, where working men drink up their weekly paychecks and then some, often fighting with other customers and creating a public nuisance for the neighborhood—then stagger home (or worse, in current times, attempt to drive home) and abuse their families. Both of these images have clearly had their real-life exemplars. Regulation of public drinking places is in part an effort to control the excesses of the saloon, while preserving the legitimate pleasures of the neighborhood tavern.

Typically, such regulation includes restrictions on operating hours, a ban on extending credit to customers, zoning restrictions to prevent taverns from operating near schools or churches or in residential areas, a requirement that food be served with the drinks, and so forth. The laws uniformly prohibit serving minors or drunks and enjoin owners/managers to maintain orderly premises—with the threat of civil penalties and license revocation for noncompliance. Indeed, tavern owners can be liable for injury and property damage caused by inebriated customers, an issue that arises most commonly in connection with auto accidents.

Mosher (1979) summarizes the current status of the law with respect to civil liability (p. 782):

> . . . 27 states and the District of Columbia impose some type of dram shop liability on servers of alcoholic beverages. . . . Most states with dram shop acts and all states recognizing dram shop liability under common law have as a matter of social policy established that commercial servers of alcoholic beverages have a broad obligation to protect the public from injuries caused by their intoxicated or underaged patrons.

The legal standard for assigning negligence (in cases that do not involve underage patrons) is a demonstration that the patron was "obviously intoxicated." Mosher points out that this standard is inordinately vague and conducive to capricious judgments and recommends that the factual inquiry in such cases be expanded to include an assessment of the "degree of care" exercised by the server—particularly with respect to the training given to employees and the host's concern for patrons' transportation arrangements. Such a reorientation may serve to enhance the preventive effectiveness of dramshop laws. At this time, in the absence of research data, the effectiveness of such laws or variations therein is entirely a matter of anecdote, *a priori* speculation, and common sense argument.

Accidents and violent crime that may result from acute episodes of intoxication in public drinking places are a central concern of on-premise control of alcohol. There is also the question of whether widespread availability of public drinking places increases the total quantity of consumption. It seems reasonable to suppose that increased availability of alcoholic beverages in restaurants, cafeterias in workplaces, sports arenas, theaters, and so forth would have an effect on per-capita consumption; generally speaking, if the practice of drinking is integrated into a wider range of day-to-day customary activities, the quantity of consumption will increase. The question of how many and what types of public places should be permitted to accommodate drinking then becomes in part an issue of public health, albeit one that can neither be readily quantified nor simply resolved. The current trend toward increases in the number and variety of drinking premises deserves attention and thoughtful analysis, for the cumulative effect on drinking practices may be substantial.

MINIMUM AGE RESTRICTIONS

While only a small fraction of the United States continues to prohibit the sale of alcoholic beverages, the prohibition of sales to one large segment of the population—youths—is currently mandated by every state. The age thresholds all lie between 18 and 21. As of 1979, 23 states set the minimum age at 18 or 19 years, 3 set the limit at 20, and 24 set the limit at 21 (12 of these, however, allowed beer sales to 18- or 19-year-olds). There was considerable flux in these legal thresholds during the 1970s: between 1970 and 1973, 24 states reduced their minimum drinking ages (Williams et al. 1975), while a number of states have raised the minimum in the last few years. These changes have provided the basis for quasi-experimental analyses of the consequences of varying minimum age restrictions.

Williams et al. (1975) performed a short-term follow-up of minimum age reductions legislated in the early 1970s in Michigan, Wisconsin, and Ontario. Douglass (1979-1980) and his colleagues performed short-term follow-up studies of minimum age reductions in Michigan, Maine, and

Vermont and a longer term follow-up for Michigan. These and other studies have focused on a single dimension of concern—drunken driving (Haddon 1979). They found consistent evidence that the age reductions resulted in an increase in the rate of auto crashes and fatalities involving youthful drivers. Williams et al. estimate that during the first year of reduced age in the three jurisdictions they studied, the number of drivers 15-20 years old who were involved in fatal crashes was about 5 percent greater than would be expected in the absence of the change.

Smart (1977a) found that in 25 states in which the drinking age was lowered, beer was the only beverage type showing a discernible increase in consumption. Douglass refined this result in his analysis of the Michigan experience, concluding that only draught beer consumption increased significantly as a result of the minimum age reduction in that state. However, Williams et al. found an increase in youthful auto fatalities in Wisconsin following a reduction in the drinking age for spirits and wine, while beer had remained constant at 18—suggesting that beer is not uniquely responsible for teenage drinking problems.

While the minimum drinking age does have an effect on alcohol consumption by youths, underage youths still drink a great deal. National surveys in the 1970s have consistently shown that over 80 percent of high school seniors have had a first drink before age 18; over one-third of high school students, including half of all students 16-17 years old, reported drinking within the past 30 days (Abelson et al. 1977, Blane and Hewitt 1977a, Johnston et al. 1979). The prohibition on sales to youths thus may reduce availability to them somewhat, but it falls far short of imposing total abstinence on this group.

Minimum age restrictions in this country reflect widely accepted beliefs that drinking tends to be more harmful for youths than for adults and that we cannot trust youths to make good decisions about when, where, and how much they should drink. While the legitimacy of this type of restriction is widely accepted, the question of precisely where the age line should be drawn remains alive in many areas. State and federal laws currently gives 18-year-olds most of the rights and responsibilities associated with adulthood, the right to purchase alcohol being the only major exception. If 18-year-olds are mature enough to vote, seek many elective offices, enter into contractual arrangements, serve in the armed services, and so forth, it would seem logically consistent to also confer the remaining symbol of adulthood, the right to drink, on this group. The response to this argument is that as a group, people aged 18-20 are extraordinarily prone to auto accidents, as well as violent crime and other forms of socially destructive activity, and it is simply foolish to exacerbate these tendencies by legalizing drinking for this age group.[4]

We do not attempt to resolve this debate, but simply to note that there is reasonable evidence that prohibition for youths does have some effect on their drinking and in particular that the choice of a minimum drinking age has a small but consistently exacerbating effect on the auto accident and fatality rates.

CONCLUSION

The common belief that alcohol control measures (government action to regulate the supply of alcohol and drinking premises) are ineffective as prevention instruments is unfounded. This belief has been engendered in part by a misunderstanding of the lessons of the Prohibition experience. There is good evidence from econometric studies that alcohol prices, as affected by excise taxation, can affect consumption levels, and probably the consequent rates of alcohol- related problems. Reductions in the minimum drinking age slightly but consistently increase auto accident involvement by younger drivers. The effects of merchandising practices, outlet density, civil liability for servers, and so forth have not been established with reliability, in part because these control mechanisms are intrinsically very difficult to study. It is possible but as yet hypothetical that the cumulative effect of a number of changes in these areas of regulation has been substantial.

[4] If the concern is centered about lowering youthful traffic accidents, one might think of raising the driving age rather than the drinking age. European countries generally have lower drinking ages and higher driving ages than the United States.

5 Shaping Drinking Practices Directly

Chapter 4 examined the potential efficacy of taxes, restrictions on availability, and the detailed regulation of stores and taverns to palliate drinking practices—primarily with respect to consumption, but also with respect to contexts. While such approaches appear to hold significant potential for controlling the alcohol problem, their influence is indirect: they produce effects on drinking practices and the shape of alcohol problems only as drinkers react in personalized, somewhat unpredictable ways to the new terms of commercial availability. Arguably, a more precise and direct approach to moderating drinking practices would be to specify "safe" and "appropriate" drinking practices based on solid facts, then use the prestige, resources, and authority of government to facilitate widespread adoption of the preferred practice. Since much drinking now occurs under private auspices, and since most on-premise drinking is heavily influenced by one's companions as well as the marketing efforts of servers, policy instruments focusing directly on private drinking practices can complement instruments regulating commercial availablility.

Governments can seek to influence collective drinking practices through instruments that operate with varying degrees of force. At one extreme, governmental authorities can embody conceptions of appropriate or tolerable drinking norms in statutes and regulations (e.g., laws prohibiting drunken driving, being drunk in public, being intoxicated while flying a plane, and drinking by minors). Clearly, these legal modes draw heavily on the moral, economic, and political resources of governments: prestige is staked on conceptions of drinking that may or may not correspond closely to commonly shared views of drinking, considerable money may be spent to enforce such laws, and the authority of the government is invoked against individual conduct.

Much less extreme are government-sponsored programs to "educate" citizens. Sometimes the programs do nothing more than provide information about drinking through labeling or mass media advertising. Other times, the programs take advantage of organizations such as schools, churches, and health maintenance organizations to disseminate the messages in more localized ways. In some cases, the programs go beyond merely providing information and seek to motivate people to moderate their drinking or actually train them in techniques for doing so. To the extent that such programs are sponsored by government and embody a specific conception of drinking, they claim the prestige and credibility of the government. The volume of resources claimed, however, can vary significantly, depending on how intensive the program is and how much of the costs can be defrayed by taking advantage of existing institutions as vehicles for the messages, motivation, or training.

At the other extreme, the government could seek to alter drinking practices through the symbolic impact of its own independent actions or statements about what constitutes safer drinking practices. In areas in which federal governmental power is largely uncontested (e.g., managing its own employees, regulating interstate transportation, and managing national parks), it could make its operations reflect a coherent and consistent idea of safe drinking practices. Compared with statutes and regulations, shaping practices through symbolic action is light and unburdensome: it operates only through the force of government's prestige and leaves individuals free to draw their own conclusions from observed policies. But these characteristics may be liabilities as well as assets: citizens may simply fail to note government's exemplary actions, draw the wrong inference, or ignore the example as unpersuasive.

The purpose of this chapter is to explore the potential of using these sorts of instruments (law; education, information, and training; and symbolic action) to shape collective drinking practices. We try to give some specific content to the general concepts of "safe" or "appropriate" drinking practices. Specifically, we look for restrictions on drinking that could command widespread support and, if more widely and enthusiastically adopted, could change the shape of alcohol problems in important areas of concern.

CONCEPTS OF SAFER AND MORE APPROPRIATE DRINKING

It is relatively easy to state and find consensus on prevention goals, so long as one keeps away from specifics. There is little controversy, scientific or political, over the goal of reducing alcohol abuse and alcoholism. When one looks more specifically at the behaviors entailed and the range of individuals potentially involved, however, the easy consensus evaporates along with the simplicity of the terms. A list of drinking practice targets that have at various times been pursued by government and/or public interest groups includes:

- eating while drinking
- drinking more diluted forms of alcohol
- drinking nonalcoholic beverages in preference to alcoholic ones
- drinking in association with recreational opportunities
- substituting sports for drinking
- not driving while drinking or after drinking
- not drinking at all by people of certain ages or ethnicities
- not drinking more than a specified amount per day
- not drinking or appearing drunk in public.

To discuss ways of achieving any such behavioral results without being specific about which one is in view cannot lead to effective results.

We have identified three aspects of drinking behavior as suitable targets of prescriptive efforts:

- driving and performing other hazardous mechanical tasks while intoxicated,
- drinking so much over a lifetime that one begins to suffer from organic damage such as cirrhosis of the liver, and
- using drunkenness as an excuse for violent or aggressive behavior.

In our view, these aspects of drinking behavior are suitable targets in part because the behavior is tied to some of the worst consequences of current drinking practices and in part because current attitudes and practices are supportive enough to allow governmental efforts but not so widely honored that such efforts would be redundant.

Some statistical and causal elaboration, focusing for illustrative purposes on the relation of alcohol to mortality, indicates what is at stake in these areas. At least 12,000, and perhaps twice this number, of the more than 100,000 accidental deaths that occur each year in the United

States may be directly attributed to the effects of drinking. In the most extensively studied area, 12,000 annual deaths are attributable to drunken driving alone (Reed, in this volume), and the majority of these deaths involve people under 30 years old.[1]

If the same proportion of other accidental deaths were attributable to alcohol as were auto fatalities, it would yield another 12,000 or so each year. This seems the maximum possible figure and may be considerably smaller (Wingard and Room 1977). The tradition of research on non-auto crashes, falls, burns, drownings, and other accidents is less rigorous and voluminous than the driving literature and cannot provide any firmer estimates.

Of the more than 30,000 annual liver cirrhosis deaths in the United States, between 20,000 and 25,000 can be attributed directly to the effects of alcohol (Ouellet et al. 1977, 1978). Nearly all of these deaths are people over 35, and most are under 65. Only about half the alcohol-caused cirrhosis deaths are drinkers who would qualify as alcoholic under clinical criteria. Some degree of risk of liver injury is actuarially present in people who sustain an average of only three to four drinks daily over a long period (Popham et al. 1976).

Finally, in regard to violence, about 10,000 murders occur each year in situations involving alcohol. The question of attribution here is quite open. It has often been taken for granted that alcohol makes people violent and that the connection is pharmacological—the textbook explanation being that alcohol "disinhibits" violent impulses. Recently, evidence has accumulated from a variety of fields suggesting that the connection between alcohol and violence is cultural rather than pharmacological. MacAndrew and Edgerton (1969) marshalled anthropological evidence of the enormous variations between cultures in behaviors while drunk, including manifestations of aggressiveness. Levine's (1977) historical study of American attitudes toward the role of alcohol in casualties and crime shows that, on the whole, colonists in this country believed alcohol made people clumsy; by and large, they did not believe it made people violent. The latter belief was an innovation of the temperance era. Stevenson's tale of *Dr. Jekyll and Mr. Hyde* captures the essence of how the 19th century came to view alcohol, as a magical substance that turned a man into a beast. Social psychological experi-

[1] It is illuminating to note that this figure is about one-quarter of all auto crash fatalities; compare the "half of all highway deaths" conventionally cited "involving" alcohol. The difference is due largely to the surprisingly high prevalence of drunken drivers at the times (late nights) when fatal crash risk is at its highest anyway because of driver fatigue, poor visibility, and lack of routine traffic stimuli. The number of crashes that would be expected even in the absence of alcohol substantially reduces the conventional figure, in a causal analysis, but it is still very large.

ments with college students using a "balanced placebo" design (four conditions: believing that one is drinking alcohol and actually drinking alcohol, believing but not drinking, drinking but not believing, neither drinking nor believing) have shown that the belief that one is drinking alcohol, not the fact of drinking alcohol, is associated with aggression. There is also the negative evidence of an extensive tradition in research of looking for physiological connections between alcohol and violence that has not produced positive findings; indeed, the disinhibition theory shelters a wide variety of essentially incompatible as well as unsubstantiated explanations (Pernanen 1976).

Nevertheless, even without a pharmacological connection, there is a strong causal connection between alcohol and violence: the culture tells people this is so, and in acting on this belief they make it come true. Research designed to sort out the specific quantitative contribution of cultural belief to alcohol-involved violence is in its infancy, but this connection requires consideration even in the absence of good epidemiological estimates.

Of course, all of the alcohol-related deaths analyzed above are simply the visible tip of the iceberg: not only tragic in themselves, but also important as signs locating the larger quantities of harm—physical, psychological, and social—that we wish to reduce. In the remainder of this chapter we look at the three modes of government influence (law, education and symbolic action) with this set of practices and corresponding problems in mind.

DRINKING PRACTICES AND THE LAW

In two major areas, governments have sought to shape drinking practices through laws backed by criminal penalties. The two areas are drunken driving and public drunkenness. These are by far the most common criminal offenses reported to be engaged in by young men (O'Donnell et al. 1976) and are the two largest categories of arrests reported to the Federal Bureau of Investigation (1979), each approximating 1.25 million arrests annually of the roughly 10 million reported. These involvements of law in the direct control of drinking practices dwarf all other efforts of government.

DRUNKEN DRIVING AND THE LAW

The law in the United States (and throughout the world) establishes clear prescriptions for drunkenness: one should not drive while drunk. The legal codes usually stipulate a specific blood alcohol content (BAC), conventionally 0.08 or 0.10, that defines drunkenness for judicial purposes. These laws reinforce public attitudes, for the great majority of the population believes that driving while drunk is wrong and dangerous (NIAAA 1978).

The question of whether the existence of these laws actually influences the rate of drunken driving and, if so, through what mechanisms is still unsettled. Some useful arguments and evidence bearing on these issues are available, however. At the outset, one might reasonably be skeptical about the value of such laws. Even though we now spend significant sums enforcing these laws and have more arrests nationwide for drunken driving than for any other offense, the probability of being arrested for drunken driving is now estimated to be roughly 1 in 2,000. Compared with the perceived risks of injury or economic losses from drunken driving (which must also operate to deter drunken driving), the risks and consequences of arrest must appear small. Moreover, one can reasonably ask if intoxicated people are generally able to calibrate risks, whether created by enforcement agencies or the laws of physics.

Given this initial skepticism about the potential value of legal sanctions in controlling drunken driving, it is somewhat startling to discover that the few careful studies in this area indicate that drunken driving *is* responsive to the perceived risk of arrest. Two bodies of evidence are especially relevant, and both focus not on drunk driving per se but on its most serious risk: fatal injury in an auto crash.

The first instance is the 1967 British Road Safety Act, carefully analyzed by Ross (1973). The Road Safety Act (RSA) defined driving with a BAC of 0.08 or higher, as determined by breath or laboratory analysis, to be an offense. Drivers were required to submit to BAC tests if asked to by police upon reasonable cause (a road accident, moving violation, erratic driving). Refusal to be tested would create the presumption of illegal intoxication. Judges were denied discretion in sentencing: a first BAC offense was to result in a mandatory 1-year suspension of the license to drive (the same statutory maximum penalty as before). The 1967 RSA thus differed from earlier custom and law (the 1962 Road Traffic Act) in three principal respects: (1) the use of BAC measurement as definitive rather than supplementary evidence of drunkenness, (2) mandatory submission to testing, and (3) mandatory sentencing linked specifically to this evidence. In addition, unprecedented publicity was generated by the government in the months before the RSA took effect, especially regarding the prospective use by police of the roadside "Alcotest" breathalyzer device, 1 million of which were purchased by the Home Office from a German manufacturer. The device was a novelty in England.

The immediate impact of the RSA was positive and dramatic. For the three months immediately following its effective date, total auto fatalities were reduced 23 percent, and the proportion of fatally injured drivers with BAC levels of 0.08 or higher (postmortem blood determination) dropped from 27 percent before passage to 17 percent the following year. However, this dramatic effect gradually diminished. The relevant time series data, adjusted by Ross for seasonal and other external variations, show a gradual decline in auto injuries from 1966 to 1967, rapidly falling when the RSA went into effect; the curve then flattened, and began to rise in 1970. As Ross concludes: "In the case at hand it appears that the benefits initially produced by the legislation had been largely canceled by the end of 1970, and a return to normal would be predicted for subsequent years" (p. 77).

The explanation proposed by Ross for this sequence of events is that the well-publicized passage of the RSA convinced many drivers that the risk of arrest and punishment when driving drunk would become much higher than it had been. They were hence deterred from drinking before driving and/or from driving after drinking. But law enforcement did not markedly increase the certainty of either detection or punishment, and, as time passed, drivers observed this and the immediate deterring impact of the RSA thus evaporated.

While this general explanation seems quite plausible, Ross's data support a richer interpretation with respect both to how drinking practices changed and to how the deterrent effect was produced. With respect to the change in behavior, Ross clearly rules out a significant overall decline in either drinking or driving. Instead, he presents evidence that the law focused its deterrent effect on a fairly narrow slice of behavior: the custom of driving to and from pubs—especially on weekend nights. In response to the new law, many habitues took to walking or drinking closer to home. This aroused a considerable outcry from several thousand pub owners; a number of less convenient establishments closed.

Similarly, a detailed analysis of the government's "massive publicity effort" reveals that the focus of the campaign was on the novel roadside breathalyzer test. While about 40 percent of adults learned the legal penalty or actual BAC limit, 99 percent could identify the breath test. It appeared that this new technology would bring about drastic changes in the activities of police and the courts, replacing the earlier workings of patrol and trial with a revolutionary scientific mechanism. In fact, there was no such result. As Ross documents, well-publicized cases soon established narrow limits to the authority of the Alcotest as evidence. It took several years to establish reliable standards for its use. Even though testing increased steadily from 1967 onward, the British police did not employ it frequently, compared with police in other countries.

If the initial dramatic effect on pub-going customs is thus attributable to anticipation that a technical innovation would transform the system of traffic surveillance and justice, that anticipation was due largely, it appears, to a somewhat superstitious belief, shared and promoted by the government, in the power of the new technology.

This added information enriches our understanding of how the law operated to produce effects, beyond the simple notion that perception of an increased risk of arrest generally deters a given behavior. It suggests that the actual changes in behavior produced may be quite narrow as citizens react to the new contingencies in ways that preserve as much of their customary activity as possible. It also suggests that deterrence works best when its threat is widely advertised and becomes the focus of much public conversation, and perhaps when the enlarged threat is tied to a technological novelty. The question of whether and how the dramatic initial results of the British Road Safety Act might have been sustained thus remains somewhat obscure.

A series of cases potentially as significant as the RSA are the Alcohol Safety Action Programs (ASAP). Between 1970 and 1977, the U.S. Department of Transportation funded 35 local sites, which varied in size but in the aggregate covered millions of drivers, in a demonstration effort to control the problem of drunken drivers. The programs cost more than $88 million. These "comprehensive, multifaceted" programs employed four categories of countermeasures: enforcement, adjudication, rehabiliation, and public information and education (National Highway Traffic Safety Administration 1979b).

For our purposes, the studies documenting and evaluating this experience have serious flaws (Zador 1976, Zimring 1977, Cameron 1979, Reed, in this volume). Serious methodological problems in the design of the program and the collection of data make it difficult to consider these program innovations as experiments. Equally important is the difficulty of isolating the separate effects of each of the program components. Since our concern is largely with the effects of enforcement and adjudication on drunken driving (including those portions of rehabilitation and education efforts that were tied to enforcement efforts), such isolation is clearly desirable.

Despite these difficulties, some useful information was developed by ASAP evaluators in their final report (Levy et al. 1978; National Highway Traffic Safety Administration 1979a,b,c,d); they attempted to meet a number of the criticisms leveled at the earlier evaluation (National Highway Traffic Safety Administration 1974). The final evaluation found that 12 of the 35 ASAP programs had produced a statistically discernible effect on nighttime auto fatalities—a good indicator of drunken driving. These 12 programs averaged a 30-percent reduction over about 3 years. This result appears broadly comparable to Ross's finding of a 23-percent reduction in this index over a similar time frame—a coincidence that lends some increased credibility to the ASAP findings. On the other hand, it should be noted that when previous optimistic evaluations by the National Highway Traffic Safety Administration (NHTSA) of the ASAP and other traffic safety programs have been reanalyzed or redone by independent investigators, much smaller or no effects were reported (Cameron 1979). At any rate, the reduction and the particular characteristics of those programs that produced it, led NHTSA to conclude that the program had been successful and, somewhat more surprisingly, that "raising the enforcement level [i.e., alcohol-related arrests, by a factor of two to three] and public awareness of that level in order to deter the social drinker" were central to producing this result (Levy et al. 1978, p. 174).

Thus, some moderately persuasive evidence exists suggesting that effectively enforced drunken driving laws will deter drunken driving and reduce accidents and fatalities associated with it. The main problem with trying to take advantage of this opportunity is that it is likely to be expensive; exactly how expensive is unclear. Using NHTSA's figures on lives saved (563) and the total costs of ASAP ($88 million) yields an estimate of roughly $156,000 per life saved (but cf. Reed, in this volume). But this simple calculation ignores many of the costs of enforcement borne by local and state agencies, the expenses of treatment programs borne by arrestees themselves, and the "cost" to innocent drivers of increased police surveillance of driving practices. Moreover, we cannot hope to save much in the way of resources by punishing more severely those whom we catch. Current research (summarized by Reed, in this volume) indicates that the prevalence of drunken driving is much more effectively influenced by changing the rate of arrests and convictions than by trying to alter the severity of punishment delivered. Our jury system, wherein many jurors identify with the defendant accused of drunken driving, prohibits the imposition of penalties out of proportion to those the public feels are appropriate. Nonetheless, drunken driving laws are probably important. At a minimum, they help sustain a widely shared disapprobation of drunken driving. They also provide an opportunity to attack a given drinking practice more aggressively if the society is willing to commit the resources, publicity, and attention necessary to make deterrence a social phenomenon rather than an abstract concept.

PUBLIC DRUNKENNESS

The second aspect of drinking practices controlled by criminal statutes is that of being drunk in public. In principle, these statutes could be justified in several different ways. Since those who become publicly drunk make themselves vulnerable to criminal attacks and are often the people experiencing the worst consequences of drinking, the laws could be understood as paternalistic efforts to discourage dangerous conduct or to allow the state to take "protective custody" of people who are in danger. Alternatively, the laws could be seen as devices for preventing drunks from harming others. To the extent that public drunkenness is related to violence, vandalism, and other forms of disorder, the laws against public drunkenness could be understood as devices to allow the police to intervene early and moderately in situations that could become dangerous—much as laws against drunken driving or the illegal carrying of weapons are assumed to operate. The laws could also be understood as devices to protect the sensibilities of people shocked by breaches in public decorum created by drunkenness.

In recent years, the laws against public drunkenness have been understood primarily in these last terms, i.e., as devices that fixed a preference about public decorum in criminal statutes and were used to stigmatize and control a minority of the population. Since this use of state power seemed suspect, a social movement developed that urged the "decriminalization" of public drunkenness. The central argument of this movement is that the real problems created by public drunkenness can be handled more effectively through extensions of civil authority and the provision of social and medical services than through the application of criminal law. To the extent that criminal laws served the purposes of protecting occasional drunks from victimization and chronic drunks from exposure and starvation, the purpose could be better served by a form of civil "protective custody" allied with social and medical services. To the extent that the laws guarded against public violence and disorder, the police could rely on statutes regulating conduct more reliably related to the worrisome offenses (e.g., assault, vandalism, etc.).

About half the states and cities of the United States have been persuaded by these arguments to "decriminalize" public drunkenness (U.S. Senate 1976). Typically, the decriminalization statutes not only eliminate criminal sanctions but also mandate that drunks be taken into protective custody and supplied with a variety of social and medical services. While these new laws have succeeded in changing the legal status of public drunkenness and much of the rhetoric about the problem, the actual handling of public drunks has changed only modestly. The police usually remain involved because they are the only agency trained and deployed in ways that bring them into contact with public drunks. Moreover, they retain explicit authority to hold a drunk in protective custody for a fixed period of time before release or transfer to a treatment facility. The social and medical services remain largely absent because the legal mandate for increased services was in few cases backed up by suitable appropriations and programmatic activities.

The effects of this liberalizing trend are unclear. There are virtually no data on how the laws regulating public drunkenness have affected the frequency and character of public violence. Some data exist (Oki et al. 1977) that suggest that neither the old laws nor the new laws have had much effect on the health and welfare of chronic public inebriates—though this may be the result of having failed to make investments in the medical and social services mandated by the reforms. In fact, the only clear effect of decriminalization seems to be the most immediate and most obvious: chronic public inebriates have, according to surveys of public impressions, become more visible on city streets (Aaronson et al. 1978, Giffen and Lambert 1978). The increase in visibility might be taken as evidence for increased frequency of public drunkenness, but this need not be so. It may be that public inebriates are now diffused more broadly throughout cities, giving the impression of greater frequency. Alternatively, it may be that chronic public drunks spend less of their time in institutions (a possibility that is consistent with the idea that episodes of drunkenness on city streets have increased but inconsistent with the idea that the number of public drunks has increased). Such effects are the natural result of increased official tolerance of public drunkenness.

There remains at least the possibility that public drunkenness has increased in frequency and that some increase in violence, vandalism, and disorder is associated with this change. We wish the research in this area were more complete and illuminating, but we suspect that the laws against public drunkenness were not operating to hold drunken violence within tight bounds all those years. It is more likely that the role of the law with regard to violence linked to alcohol involves the acceptance, by judges, juries, or other participants in the legal process, of the cultural presumption that alcohol causes violence and therefore serves to excuse it. In the absence of a body of studies on the treatment of drunkenness in this connection, we are unable to extend this analysis into the question of what might be done in this area. (Chapter 6 includes a discussion of public drunkenness from an environmental perspective).

EDUCATION, INFORMATION, AND TRAINING PROGRAMS

A second way to directly influence the drinking practices of individuals is through education, information, and training programs. To a degree, of course, such efforts must be seen as important complements to legal approaches. Laws are effective in shaping conduct only if they are promulgated, a function that is at least in part one of information and education. Moreover, the language of a law has educational value: it instructs people about what the society considers desirable as well as threatens them with official action if they fail to comply. Finally, intensive education and training programs are sometimes used as alternative sanctions for violations of the legal requirements. Thus, education, information, and training are often intertwined with laws.

They can also operate as instruments separate from the law. They can seek to persuade and advise people in areas in which society is reluctant to establish legal restrictions. In fact, in a democratic society, educational programs operating apart from the law enjoy great prestige and legitimacy, despite the scanty evidence of their effectiveness. In any case, it is this advisory form of educational effort that is the focus of this section. Specifically, we look at three kinds of programs: educational programs directed at youths through educational institutions (primarily public schools); mass media information campaigns; and community-based health information and training programs.

SCHOOL EDUCATION PROGRAMS

Alcohol education in the public schools was originally a creature of the temperance movement. By the late 19th century, state authorization and implementation of alcohol education programs was almost universal. The character and tone of this education were consistent with prevailing views at the time and were strongly antialcohol. While some attention was paid to the adverse health consequences of drinking, the dominant focus was on alcohol's disastrous effects on character, family life, and the fabric of society. Ruin was predicted for the vast majority who drank at all. With repeal and the early beginnings of the alcoholism movement in the 1930s and the 1940s, the approach shifted to a more health-oriented, factual presentation, but the emphasis was still on the negative consequences of drinking. By the late 1950s and the early 1960s, alcohol education had become less judgmental regarding alcohol use and more colorless, scanty, and limited. During the 1960s the emphasis shifted gradually from alcohol education toward alcoholism education. This carefully neutral approach presented alcoholism as a disease while avoiding alarmist interpretations of the small but quite real risks of alcoholism for the individual drinker.

In the 1970s, attention to alcohol education as a device for controlling the alcohol problem increased markedly, as a result of at least three factors: (1) the establishment of NIAAA with a broad mandate for controlling the alcohol problem; (2) the heavy reliance on drug education for coping with the drug problem; and (3) the "discovery" of significant levels of teenage drinking. Using such approaches as "values clarification," "enhancement of self-concept," "training in coping skills," and "peer counseling," the approach focused on teaching individuals strategies to cope with a variety of life problems—not just alcohol and drinking. This emphasis appears in the "responsible drinking" campaign promoted by Morris Chafetz as founding director of the NIAAA. The alliance of the National Clearinghouse for Alcohol Information in NIAAA and the network of prevention coordinators within state alcoholism authorities resulted in a proliferation of new educational materials as well as a marked increase in attention to evaluation and research. Whereas 15 years ago alcohol education was a desultory part of the high school health curriculum, there are now numerous alcohol-centered programs sponsored by NIAAA's Prevention Division, the National Institute on Drug Abuse, the U.S. Department of Education, and a number of private organizations. Replication and evaluation of projects in anticipation of national dissemination are now under way (NIAAA 1979).

The ultimate effects of these educational programs remain unclear. The data that would allow evaluation of the programs have become more voluminous and sophisticated, but remain inconclusive (Staulcup et al. 1979, Wittman 1980). Still, it is possible to reflect a bit on the strategic orientation of these programs in the light of what we do know about drinking practices.

In general, we think that the ambitions of school-based alcohol education programs are too grand. They are seeking to produce broad and durable changes in the orientations of students toward drinking: in effect, to innoculate the student population against future drinking problems. Moreover, they are relying on a minor part of the curriculum in the schools to produce this effect. Given what we know about drinking practices, this strategy seems overly ambitious. It is known, for example, that drinking practices change dramatically over a person's life. Consequently, the idea that we might fix lifetime drinking practices within certain parameters by educating people early evokes skepticism. Similarly, it is known (Blane 1976) that most drinking behavior is learned at home, and that students listen with only half an ear to teachers when they raise questions about life-style. Consequently, the public schools may not be the most powerful institutional base from which to launch an alcohol-prevention education program. For these reasons, we believe that it is inappropriate to define the objectives of school-based education programs as the innoculation of students against future alcohol problems.

A different objective might be effectively pursued through school-based programs. Significant problems are associated with the significant amount of drinking done by young people of high school and college age (e.g., traffic accidents, fights, and diminished classroom performance due to drunkenness on school nights, in class, or instead of class). It would be valuable and feasible for education programs to focus on the *current* problems of youthful drinkers without reference to the question of whether the training efforts shaped long-term practices. It could conceivably be useful to teach youths that they are responsible for violent and vicious behavior whether drunk or sober—that it is a matter of responsible behavior, not just responsible drinking. In effect, high schools and universities should adopt the objective of shaping drinking practices for the student populations while they are of school age and should register successes if they succeed in reducing drinking and problems associated with drinking for their current populations. This reorientation may not require much change in current approaches and materials, but it would certainly require replacing the prevailing vague objectives of school-based alcohol education programs with clear behavioral goals. In any case, it would make a big difference in how we evaluate these programs and in the urgency that the schools and students may feel in coping with the problems.

MASS MEDIA INFORMATION CAMPAIGNS

Typically, education for adults is viewed primarily in terms of mass media campaigns. The publicists of the temperance movement were masters of media campaigns. After repeal, media efforts along temperance lines for the most part disappeared. The alcoholism movement at this time invested little effort in the mass media to ain acceptance of the disease concept. Since the mid-1960s, however, interest has greatly increased, as has investment in media campaigns.

The consensus of those who have surveyed the potential of using mass media for reducing alcohol problems parallels our judgment on school-based education programs. There is very little hard evidence pertaining to the impact of these programs, with some recent, notable exceptions. The criticisms are numerous: poorly designed evaluation, no evaluation, outcomes limited to changes in attitudes or information with little attempt to connect these outcomes with behavior. Blane and Hewitt (1977b, p. 32) speak for many reviewers in concluding that: "the effects of public education are largely limited to increasing knowledge and reinforcing established attitudes or behavior patterns." Blane, in another article (1976, p. 357), notes that "The effects of public information and education programs on alteration of drinking patterns is indeterminant, although it is generally felt to be slight. Research evidence is almost totally lacking; all that exists are bits and pieces of information that are no more than suggestive." Moreover, on the basis of a combination of general empirical findings, common sense, and logic, Wallack (1980) makes a persuasive case that noticeable behavioral changes would be unlikely as a result of mass media campaigns alone.[2]

The future of mass media campaigns in controlling the alcohol problem is not likely to be as an independent instrument but as a program undertaken in conjunction with something else: as a device for increasing the salience of a new legal restriction as in the British Road Safety Act, as reinforcement for ideas presented in school-based programs, or as a way of triggering interest in programs that can provide training to individuals who want to change these current practices. It is to this last idea that we next turn.

COMMUNITY-BASED HEALTH INFORMATION AND TRAINING

Perhaps the most promising new approach in the area of education and training is a program that combines carefully designed mass media campaigns with personalized behavioral training provided through schools, volunteer associations, or health maintenance organizations. The most widely known (but by no means the only) example of such a program is the Three Community Study (TCS) of the Stanford Heart Disease Prevention Project.

In the area of education and training the TCS is an example parallel to the study of the Road Safety Act in drunken-driving deterrence; we present a detailed review in order to give readers a more concrete idea of the genre. The TCS was intended to determine whether state-of-the-art programming of mass media in a community and intensive training of a segment of a community would be effective in modifying behaviors known to contribute to the risk of heart disease (Hochheimer, in this volume; Maccoby 1980; Farquhar et al. 1977). Two towns of about

[2] Despite the enormous expenditure for advertising different brands and types of beverages, "there are few reports of empirical studies of the effects of alcohol advertising and no clear evidence that advertising affects consumption" (Ogburne and Smart 1980, p. 293). It is generally thought that the main effect of commercial advertising is to alert the public to new brands, in competition with older ones, and conversely to protect or expand the market shares of established brands. The available scientific evidence is too sparse to permit us any extended discussion of the effects of advertising policies. Nevertheless, important issues of principle are involved in such policies, which may serve as barometers of the influence of different governing ideas.

13,000 in central California were selected for public health education, and a third town served as a control. A random probability sample of adults (age 35-59) in all three communities completed interviews about their knowledge of heart disease and about behaviors and life-styles affecting their risk of heart attack. The participants also provided measurements of serum cholesterol, triglyceride, weight, and blood pressure.

After the initial survey, educational campaigns via mass media were started in the two towns. Specifically, the programs advocated dietary changes (a reduction in animal fats, cholesterol, sugar, alcohol, and salt and an increase in fiber), giving up cigarettes, a return to ideal weight, and a program of regular exercise. Most of the behavior changes that were advocated involved not only motivation to change (almost all adult cigarette smokers know that smoking is dangerous to their health and the great majority have tried to stop smoking) but also the learning of new information, the acquisition of new skills, and the practice of techniques for maintaining these skills.

A multimedia program, developed in the Department of Communication at Stanford University, was carefully orchestrated to achieve a series of objectives. With the cooperation of local broadcasting and newspaper outlets, messages were delivered through a variety of media. The sponsors of TCS reasoned that televised and radio spots of 10 to 60 seconds' duration, if widely viewed many times, could get messages attended to very well; mass media might not succeed in telling people what to think, but they could do an excellent job of telling people what to think about (Cohen 1963). There were some 50 TV spots and more than 100 radio spots. In addition, more than three hours of television programming, many hours of radio programming, weekly newspaper columns, newspaper articles and advertisements, and direct mail were involved. In one town, a group of people found to be at high risk of cardiovascular disease were selected at random to receive an additional 14-week program of intensive instruction. They received home counseling or took part in group classes led by university physicians, health educators, and graduate students trained in health education.

Surveys and medical examinations were undertaken after one, two, and three years of the campaign. As might be expected, those receiving intensive training in addition to the media campaign showed the sharpest initial reduction in risk. By the end of two years, however, the town receiving health messages through the media only had caught up with the community including the intensive instruction group. When overall risk of heart disease was calculated, participants showed reductions of between 16 and 18 percent after two years. In the town that received no educational campaign, the average risk had increased by 6.5 percent.

There was some retrogression during the third year in the mass-media-only town when educational programming was sharply curtailed, but not in the mass-media town that included intensive instruction. Apparently the supplemental value of face-to-face instruction had more staying power.

Certain detailed findings are of particular interest. Smoking cessation was not achieved to any great extent in the mass-media-only town, but 50 percent of those receiving intensive instructions (versus 0 to 15 percent of the high-risk controls) had quit smoking after three years. Dietary cholesterol levels decreased by more than 30 percent in both experimental communities (versus 10 percent in the control). Changes in weight were not achieved by either method (see Stunkard and Penick 1979); in contrast, however, the control sample gained weight.

The TCS is only a single study of community-based health training practices. Extensions and replications now under way will improve the breadth and reliability of the data base. One extension of the intensive instruction method using peer-group trainers has been put into effect in one of two matched California junior high schools (McAlister et al. 1980). The students who were trained to resist social pressures toward substance use were reporting cigarette, marijuana, and alcohol consumption rates less than half those in the control school, nearly two years after training was initiated. The authors note that (p. 721): "The exact processes through which this apparent effect has been produced are uncertain. Our impression is that the program created generally negative attitudes about smoking at least as much as it actually taught skills for resisting pressures to smoke." The attitude change, it should be emphasized, was an *effect* of the training, not its focus.

In conclusion, there is now some accumulating evidence, from TCS and other efforts, that health information and training programs can be effective in reducing health-damaging activities involving overconsumption or poorly managed consumption. There are, however, many questions still to be resolved. Efforts are still in an experimental stage, and there is a substantial difference between experimental innovative programs (and closely associated replications), on one hand, and widely diffused "mass production" equivalents, on the other. The step from research-based implementation by highly committed originators to routine dissemination by initially indifferent or skeptical organizations and operatives is a long one whose outcome cannot be ensured. Nevertheless, the experiments are promising in an area in which there has been a dearth of empirically based encouragement for many years. Moreover, an institutional base that could be exploited in implementing such programs is now being expanded. That is the network of health maintenance organizations springing up around the country. Such organizations are likely to be interested in and capble of providing suitable training programs at relatively low cost.

CONCLUSIONS ABOUT EDUCATION, INFORMATION, AND TRAINING PROGRAMS

Reflecting on this accumulated experience with education, information, and training programs, and considering what distinguishes relatively promising efforts from less successful ones, we can draw several tentative conclusions.

First, it is important to define the behavior that is the target of the educational effort as concretely as possible. Otherwise, it is impossible to design or evaluate a program effectively.

Second, the emphasis should be on teaching specific new knowledge, acquiring new skills, and practicing techniques for maintaining the new skills. Education, information, and training are intended as tools of informed choice. They should be viewed as ways of expanding individual and community capacities to achieve desired ends, rather than as ways of shaping, changing, or clarifying these ends. Education may have this latter effect, but it is not the sole or main focus of attention.

Third, the programs that have been successful have adopted an experimental approach and have drawn heavily on the professional knowledge and skills of people trained in behavioral and communication sciences. This professional and experimental orientation has not been characteristic of past education efforts, which have relied on the expertise of commercial advertisers and the enthusiasm of a wide variety of local volunteer groups. Such operations have generally failed in the detailed planning, precise execution, and frequent evaluations that seem essential to fielding an effective training program.

Thus, we think there is potential in these areas, but it does not lie where we have commonly looked. It is not exclusively in the schools or in mass media advertising. It may be in information and training programs sponsored by universities and health maintenance organizations focusing on the health risks of some drinking practices and teaching techniques for modifying personal drinking habits.

SETTING A GOOD EXAMPLE: A NOTE ON SYMBOLIC EFFECTS OF GOVERNMENT ACTIONS

As Mosher and Mottl document (in this volume), there are a surprising number of jurisdictions, populations, and spheres of work and leisure in which the federal government retains formal control over the supply of alcohol, drinking practices, or products or facilities directly relevant to drinking problems. In certain areas the government has chosen to clearly recognize alcohol as an intoxicating, controlled substance—that is, a drug. Two examples are the prohibition on the presence or sale of alcoholic beverages in virtually all federal work sites and the establishment of the National Institute on Alcohol Abuse and Alcoholism with responsibility both to treat alcoholism and to work at the prevention of alcohol abuse and alcoholism. In other areas, the federal government has treated alcoholic beverages as commercial goods that provide legitimate business opportunities for entrepreneurs (e.g., in the granting of small business loans to fledgling package stores). In still other areas, such as its management of most national parks, the federal government has chosen to fit its alcohol policies to local views.

It can be argued that a reasonably consistent approach to alcohol throughout the federal government's jurisdiction would be a desirable component of any government-sponsored program to shape drinking practices. This argument is appealing in institutional terms because it seems easier to establish a coherent line within one level of government than to operate through the myriad state and local agencies that write laws and regulations, enforce them, mount preventive training programs, and so on. Not only are there fewer agencies at the federal level, but one can rely on NIAAA to provide a natural and accepted institutional focus for such an effort. Moreover, it seems conceivable that there would be substantial benefits from striking a consistent position within the federal government. The terrain covered by the federal government is large enough to suggest that the *direct* effects of federal efforts would not necessarily be small. In addition, the symbolic weight of the federal government's position might play an important role in reinforcing individual decisions about drinking practices or mobilizing institutions at other levels of government and in the private sector to move in similar directions. In essence, if the weight of the federal government could swing in a consistent, coherent direction, some broad social momentum could be developed even though many of the relevant agencies lie outside the direct control of the federal government.

This argument seems plausible, but one can be skeptical about how easy it would be to develop a coherent line throughout the federal domain and how much this would matter to the positions of private individuals and other agencies concerned with the problem. Probably the strongest argument for undertaking the arduous bureaucratic task of coordinating alcohol policy throughout the federal government is not for the symbolic effects, but for the material effects. Enough drinking behavior is within the direct reach of the federal government to make it worthwhile to manage this drinking, quite apart from any symbolic effect.

CONCLUSION

We began this chapter by rehearsing the substantial normative and pragmatic objections to the general idea that government might be able to shape drinking practices directly. We conclude with some guarded optimism about the potential for some carefully designed efforts to accomplish this purpose in selected areas.

In general, we recognize that the law is a clumsy and expensive instrument to wield in shaping private drinking practices, but in one area we think it is feasible and desirable to do so: the regulation of drunken driving. The problems associated with drunken driving are large enough and involve enough innocent third parties to warrant legal intervention. Moreover, institutions exist that can enforce such laws more aggressively and are backed by a consensus against drunken driving. Major barriers to more aggressive action in this area seem to be some confusion about the nature of the offense and ambiguity about institutional leadership. Ironically, the confusion about the offense has been created partly by mass media campaigns against drunken driving. Ads suggest that any level of drinking is dangerous in combination with driving. This causes many people to assume that the criminal offense is simply having a drink and then driving—something that many people have done. Hence, they are inclined to be lenient with accused offenders. Actually, the offense is *drunken* driving, and the standard for drunkenness is usually set far above the levels of intoxication that most people reach. If people (particularly jurors) knew just how drunk one had to be to be charged with the offense, it is conceivable that they would be more willing to convict, because the legal level of drunkenness really does establish a deviant level of drinking. The confusion about institutional leadership exists because NHTSA and NIAAA share federal jurisdiction in this area, and virtually all of the enforcement capacity is at state and local levels.

Education programs seem attractive because they generally appear to be less intrusive, more flexible, and more precise, but there is little direct evidence indicating that they can succeed, particularly on a large scale. We are especially skeptical of the idea that adults or youths in high school and college can be told things about drinking that will then keep their drinking within safe limits for the whole of their lives. We do think that educational institutions might be able to shape current drinking practices in safer directions, by talking about current dangers and risks and providing some techniques for improving practices if people are so inclined. More generally, we think that the key to successful education programs is to combine (1) well-designed local mass media campaigns that increase public salience and private concern about drinking practices with (2) low-keyed programs offering simple behavioral training techniques to those who are motivated to change their drinking practices. In particular, we think programs focusing on the health consequences of too much or too reckless drinking mounted by local hospitals and health maintenance organizations might succeed in reducing some of the adverse consequences of drinking. In undertaking such programs it is very important to understand that the objective is to change specific bits of behavior and also that we are fundamentally dependent on the preexisting motivation of people to produce these changes. The task is thus to make it relatively easy and unembarassing for people to learn how better to understand and control their drinking.

Finally, we think it is desirable for the federal government to review its programs in a variety of areas and ensure that its actions are consistent with its position with respect to drinking. This is important primarily because the cumulative impact of the various areas of federal influence might be substantial and also because the symbolic effect of doing so might stimulate other institutions to move in similar directions.

We note that actions of the type suggested here always carry with them significant social risks. In a society as large and heterogeneous with respect to drinking practices as ours is, it is difficult to establish any kind of boundary around appropriate drinking practices. Those whose practices are now well within the boundary may feel outraged that the government could be so irresponsible as to license a kind of drinking that seems very reckless to them. Those whose practices are outside the boundary may feel indignant that the government is discouraging conduct that seems quite safe to them and may complain that the government is interfering. Moreover, to make matters worse, whenever the government draws lines with respect to drinking practices, many third parties are mobilized and feel entitled to disapprove of and comment on practices that are outside those lines. This can exacerbate social conflict over drinking and produce important consequences even if the level and character of drinking does not change. It is for these reasons that we considered the substantive changes in drinking practices simultaneously with the instruments. We believe there is sufficient consensus about drunken driving and heavy cumulative consumption to make these suitable subjects for the law and for community-based health training intervention, respectively; we remain uncertain about how we might proceed to discourage drunken violence.

6 Reducing Environmental Risk

In the two preceding chapters, strategies were outlined that sought to influence the supply of alcohol, drinking premises, and drinking practices—and thereby to reduce some of the problematic consequences of drinking. These discussions centered on ways to diminish the quantity of consumption or the pairing of intoxication with certain activities, by using taxation, the law, training, and so forth. The focus in this chapter shifts from the management of drinking practices as such to ways of modifying environments so that when drinking or drunken activities occur, they are less likely to cause or exacerbate damage.

We are interested in this approach because even the most strenuous efforts that might be taken to reduce hazardous consumption or episodes of drinking cannot realistically be expected to eliminate or radically curtail them. Therefore it is sensible to investigate an additional set of mechanisms that will extend to people and events unaffected by other prevention policies. This approach—manipulating the environment so that risks of harmful effects are reduced even in the presence of drunken behavior—has been characterized as "making the world safe for drunks," a characterization that tends to evoke moral uneasiness or even outrage in some quarters, rooted in a long-standing view of drunkenness as moral error that should get its just deserts. This view is sufficiently widespread and seductive that it must be explicitly countered if we are to contemplate the use of this strategy.

The approach can be more accurately, if more awkwardly, described as "making the world safer for, and from, people who are affected by alcohol intoxication or other impairments." This elaborate description is meant to emphasize two important features. First, the world can be made safer not only for but also from drunkenness. When a person drinks too much, whether frequently or only once in a lifetime, hazards may be created not only for the drinker, but also for many others with whom the drinker has contact. To give a dramatic example, "dead man" controls, which automatically stop a train run by an intoxicated engineer who has passed out, may not only save the engineer's life, but also the lives of many others. On a more mundane level, passive restraints in cars protect drunken drivers as well as "innocent" passengers and drivers in other cars so equipped.

Second, impairments due to drunkenness are not unlike those due to fatigue, absent-mindedness, anger, previous minor injuries, etc. They are routine matters, and the world is in many ways arranged to keep them from causing further harm. When drunkenness enters the equation, however, the normal calculus of risks is revised or suspended, because drunkenness is seen as a moral issue, implying intentionality with respect to any consequences. For example, statistics on alcohol's involvement in casualties other than auto accidents are not collected or kept; this is apparently due, at least in part, to avoidance of the moral loading that the involvement of drunkenness would put on these situations. Whatever the moral evaluation, it seems clear to us that information about drunkenness as a cause of accidents would be a significant aid in detecting potentials for "human error" and making products and structures safer from its results. Thus, the benefit of this approach is not limited by the degree to which alcohol is involved in particular hazardous consequences.

If there were a moral justification for "making drunks pay" for their errors, the debtors would include everyone who has ever had "a bit too much," which includes a substantial majority of the adult U.S. population. As indicated in chapter 2, while very heavy drinkers, including the minority who are definitely alcoholics, account for a disproportionate share of drunken episodes, the far larger numbers of less heavy drinkers account for the majority of such episodes. In any case, a "let them pay" attitude clashes with long-standing medical and public health values. Just as a private doctor does not refuse to care for a patient because the illness or disability is the patient's fault, there is a public health commitment to limiting morbidity and mortality irrespective of other moral issues. We do not withhold public funds from the treatment of heart disease in smokers or obese people, or discourage the spread of cardiopulmonary resuscitation training or blood pressure testing, on the grounds that many or even most of the people at risk may contribute to their own problems by dietary indulgence and deficient physical discipline.

The rationale that a "let them pay" approach acts as a deterrent to drinking is analogous to banning public clinics for treatment of venereal diseases on the basis that they encourage promiscuity. We do not make policy, deliberately or by default, that syphilis is a just and useful punishment for the sins of fornicators and adulterers—even in jurisdictions in which their acts may be crimes. In general, our national policy is to enhance public health without special regard to the morality or even legality of contributary actions. We do not passively permit deaths and injuries that can be prevented.

Assuming that these points neutralize the main ideological obstacles to an approach focused on minimizing the harmful consequences of drinking, how can it be done? As mentioned earlier, this chapter is primarily concerned with environmental modifications based on the premise that hazardous drinking takes place. The range of possible modifications is extensive, taking account of both the psychobiological and cultural properties of alcohol.

There are two principal areas of such modifications: first, ones that try to reduce the likelihood of traumatic injuries from mechanical causes, either by making the physical environment more forgiving or by placing social "cushions" around the drinker; second, ones that try to reduce damage wrought by interpersonal events, either by physically isolating the drinker from certain hostile reactions or by trying to reduce the intensity of such hostility.

DAMAGE IN THE PHYSICAL ENVIRONMENT

PHYSICAL SAFETY MEASURES

Alcohol intoxication beyond a minimum level makes one clumsy and inattentive, and it is therefore directly implicated in a certain proportion of casualties due to ineptness in dealing with the environment: burns, drownings, falls, and other momentum injuries, notably motor vehicle accidents. These alcohol-related problems can thus be reduced by making the created physical environment more tolerant of inept (including drunken) behavior. If everyday materials are made less likely to burn, cut, trip, or gash people who are temporarily operating below normal levels of efficiency, the rate of alcohol-related casualties will decline. Since alcohol-related casualties tend to be concentrated in the most severe categories— deaths and permanent disabilities—safety improvements will affect alcohol-related injuries most dramatically.

Our society has chosen to give substantial attention to making the environment safer through such public and private agencies as the National Safety Council, the Occupational Safety and Health Administration, the Underwriter's Laboratories, the Consumer Product Safety Commission, and a myriad of others. Traditionally, safety standards have been set assuming adults are alert and active—and, in many situations (including driving), trained in specific skills needed in the situation. Gradually, it has been recognized that engineering standards should take more realistic account of "human factors," including the likelihood of occasional failures of people to live up to the ideal of an alert, active, and skillful adult. So far, such factors have primarily been taken into account in policy making for children and for the physically handicapped or enfeebled. Virtually no attention has been paid to those handicapped by drunkenness or other mental states seen as voluntarily assumed. This gap in official policy contrasts with the great deal of unofficial action by drinkers and their associates to arrange circumstances so as to minimize damage around drinking.

Alcohol-related accidents are important contributors to morbidity and mortality statistics, particularly in younger age groups, in which death or disablement affects many years of expected life. One-half of all deaths among people aged 15-24, and one-third of all deaths among people aged 25-34, occur as a result of accidental injury (National Center for Health Statistics 1977); the overall death count from accidental injury was 103,000 in 1979 (National Center for Health Statistics 1980), about evenly divided between road crashes and other types of fatalities.

The most important class of technological devices for the prevention of traumatic deaths related to alcohol is probably passive restraints (such as air bags) for auto passengers. Due to the nearly universal practice (in the United States) of routinely determining the BAC of decedents in auto fatalities and the funding of research to develop comparison samples of drivers (Borkenstein et al. 1974, Farris et al. 1976, Holcomb 1938, McCaroll and Haddon 1962, Perrine et al. 1971), we are able to estimate with some confidence that there are about 12,000 deaths due to drunken driving annually in the United States (Reed, in this volume). A large fraction of these would probably be prevented by sound passive restraint systems.

This is not the only automobile safety technology possible, nor is auto safety the only area in which alcohol contributes to traumatic injury. Mosher and Mottl (in this volume) indicate a series of fireproofing technologies that have not been used by manufacturers of home furnishings and cigarettes, despite the involvement of these ignition sources in many residential fires. The connection of alcohol-aided fatigue and carelessness in these fires is well documented (Aarens et al. 1977), although there are insufficient data to estimate as narrow a range of attributable deaths, injuries, or property damages as can be done with auto accidents.

There is in fact a welter of potentially hazardous environments and circumstances, and there is a corresponding array of safety and other relevant agencies involved or potentially involved in this strategy. This variety poses a problem for public policy: how can these many opportunities to favorably affect the rate of alcohol-induced injury be detected and evaluated for efficient returns?

In other circumstances, this type of problem has been handled by making a "watchdog," oversight, or coordinating unit responsible for energizing and prodding the relevant agencies to take account of the specific issues that reach across jurisdictions for which consumer protection in regard to alcohol is relevant. Thus, for example, the National Transportation Safety Board has jurisdiction with respect to safety in the regulatory territory of a variety of other agencies, from the Coast Guard to the Federal Railway Administration. In a different mode, the White House Interdepartmental Task Force on Women has served as a coordinating center across all federal agencies for issues of concern to women as an interest group. What is needed, either as a specific function of the National Institute on Alcohol Abuse and Alcoholism or in some other form, is an institutional focus on environmental modifications to reduce the hazards of drunkenness.

The first item on the agenda for such an effort would be to bring the same attention given to BAC in auto fatalities to other accident reporting systems. An attempt to build a comprehensive overview of the role of alcohol in traumatic death and injury at the national level would be quite valuable (cf. Haberman and Baden 1978).

The fact that a given sequence of events or product design features is found to be important in producing or reducing injury does not answer the question of what course of government action is appropriate. To use the example of passive restraint technologies in automobiles, it is possible to promote their use by the following:

- educating consumers to buy or demand them;
- providing research and development grants to improve them;
- differentially taxing autos that are not equipped with them;
- requiring that manufacturers make one or more of them available as optional equipment; or
- requiring every automobile sold or licensed to be furnished with one of them.

This list is neither exhaustive nor mutually exclusive; it serves as a reminder that there are no automatic connections between knowledge about damage and subsequent government action. Moreover, the cost of even the least expensive action may be so high as to make doing nothing an important option. An agency broadly interested in the role of alcohol-induced and other impairments in accidental injury, by fostering the development of more suitable accident reporting systems, could provide a focus for investigating the value of alternative policies suggested by this information.

SOCIAL SAFETY MEASURES

Apart from the emphasis on consumer product safety, the social environment can be manipulated to decrease the exposure of drinkers to alcohol-induced casualties. The intervention of third parties to keep the intoxicated individual from performing potentially hazardous acts (e.g., saving an intoxicated friend from having to drive home from a party; training bartenders to carry out a similar function for their customers; routine checks by supervisors of personnel reporting for work in situations in which their job entails a potentially dangerous task[1]) are examples of alterations in the social environment aimed at avoiding the consequences of drinking. They are, in essence, strategies to persuade those in potential settings for drunkenness to take active steps to insulate the drinker from harm, and thus to put a social cushion between drinkers and potentially hazardous environments. These measures do not seek to basically alter or interfere with drinking or drunkenness—but to limit certain consequences that are generally aversive or at best ambiguously regarded.

[1] This approach undoubtedly raises serious issues of civil liberty, but the pros and cons of the "intrusion" must be weighed. Checks can be applied in an objective and nondiscriminatory manner, using, for example, a breathalyzer for all personnel with certain responsibilities. Although the check-in may be regarded as odious, the occupational safety rationale should make it more acceptable. Clearly it would be a message about the hazards of certain jobs, a warning to those who know that they drink excessively, and a communication that drunkenness will not be covered up or implicitly condoned in jobs in which certain risks are involved. Specific identification of such jobs is not a task that this panel has undertaken, but categories such as air traffic and flight control, missile guidance, heavy crane operation, surgery, and medical intensive care come to mind. More generally, 36 percent of the labor force in 1971 was engaged in work involving a fairly complex relationship to "things" (i.e., machines, tools, equipment, products), and 15 percent of all the occupations listed in the *Dictionary of Occupational Titles* involve working under hazardous conditions (U.S. Department of Labor 1977; see Miller et al. 1980 for further specification and analysis).

DAMAGE IN THE SOCIAL ENVIRONMENT

VICTIMIZATION

Associated with its physiological effect of producing clumsiness, alcohol makes people vulnerable to harm or exploitation. The least arguable connection between alcohol and crime is that drunks are easy and traditional prey for criminal harm and exploitation. There is a specific and ancient criminal profession—the jackroller—that specializes in mugging drunks. The tendency of American culture to enclave public drunkenness (and, for that matter, tourist and convention activities) conveniently concentrates victims of the jackroller and allied trades in particular city districts. Victimization of the drunk, of course, extends beyond skid row inhabitants into all social strata.

Despite the evidence that drunkenness leads to victimization, there has been little public discussion of the issue or of its implications for public policy, primarily because of the morality and culpability issues seen as attached to drunkenness. The substantial recent literature on victimization makes little or no reference to drunkenness of victims. Much of this literature is ideologically oriented toward increasing social assistance and benefits for victims of crimes, and it may be that drunkenness is seen as detracting from the deservingness of victims (Miers 1978). When San Francisco police initiated a decoy program to catch muggers, in which police posed as semicomatose drunks with wallets half-visible, commentary in local newspapers tended to depict this as a waste of police resources, and there was even some sympathy for the criminals thus caught as somehow being victims of entrapment.

While there is clearly a strong association of drunkenness and victimization, it is poorly quantified—it is not even included in such general accountings of the costs of drinking as the Berry and Boland (1977) study—and there has been no study of the possibilities of protecting the environment of drunkenness against criminal activity. This seems a potentially fruitful area for study under the auspices of the National Institute of Justice and the National Institute on Alcohol Abuse and Alcoholism.

PUBLIC DRUNKENNESS AND THE URBAN ENVIRONMENT

By definition, public drunkenness is a problem that is in part a matter of environment; private drunkenness is generally not in itself illegal. One potential strategy for diminishing the problem of public drunkenness is thus to rearrange the environment of drunkenness so that it is inconspicuous or out of public view. This strategy is, of course, widely pursued by individual drinkers and their associates; it is also clearly available as an instrument of social policy. In core areas of American cities, however, the trend in the last 35 years has been one of urban renewal, demolishing the traditional institutions of skid row, often forcing the habitues into the street, with the intention of obliterating any haven for impoverished drunks in the city. In many cities, it was during the heyday of urban renewal that arrests for public drunkenness reached their peak.

Social policy toward skid row drunkenness in the past has mainly tried to treat the problem out of existence. Despite the many humane efforts and arrangements pursued along this line, it cannot be said to be a policy that solves the problem. On a local, not federal, level, there have been scattered attempts at environmental solutions. Reinventing the older tradition of municipal lodging houses, San Francisco for a time used federal Model Cities money to run a "wet hotel," i.e., a residential hotel in which the poor, like the rich, had the right to drink on the premises. An approach tempered by the modern alcoholism perspective would modify the "wet hotel": after an initial detoxification (affording an opportunity to begin a healing process, physically and emotionally), those desiring to continue treatment could do so on a voluntary basis in a setting of their choice, while those opting to resume drinking could do so, in the hotel if they wished, with access to an ongoing possibility of care.

DEEMPHASIZING HOSTILITY

The collective informal social norms that largely govern drinking behavior in the United States tend to operate within wide margins of tolerance. Typically, the broad post-repeal accommodation of alcohol in this society and the fact that most adult Americans have personal exposure or experience with being drunk make most people hesitant either to intervene directly, or too roundly to criticize, the drunken behavior of others. In short, the typical reaction to intoxication in this country tends to be underreaction, especially if we consider the known health effects of excessive consumption.

There are occasions, however, when the range of tolerance becomes too narrow. This occurs when evidence of past or present drinking behavior is used as an excuse to treat people who drink, or some segment of drinkers, as though they were not deserving of their full civil liberties. Where unwarranted hostility to some people's drinking, intoxication, or past drinking problems leads to denial of their normal opportunities for work or leisure, an effort to defuse or deemphasize this hostility is well worth consideration.

It can be argued that the reduction of an alcohol problem by deemphasis in fact occurred in the course of the 1960s in the United States. Well before the promulgation in 1971 of the model Uniform Alcoholism Intoxication Treatment Act by the National Conference on Uniform State Laws, police in a number of cities were becoming more tolerant of street drinkers, and this tolerance was reflected in a declining arrest rate (Federal Bureau of Investigation, annual). It can be argued that the still-high rate of arrests for this offense reflects an intolerance, not so much on the part of the public at large as on the part of local merchants. Of course, such a policy of deemphasis poses sharp value conflicts—for instance, between the potential benefit to health from compulsory treatment versus the deprivation of liberty involved in such benefits.

In addition to public drunkenness, there are three other areas in which deemphasis may prove to have advantages, although empirical evidence is scarce: teenage drinking, regional differences, and recovery from alcoholism. Teenage drinking has been widely publicized in our society in the past decade. Some students of alcohol policy have suggested that this problem has been exaggerated by governmental agencies and that it has diverted public attention from problems of adult drinking (Chauncey 1980). Certainly the criteria used to discern drinking problems among teenagers are far less tolerant of drunkenness than the criteria used for adults (NIAAA 1978). It is notable that the level of public concern with teenage drinking has risen appreciably in the 1970s, yet there is substantial evidence that the national quantity and pattern of teenage drinking has remained about the same from 1965 to the present (Blane and Hewitt 1977a, Abelson et al. 1977, Johnston et al. 1979; there may have been significant local departures from this trend: see, e.g., Blackford 1977). It is doubtful that the increased level of concern has contributed to the prevention of teenage drinking as such, although it may have been instrumental in the reversal of statutes lowering the minimum drinking age in several states.

Rates of drinking are disproportionately distributed throughout the country—that is, the southern prairie and mountain areas are relatively dry, and the northeast and coastal regions are much wetter. Hence, policies that focus public attention on drinking may be inappropriate and may needlessly arouse concern in the drier regions. It has been found that, particularly in the traditionally dryer parts of the country, some drinkers report family or other complaints about relatively light patterns of drinking (Cahalan and Room 1974). Whether these reports are indicative of a serious problem of overreaction is not now known.

The third area in which strategies of deemphasis might prove valuable is occupational opportunities for recovered alcoholics. Irrespective of current drinking status, recovered alcoholics often face discrimination when seeking employment and quickly learn to conceal their past drinking history. A similar concern about exposure and its impact on their jobs, reputations, etc. contributes to the reluctance of some otherwise interested alcoholics to seek treatment or hospitalization. Alcoholism treatment was not covered, until recently, by the majority of health insurance plans.

In the past five years, however, alcoholics have gained certain legal protections. Legislation forbidding discrimination against them in housing and employment, legislation regulating the confidentiality of the medical records of alcoholic and drug abuse patients, and legislation providing Blue Cross/Blue Shield coverage are now on the books.[2]

Government policy on alcohol in the last 10 years or so has tilted very firmly toward the identification of an ever-broadening range of behaviors as problems and toward an increased likelihood that the problems will be identified as alcohol problems. The current practice of raising the threshold of people's worries about their own and others' drinking is often taken for granted as a policy imperative, without recognizing that there are potential costs as well as benefits from such an approach. To define more and more drinking behaviors as unacceptable may reduce somewhat the rate of those behaviors; it also tends to make deviants of those who persist in the behaviors, which accelerates the development of an isolated subculture at odds with the larger society, requiring increased attention of legal agencies to the reduction of this conflict.

CONCLUSION

Public concern about alcohol problems has centered on two specific areas. First is the recurrent default of major social roles—in the family and at work—that accompanies long-term loss of control over drinking behavior, which is the existential experience at the heart of the disease concept of alcoholism. Second is drunken driving, which is seen to overlap with the problems of alcoholism as well as to constitute an issue in its own right.

The emphasis in both of these areas has been on influencing the drinker—getting the alcoholic to stop drinking and the drunken driver to not pair the two activities. Whatever progress may be made, prevention and recovery efforts directed at these problems will not eliminate

[2] The panel did not investigate and takes no position regarding recent changes in health insurance coverage for alcoholism treatment.

either of them. Recognizing this reality, the approach suggested in this chapter is to consider how these and related effects of hazardous drinking can be lessened in severity: to examine ways of modifying humanly built, humanly formed environments to make them safer for, and from, drunkenness and other impairments.

In the area of accident prevention, it is clear that alcohol intoxication is a prominent form of human factor error in the most severe categories of injury. Alcohol intoxication has been investigated in depth only in the area of road accidents, although a number of limited studies (see Aarens et al. 1977, Mosher and Mottl, in this volume for detailed bibliography and review) have suggested that some other accident categories may have as high or higher overall rates of alcohol involvement. Further inquiry into these areas would be greatly facilitated by systematic attention to BAC in the reporting of accidents by various agencies. These reports ought to be monitored to discover information useful to product design and public policy formation. A marriage of technological safety innovations specifically adapted to human factors, with accident reporting systems that attend to BAC as a major indicator of impairment, seems to us a quite promising strategy. Depending on cost and acceptability, informed choices could then be made among a variety of ways of achieving safer environments.

Accident prevention by increased social buffering of drunkenness is another area of potential intervention. These strategies are similar to a number of those discussed in chapter 5, but they differ in that the focus is not on the drinker as such but on third parties who may be in a position to help prevent drinkers from engendering serious damages.

Moving from the area of traumatic accidents to the realm of damaging societal reactions, we find that the data, although much scarcer, are suggestive of important policy issues. The notion that alcohol is a cause of violent attacks has been discussed in chapter 5, where the cultural rather than the strictly pharmacological basis of this connection has been stressed. In looking at alcohol as contributory to victimization, we find a similar phenomenon, that the undoubted impairment of functioning induced by intoxication is converted into risk of victimization by a series of sociocultural arrangements, arrangments that are subject to change. The recent shifts in policy dealing with public intoxication are exemplary of such changes.

We argued previously that drinking practices are indigenous, collective complexes that have long histories and that they are both autonomous and variously resistant to the efforts of government to change them quickly. Nevertheless, some actions of government can affect drinking practices, in predictable, important, and even desirable ways.

All of this applies as well to the range of economic and other practices that shape the safety of environments in which drinking and intoxication occur. While government agencies such as NIAAA need to act forcefully if they are to have any impact, they need not batter blindly at one side of a problem and ignore any undesirable results that may be created at its other side. For this reason we think it is important to improve current knowledge about how increased sensitivity to alcohol problems may have negative effects, sobering reminders that sometimes problems are better approached by dealing with symptoms as they occasionally occur, rather than raising alarms at any indication that possible causes are present.

Summary

Alcohol problems have been a part of this society since colonial times, and they have left profound marks in the nation's consciousness, including two amendments to the Constitution. There is every reason to think that drinking and its effects for good and ill are here to stay for the foreseeable future. The permanence of drinking and its effects as a feature of society is a necessary premise for any realistic analysis—but permanence does not necessarily mean unresponsiveness. Problems may always remain, yet they may grow larger or smaller, change shape or focus, depending on the measures taken or not taken to affect them. These measures, in turn, reflect values that people place on alcohol use, values that are diverse and, at various points, conflicting.

Public efforts to control the making, selling, use, and effects of beverage alcohol have waxed and waned over the 200 years of U.S. history. On balance, it is clear that a *laissez faire* approach has not prevailed, and there are sound reasons why it has not. The Prohibition era, however, left a sour taste on the national palate, inducing not only a decisive rejection of radical measures of control, but also inclining people toward the belief that no measure of prevention can have a beneficial result. That belief is not sustained by the evidence.

In the first part of this report, we provide a picture of alcohol use and its effects drawn in a broad perspective, with dimensions extending historically, conceptually, and demographically. The historical review points out how a few simple governing ideas have dominated and defined the problem: the idea of normal drinking customs, of alcohol as an addicting poison, and of alcoholism as a disease. The hoary parable of the elephant and the blind men seems to apply, except that in this case the men are not simply groping for descriptions, they are outfitting the beast.

The conceptual model of alcohol problems provided in chapter 2 is meant to escape this tendency of each model to provide too few moving parts, to simplify too greatly. The basic concepts ought to have enough degrees of freedom to see why and where any single plan might go awry and to permit one to consider the use of a number of different levers.

In defining alcohol problems, we began with the set of effects that assume social significance. These include health, economic, and social effects. These effects are differentiated among drinkers, their intimates, and others in society. Moreover, in looking at how these effects emerge from drinking practices, we found it useful to distinguish among degrees of intoxication (how drunk), frequency of intoxication (how often), and cumulative consumption (how much altogether); the contexts and activities that were commonly paired with intoxication (where, when, while doing what); and the general level of risks built into the existing social and physical environment of drinkers. These distinctions are important partly because they contribute differentially to the significant social effects, but even more importantly because each of these features could become a separate avenue for seeking to alter the size and shape of alcohol problems.

When the magnitude and distribution of this complex of effects are examined in the light of our understanding of how they are produced, we discover something unexpected about the location of alcohol problems in the drinking population. As one might expect, a small fraction of drinkers—the heaviest ones—carry a disproportionate amount of the consequences of drinking. But this proportion does not actually account for most of the problems. Even though the *rate* of drinking problems per person among non-heavy drinkers may be much lower than for heavy drinkers, the sheer number of such people is very large. Therefore, substantial fractions of alcohol problems—the major parts of many particulars—are not isolated in a small fraction of the population. These problems are broadly based and cannot be effectively approached except through broad, general measures of the sort that we have called *prevention policies*. If these measures are to be general, they must of necessity be relatively nonpersonal, because of the enormous costs—economic and political—that would arise from trying to tailor such a mass activity to the variable personal characteristics of tens of millions of people. Wide individual variation is a bedrock reality of drinking behavior, but public prevention policies based on individual attention are well beyond the practical limits of current resources.

Beyond the characteristics of being general and nonpersonal, prevention policies do not conform to a single model. For our purposes, we divided prevention policies into three broad classes according to their proximate targets: first, those instruments directed at the supply of alcohol and places to drink it; second, those aimed at the drinking practices of consumers once they have alcohol; and finally those that seek to alter characteristics of the social and physical environment involved in producing certain consequences of drinking (e.g., automobiles in drunken driving, employers' attitudes toward recovered alcoholics).

In the second part of this report we reviewed the evidence regarding these instruments and came to the following conclusions:

1. Regulating the supply of alcohol, through such mechanisms as licensing, limited prohibitions (age, location), and relatively high taxation have a long history in the United States and elsewhere. The machinery of implementation for these measures is available in local, state, and federal Alcoholic Beverage Control (ABC) authorities. The trend in the last 25 years has been a general relaxation of restrictions on the alcohol market, including lower legal purchasing age minima; gradual expansion of outlet numbers, types, and hours; and substantial reductions in real tax rates. This trend has been commensurate with increases in alcohol consumption.

The evidence about regulation of supply induces caution in regard to continued relaxation of controls. Most studies have been of single effects, and it is probable that these effects are cumulative. The evidence is most extensive and uniform in regard to the effects of taxation. It shows that taxes affect prices, prices affect the quantity of consumption, and the quantity of consumption affects the health and safety of drinkers. An increased tax on alcoholic beverages has the particular effect of improving the chronic health picture (as indexed by liver failure) of the heavier drinkers—who are, it can be added, paying most of the tax increase. Therefore we see good grounds for incorporating an interest in the prevention of alcohol problems into the setting of tax rates on alcohol.

2. Drinking practices can be affected principally by law and by education. The use of law to reduce drunken driving has received the most careful study. But legal action does not just mean passing stiffer penalties. The crucial elements appear to be letting the public know police are bent on enforcing the law and increasing police surveillance of night-

time traffic patterns, in which most alcohol-induced highway deaths occur. Such programs appear to have had measurable short-run effects in reducing the number of fatal alcohol-related crashes. Paying for levels of enforcement sufficient to sustain these effects, including the effort to keep these programs salient in the public eye, is apt to be costly.

Education, information, and training to reduce alcohol problems have had a checkered history. In recent years, curricular materials about alcohol use and consequences have improved considerably in their balance and accuracy. But good tools alone do not build bridges—skilled labor and clear plans are necessary as well. As yet, alcohol education has been short on both, and there is little encouragement to be gotten from previous efforts. Nevertheless, a series of new developments in the field of health education combine mass communication principles to increase the salience of a given issue with relatively nonpersonal training techniques designed to help people change their behavior in ways they think are desirable. There is sufficient promise in these new efforts to warrant investment in experimental alcohol training on this model, understanding that, at this stage, collection of knowledge is the primary goal to be sought in such efforts. The feasibility and costs of managing these programs on a large scale remain uncertain.

3. The strategy of environmental intervention in the consequences of drinking follows from the observation that certain ill effects of drinking occur only when particular physical or sociocultural contexts are combined with drunkenness. In many of these cases, altering certain features of the context will reduce the ill effect. This is so especially when drunkenness, like fatigue, reduces alertness in situations in which human factors engineering would help protect against human error. Automobiles are the best-studied example and passive restraint technology the most promising innovation. But auto crashes cause less than half of the fatal accidental injuries sustained in the United States annually. The systematic extension of blood alcohol content testing to reporting systems in other areas of accident and safety research would help considerably in reducing the range of uncertainty about the size of alcohol problems in different areas and would permit concentration on preventive measures that are most likely to have significant effect.

In developing and applying the prevention perspective, we have been struck by, and had to resist most forcibly, the tendency to think about policy in terms of opposed pairs: dry versus wet, prohibition versus unlimited access, treatment versus prevention, good drinking versus bad drinking. A clear-eyed examination of current policies shows them to be juxtapositions of different governing ideas. In such an architecture of compromise, it is often the case, as Mies van der Rohe has said, that "God is in the details." We are convinced that the regulation of supply, legal and pedogogical approaches to drinking practices, and interventions in the environment mediating between drinking and certain of its consequences, represent valid approaches with promise for sustained improvement. Each detailed element will fail or succeed only as it is implemented properly and thoroughly; tactics that are undertaken as part of a broad and coordinated approach are more likely to be effective than ones undertaken in isolation.

The building of effective policy thus needs a general vision as well as a fine hand. The vision that we propose is threefold:

- Alcohol problems are permanent, because drinking is an important and ineradicable part of this society and culture.
- Alcohol problems tend to be so broadly felt and distributed as to be a general social problem, even though they are excessively prevalent in a relatively small fraction of the population.
- The possibilities for reducing the problem by preventive measures are modest but real and should increase with experience; they should not be ignored because of ghosts from the past.

References

Aarens, Marc, Tracy Cameron, Judy Roizen, Ron Roizen, Robin Room, Dan Schneberk, and Deborah Wingard (1977) *Alcohol, Casualties, and Crime*. Final Report, Alcohol, Casualties, and Crime Project, NIAAA Contract No. ADM-281-76-0027. Berkeley, Calif.: Social Research Group (reprinted August 1978).

Aaronson, D. E., C. T. Dienes, and M. C. Musheno (1978) Changing the public drunkenness laws: The impact of decriminalization. *Law and Society Review* 12(3):405-436.

Abelson, Herbert I., Patricia M. Fishburne, and Ira H. Cisin (1977) *National Survey on Drug Abuse: A Nationwide Study—Youth, Young Adults, and Other People*. Volume 1 (Main Findings). DHEW Publication No. (ADM) 78-618. Rockville, Md.: National Institute on Drug Abuse.

Armor, David J., J. Michael Polich, and Harriett Stambul (1976) *Alcoholism and Treatment*. Report R-1739 NIAAA. Santa Monica, Calif.: Rand Corp. (Published by John Wiley, 1978).

Babor, T. F., J. H. Mendelson, I. Greenberg, and J. Kuehnle (1978) Experimental analysis of the "happy hour"; Effects of purchase price on alcohol consumption. *Psychopharmacology, Berl.* 58:35-41.

Beauchamp, Dan E. (1973) Precarious Politics: Alcoholism and Public Policy. Unpublished doctoral dissertation, Department of Political Science, Johns Hopkins University, Baltimore.

Berry, Robert E., Jr., and James P. Boland (1977) *The Economic Cost of Alcohol Abuse*. New York: The Free Press.

Blackford, L. (1977) *Summary Report. Surveys of Student Drug Use, San Mateo, CA: Alcohol, Amphetamines, Barbiturates, Heroin, LSD, Marijuana, Tobacco; Trends in Levels of Use Reported by Junior and Senior High School Students 1968-1976*. San Mateo, Calif.: San Mateo County Department of Public Health and Welfare.

Blane, Howard T. (1976) Education and the prevention of alcoholism. Pp. 519-578 in Benjamin Kissin and Henri Begleiter, eds., *The Biology of Alcoholism, Volume 4: Social Aspects of Alcoholism*. New York: Plenum Press.

Blane, Howard T., and Linda E. Hewitt (1977a) *Alcohol and Youth: An Analysis of the*

Literature 1960-1975. Publication No. PB-268 698. National Technical Information Service: Springfield, Va.

Blane, Howard T., and Linda E. Hewitt (1977b) Mass Media, Public Education and Alcohol: A State of the Art Review. Rockville, Md.: National Institute on Alcohol Abuse and Alcoholism.

Boland, Bradley, and Ronald Roizen (1973) Sales slips and survey responses: New data on the reliabilty of survey consumption measures. *Drinking and Drug Practices Surveyor* 8:5-10.

Borkenstein, R. F., R. F. Crowther, R. P. Shumate, W. B. Ziel, and R. Zylman (1974) *The Role of the Drinking Driver in Traffic Accidents. Blutalkohol* 11 (Supplement 1). (Originally prepared in 1964).

Bruun, Kettil (1969) The actual and the registered frequency of drunkenness in Helsinki. *British Journal of Addiction* 64(1):3-8.

Brunn, Kettil (1971) Implications of legislation relating to alcoholism and drug dependence. Pp. 173-181 in C. G. Kiloh and D. S. Bell, eds., *Proceedings, 29th International Congress on Alcoholism and Drug Dependence*. Butterworths.

Bruun, Kettil, Griffith Edwards, Martti Lumio, Klaus Mäkelä, Lynn Pan, Robert E. Popham, Robin Room, Wolfgang Schmidt, Ole-Jørgen Skog, Pekka Sulkunen, and Esa Österberg (1975) *Alcohol Control Policies in Public Health Perspective*. New Brunswick, N.J.: Rutgers University Center of Alcohol Studies.

Bureau of the Census (1975) *Historical Statistics of the United States: Colonial Times to 1970. Bicentennial Edition.* Washington, D.C.: U.S. Department of Commerce.

Cahalan, Don, and Ira H. Cisin (1968) American drinking practices: Summary of findings from a national probability sample. II. Measurement of massed versus spaced drinking. *Quarterly Journal of Alcohol Studies* 29:642-656.

Cahalan, Don, Ira H. Cisin, and Helen M. Crossley (1969) *American Drinking Practices*. New Brunswick, N.J.: Rutgers University Center of Alcohol Studies.

Cahalan, Don, and Robin Room (1974) *Problem Drinking Among American Men*. New Brunswick, N.J.: Rutgers University Center of Alcohol Studies.

Cameron, Tracy (1979) The impact of drinking-driving countermeasures: A review and evaluation. *Contemporary Drug Problems* 8(4):495-566.

Cappell, Howard, and C. Peter Herman (1972) Alcohol and tension reduction: A review. *Quarterly Journal of Studies on Alcohol* 33(1):33-64.

Chauncey, Robert L. (1980) New careers for moral entrepreneurs: Teenage drinking. *Journal of Drug Issues* 10(1):45-70.

Clark, Norman H. (1976) *Deliver Us From Evil: An Interpretation of American Prohibition*. New York: Norton.

Clark, Walter B., and Don Cahalan (1976) Changes in problem drinking over a four-year span. *Addictive Behaviors* 1:251-259.

Clark, Walter B. (1977) Contextual and Situational Variables in Drinking Behavior. (Mimeograph) Publication No. F60. Social Research Group, University of California, Berkeley.

Clark, Walter B., and Lorraine Midanik (1980) Results of the 1979 National Survey. (Mimeograph). Social Research Group, University of California, Berkeley.

Cohen, Bernard C. (1963) *The Press, The Public, and Foreign Policy*. Princeton, N.J.: Princeton University Press.

deLint, Jan (1962) Pathological Wine Consumption in Toronto: A Study of Its Definition for Sociological Analysis. M.A. Thesis. Department of Sociology, University of Toronto.

deLint, Jan, and Wolfgang Schmidt (1976) Alcoholism and mortality. Pp. 275-305, in

Benjamin Kissin and Henri Begleiter, eds., *The Biology of Alcoholism, Volume 4: Social Aspects of Alcoholism*. New York: Plenum Press.

DISCUS (1977) *Distilled Spirits Industry 1976 Annual Statistical Review*. Washington, D.C.: Distilled Spirits Council of the United States.

Douglass, Richard L. (1979-1980) The legal drinking age and traffic casualties: A special case of changing alcohol availability in a public health context. *Alcohol Health and Research World* 4(2):101-117.

Farquhar, J. W., N. Maccoby, P. Wood, J. K. Alexander, H. Breitrose, B. Brown, W. Haskell, A. McAlister, A. Meyer, J. Nash, and M. Stern (1977) Community education for cardiovascular health. *Lancet* 1:1192-1195.

Farris, R., T. B. Malone, and H. Lilliefors (1976) *A Comparison of Alcohol Impairment in Exposed and Injured Drivers*. Report prepared for National Highway Traffic Safety Administration, Contract No. DOT HS-4-00954. Alexandria, Va.: Essex Corporation.

Federal Bureau of Investigation (Annual) *Uniform Crime Reports for the United States*. Washington, D.C.: U.S. Government Printing Office.

Federal Bureau of Investigation (1979) *Uniform Crime Reports: Crime in the United States, 1978*. Washington, D.C.: U.S. Government Printing Office.

Fillmore, Kaye M. (1974) Drinking and problem drinking in middle age: An exploratory 20-year follow-up study. *Quarterly Journal of Alcohol Studies* 35(3):819-940.

Fillmore, Kaye M. (1975) Relationships between specific drinking problems in early adulthood and middle age: An exploratory 20-year follow-up study. *Quarterly Journal of Alcohol Studies* 36(7):882-907.

Fillmore, Kaye, Selden Bacon, and Merton Hyman (1977) Alcohol Drinking Behavior and Attitudes: Rutgers Panel Study. Report prepared for the National Institute on Alcohol Abuse and Alcoholism, Contract No. ADM 281-76-0015. June.

Fitzgerald, J. L., and H. A. Mulford (1978) Distribution of alcohol consumption and problem drinking; Comparison of sales records and survey data. *Journal of Studies on Alcohol* 39(5):879-893.

Fosdick, Raymond, and A. Scott (1933) *Toward Liquor Control*. New York: Harper and Brothers.

Gastineau, Clifford F., William J. Darby, and Thomas B. Turner, eds. (1979) *Fermented Food Beverages in Nutrition*. New York: Academic Press.

Giffen, P. J., and S. Lambert (1978) Decriminalization of public drunkenness. Pp. 395-440 in Yedi Israel, Frederick B. Glaser, Harold Kalant, Robert E. Popham, Wolfgang Schmidt, and Reginald G. Smart, eds., *Research Advances in Alcohol and Drug Problems, Volume 4*. New York: Plenum Press.

Haberman, Paul W., and Michael M. Baden (1978) *Alcohol, Other Drugs and Violent Death*. New York: Oxford University Press.

Haddon, William Jr. (1979) Options for Prevention of Motor Vehicle Injury. (Mimeograph) Washington, D.C.: Insurance Institute for Highway Safety.

Harford, Thomas C., Douglas A. Parker, Charles Pautler, and Michael Wolz (1979) Relationship between the number of on-premise outlets and alcoholism. *Journal of Studies on Alcohol* 110(11):1053-1057.

Hennekens, Charles, Bernard Rosner, and Debrah S. Cole (1978) Daily alcohol consumption and fatal coronary heart disease. *American Journal of Epidemiology* 107(3):196-200.

Holcomb, Robert L. (1938) Alcohol in relation to traffic accidents. *Journal of the American Medical Association* 3(12):1076-1085.

Hu, Tun-Yüan (1950) *The Liquor Tax in the United States, 1791-1947*. New York: Columbia University Press.

Hyman, Merton H., Marilyn A. Zimmerman, Carol Gurioli, and Alice Helrich (1980) *Drinkers, Drinking, and Alcohol-Related Mortality and Hospitalizations: A Statistical Compendium*. New Brunswick, N.J.: Rutgers University Center of Alcohol Studies.

Institute of Medicine (1980) *Alcoholism, Alcohol Abuse and Related Problems: Opportunities for Research—Report of a Study*. Washington, D.C.: National Academy of Sciences. July.

Jellinek, E. M. (1947-1948) Recent trends in alcoholism and alcohol consumption. *Quarterly Journal of Studies on Alcohol* 8:1-42.

Johnson, Paula, David J. Armor, Suzanne Polich, and Harriett Stambul (1977) U.S. Adult Drinking Practices: Time Trends, Social Correlates and Sex Roles. A working note prepared for the NIAAA, Contract No. ADM-281-76-0020. Rand Corporation, Santa Monica, Calif.

Johnston, Lloyd D., Jerald G. Bachman, and Patrick M. O'Malley (1979) *Drugs and the Class of '78: Behaviors, Attitudes, and Recent National Trends*. DHEW Publication No. (ADM) 79-877. Rockville, Md.: National Institute on Drug Abuse.

Joint Committee of the States to Study Alcoholic Beverage Laws (1973) *Alcoholic Beverage Control: Administration Licensing and Enforcement, An Official Study*. (Revised and updated by Benjamin W. Corrado, Consultant). Washington, D.C.: The Joint Committee of States to Study Alcoholic Beverage Laws.

Keller, Mark, and Carol Gurioli (1976) *Statistics on Consumption of Alcohol and on Alcoholism*. New Brunswick, N.J.: Journal of Studies on Alcohol, Inc.

Klatsky, A. J., G. D. Friedman, and A. B. Siegelaub (1978) Alcohol use, myocardial infarction, sudden cardiac death, and hyptertension. *Alcoholism: Clinical and Experimental Research* 3(1):33-39.

Ledermann, Sully (1956) *Alcool—Alcoolisme—Alcoolisation*. Données scientifiques de caractère physiologique, économique et social, Institut National d'Études Démographiques, Travaux et Documents, Cahier N° 29. Paris: Presses Universitaires de France.

Levine, Harry G. (1977) Colonial and Nineteenth Century American Thought About Liquor as a Cause of Crime and Accidents. (Mimeograph). Publication No. E48. Social Research Group, University of California, Berkeley.

Levine, Harry G. (1978) The discovery of addiction: Changing conceptions of habitual drunkenness in America. *Journal of Studies on Alcohol* 39(1):143-177.

Levine, Harry G. (1980) The Committee of Fifty and the Origins of Alcohol Control. Publication No. F129. Social Research Group, University of California, Berkeley.

Levy, Paul, Robert Voas, Penelope Johnson, and Terry M. Klein (1978) An evaluation of the Department of Transportation's alcohol safety action projects. *Journal of Safety Research* 10(4):162-176.

MacAndrew, Craig, and Robert B. Edgerton (1969) *Drunken Comportment: A Social Explanation*. Chicago: Aldine.

Maccoby, Nathan (1980) Education for Alcohol Abuse Prevention. Working paper prepared for Panel on Alternative Policies Affecting the Prevention of Alcohol Abuse and Alcoholism, National Research Council.

Mäkelä, Klaus (1971) Concentration of alcohol consumption. *Scandinavian Studies in Criminology* 3:77-88.

Mäkelä, Klaus, and Esa Österberg (1979) Notes on analyzing economic costs of alcohol use. *Drinking and Drug Practices Surveyor* 15:7-10.

Mass Observation (1970) *The Pub and the People*. 2nd edition. London: Seven Dials Press.

McAlister, Alfred, Cheryl Perry, Joel Killen, Lee Ann Slinkard, and Nathan Maccoby (1980) Pilot study of smoking, alcohol, and drug abuse prevention. *American Journal of Public Health* 70(7):719-721.

McCarrol, James R., William Haddon, Jr. (1962) A controlled study of fatal automobile accidents in New York City. *Journal of Chronic Diseases* 15:811-826.

Medicine in the Public Interest, Inc. (1979) *The Effects of Alcoholic-Beverage-Control Laws*. Washington, D.C.: Medicine in the Public Interest.

Mello, Nancy (1972) Behavioral studies of alcoholism. Pp. 219-291 in Benjamin Kissin and Henri Begleiter, eds., *The Biology of Alcoholism. Vol. 2: Physiology and Behavior.* New York: Plenum Press.

Midanik, Lorraine (1980) The Validity of Self-Reported Alcohol Consumption and Alcohol Problems: A Literature Review. (Mimeograph) Social Research Group, University of California, Berkeley.

Miers, David (1978) *Responses to Victimization*. Abingdon, England: Professional Books.

Miller, Ann R., Donald J. Treiman, Pamela S. Cain, and Patricia A. Roos, eds. (1980) *Work, Jobs, and Occupations: A Critical Review of the Dictionary of Occupational Titles*. Committee on Occupational Classification and Analysis, National Research Council. Washington, D.C.: National Academy Press.

Mosher, James (1979) Dram shop liability and the prevention of alcohol-related problems. *Journal of Studies on Alcohol* 40(9):773-798.

Mosher, James (1980) Alcoholic Beverages as Tax Deductible Business Expenses: An Issue of Public Health Policy and Prevention Strategy. (Mimeograph) Social Research Group, University of California, Berkeley.

National Center for Health Statistics (1977) *Vital and Health Statistics of the U.S.* Washington, D.C.: U.S. Government Printing Office.

National Center for Health Statistics (1980) *Monthly Vital Statistics Report: Provisional Statistics*. DHEW Publication No. PHS 80-1120, Vol. 29(1), April 9.

National Highway Traffic Safety Administration (1974) *Alcohol Safety Action Projects—Evaluation of Operations, Vol. II*. Publication No. DOT 800-874. Washington, D.C.: U.S. Department of Transportation.

National Highway Traffic Safety Administration (1979a) *Alcohol Safety Action Projects Evaluation Methodology and Overall Program Impact*. Publication No. DOT HS 803-896. Washington, D.C.: U.S. Department of Transportation.

National Highway Traffic Safety Administration (1979b) *Results of National Alcohol Safety Action Projects*. Publication No. DOT HS 804-033. Washington, D.C.: U.S. Department of Transportation.

National Highway Traffic Safety Administration (1979c) *Alcohol Safety Action Projects Evaluation of Operations: Data, Tables of Results, and Formulation*. Publication No. DOT HS 804-085. Washington, D.C.: U.S. Department of Transportation.

National Highway Traffic Safety Administration (1979d) *Summary of National Alcohol Safety Action Projects*. Publication No. DOT HS 804-032 Washington, D.C.: U.S. Department of Transportation.

National Institute on Alcohol Abuse and Alcoholism (1974) *Second Special Report to the U.S. Congress on Alcohol and Health, June 1974, from the Secretary of Health, Education, and Welfare*. USDHEW Publication No. HSM-72-9099. Washington, D.C.: U.S. Government Printing Office.

National Institute on Alcohol Abuse and Alcoholism (1978) *Third Special Report to the U.S. Congress on Alcohol and Health, from the Secretary of Health, Education, and Welfare, June 1978*. USDHEW Publication No. (ADM) 78-569. Washington, D.C.: U.S. Government Printing Office.

National Institute on Alcohol Abuse and Alcoholism (1979) Prevention Model Replication Program of the National Institute on Alcohol Abuse and Alcoholism: Introduction and Project Summaries. (Mimeograph) NIAAA, Rockville, Md.

National Institute on Alcohol Abuse and Alcoholism (1981) *Fourth Special Report to the*

U.S. Congress on Alcohol and Health from the Secretary of Health and Human Services (Preprint copy). Rockville, Md.: NIAAA.

Niskanen, W.A. (1962) *The Demand for Alcoholic Beverages: An Experiment in Econometric Method*. Publication No. P-2583. Santa Monica, Calif.: Rand Corporation.

O'Donnell, John A., Harold L. Voss, Richard R. Clayton, Gerald T. Slatin, and Robin G.W. Room (1976) *Young Men and Drugs--A Nationwide Survey*. National Institute on Drug Abuse Monograph Series No. 5. Washington, D.C.: U.S. Government Printing Office.

Ogborne, Alan C., and Reginald G. Smart (1980) Will restrictions on alcohol advertising reduce alcohol consumption? *British Journal of Addiction* 75:293-296.

Oki, G., G. Rankin, N. Giesbrecht, P. J. Giffen, and S. Lambert (1977) Some physical findings in a sample of chronic public inebriates following decriminalization. Toronto, Ontario: Addiction Research Foundation.

Ornstein, Stanley I., (1980) The control of alcohol consumption through price increases. *Journal of Studies on Alcohol* 41(a):807-818.

Ornstein, Stanley, and D. Hanssens (no date) An Economic Analysis of the Relationship of Alcohol Control Laws to the Consumption of Alcoholic Beverages. Graduate School of Management, University of California, Los Angeles.

Ornstein, Stanley, and D. Levy (no date) Price and Income Elasticities of Demand for Alcoholic Beverages. Graduate School of Management, University of California, Los Angeles.

Ouellet, Barbara L., Jean-Marie Romeder, and Jean-Marie Lance (1977) *Premature Mortality Attributable to Smoking and Hazardous Drinking in Canada. Volume I: Summary*. Ottawa, Canada: Long Range Health Planning Branch, Department of National Health and Welfare.

Ouellet, Barbara L., Jean-Marie Romeder, and Jean-Marie Lance (1978) *Premature Mortality Attributable to Smoking and Hazardous Drinking in Canada. Volume II: Detailed Calculations*. Ottawa, Canada: Long Range Health Planning Branch, Department of National Health and Welfare.

Partanen, Juha (1975) On the role of situational factors in alcohol research: Drinking in restaurants vs. drinking at home. *The Drinking and Drug Practices Surveyor* 10:14-16.

Pernanen, Kai (1976) Alcohol and crimes of violence. Pp. 351-444 in Benjamin Kissin and Henri Begleiter, eds., *The Biology of Alcoholism, Volume 4: Social Aspects of Alcoholism*. New York: Plenum Press.

Perrine, M. W., J. A. Waller, and L. S. Harris (1971) *Alcohol and Highway Safety: Behavioral and Medical Aspects. Final Report*. Washington, D.C.: National Highway Traffic Safety Administration.

Plaut, Thomas (1967) *Alcohol Problems: A Report to the Nation*. New York: Oxford University Press.

Poikolainen, Kari (1977) *Alcohol Poisoning Mortality in Four Nordic Countries*. Forssa: Finnish Foundation for Alcohol Studies.

Polich, J. Michael, and B. R. Orvis (1979) *Alcohol Problems: Patterns and Prevalence in the U.S. Air Force*. Santa Monica, Calif.: Rand Corporation.

Polich, J. Michael, David J. Armor, and Harriet B. Braiker (1980) *The Course of Alcoholism: Four Years After Treatment*. Santa Monica, Calif.: Rand Corporation.

Popham, Robert E., and Wolfgang Schmidt (1981) Words and deeds: The validity of self-report data on alcohol consumption. *Journal of Studies on Alcohol* 42(3):355-358. In the same volume: comments by J. deLint, Merton M. Hyman, H. A. Mulford, J. L. Fitzgerald, and Henry Weschler.

Popham, Robert E., Wolfgang Schmidt, and Jan deLint (1976) The effects of legal restraint on drinking. Pp. 579-625 in Benjamin Kissin and Henri Begleiter, eds., *The Biology of Alcoholism, Volume 4: Social Aspects of Alcoholism*. New York: Plenum Press.

Popham, Robert E., Wolfgang Schmidt, and Jan deLint (1978) Government control measures to prevent hazardous drinking. Pp. 239-266, in James A. Ewing and Beatrice A. Rouse, eds., *Drinking*. Chicago: Nelson-Hall.

Regan, T. S., and P. O. Ettinger (1979) Alcohol and the heart. Pp. 259-274 in J. W. Hurst, ed., *The Heart: Update I*. New York: McGraw Hill.

Room, Robin (1971) Survey vs. sales data for the U.S. *Drinking and Drug Practices Surveyor* 3:15-16.

Room, Robin (1974) Governing images and the prevention of alcohol problems. *Preventive Medicine* 3:11-23.

Room, Robin, and James Mosher (1979-1980) Out of the shadow of treatment: A role for regulatory agencies in the prevention of alcohol problems. *Alcohol Health and Research World* 4(2):11-17.

Rorabaugh, William J. (1979) *The Alcoholic Republic: An American Tradition*. New York: Oxford University Press.

Ross, H. Laurence (1973) Law, science, and accidents: The British Road Safety Act of 1967. *The Journal of Legal Studies* 2(1):1-78.

Schmidt, Wolfgang, and Robert E. Popham (1975) Heavy alcohol consumption and physical health problems: a review of the epidemiological evidence. *Drug and Alcohol Dependence* 1:27-50.

Seeley, John R. (1960) Death by liver cirrhosis and the price of beverage alcohol. *Canadian Medical Association Journal* 83:1361-1366.

Simon, J. L. (1966) The price elasticity of liquor in the U.S. and a simple method of determination. *Econometrica* 34(1);193-205.

Skog, Ole-Jørgen (1971) *Alkoholkonsumets fordeling i befolkningen* (The Distribution of Alcohol Consumption in the Population). Oslo, Norway: National Institute for Alcohol Research.

Smart, Reginald G. (1977a) Changes in alcoholic beverage sales after reduction in the legal drinking age. *American Journal of Drug Alcohol Abuse* 4(1):101-108.

Smart, Reginald G. (1977b) The relationship of availability of alcoholic beverages to per capita consumption and alcoholism rates. *Journal of Studies on Alcohol* 38(5):891-896.

Staulcup, Herbert, Kevin Kenward, and Daniel Frigo (1979) A review of federal primary alcoholism prevention projects. *Journal of Studies on Alcohol* 40(11):943-968.

Stunkard, Albert M., and Sydnor B. Penick (1979) Behavior modification in the treatment of obesity: The problem of maintaining weight loss. *Archives of General Psychiatry* 36:801-806.

Sulkunen, Pekka (1976) Drinking patterns and the level of alcohol consumption: An international overview. Chapter 4, pp. 223-281, in R. J. Gibbons et al., eds., *Research Advances in Alcohol and Drug Problems, Volume 3*. New York: John Wiley.

Sulkunen, Pekka (1978) *Developments in the Availability of Alcoholic Beverages in the EEC Countries*. Helsinki: Social Research Institute of Alcohol Studies.

United States Department of Labor (1977) *Dictionary of Occupational Titles*. Fourth Edition. Washington, D.C.: U.S. Government Printing Office.

United States Senate (1976) *Hearings before the Subcommittee on Alcoholism and Narcotics of the Committee on Labor and Public Welfare*. Washington, D.C.: U.S. Government Printing Office.

Wallack, Lawrence M. (1980) Assessing effects of mass media campaigns: An alternative perspective. *Alcohol Health and Research World* 5(1);17-29.

Walsh, Brendan M., and Dermot Walsh (1970) Economic aspects of alcohol consumption in the Republic of Ireland. *Economic and Social Review* 2(1):115-138.

Warburton, Clark (1932) *The Economic Results of Prohibition*. New York: Columbia University.

Wiener, Carolyn (1980) *The Politics of Alcoholism: Building an Arena Around a Social Problem.* New Brunswick, N.J.: Transaction Books.

Wilkinson, Rupert (1970) *The Prevention of Drinking Problems: Alcohol and Cultural Influences*. New York: Oxford University Press.

Williams, A. F., R. F. Rich, P. L. Zador, and L. J. Robertson (1975) The legal minimum age and fatal motor vehicle crashes. *Journal of Legal Studies* 4(1):219-239.

Wingard, Deborah, and Robin Room (1977) Alcohol and home, industrial, and recreational accidents. Pp. 39-119 in Marc Aarens, Tracy Cameron, Judy Roizen, Ron Roizen, Robin Room, Dan Schneberk, and Deborah Wingard (1977) *Alcohol, Casualties, and Crime*. Final Report, Alcohol, Casualties, and Crime Project, NIAAA Contract No. ADM-281-76-0027. Berkeley, Calif.: Social Research Group (reprinted August 1978).

Wittman, Friedner D. (1980) Current Status of Research Demonstration Projects in the Primary Prevention of Alcohol Problems. (Mimeograph) Social Research Group, University of California, Berkeley.

World Heath Organization (1980) *Problems Related to Alcohol Consumption. Report of a WHO Expert Committee.* Technical Report Series 650. Geneva: World Health Organization.

Zador, Paul (1976) Statistical Evaluation of the Effectiveness of "Alcohol Safety Action Projects." *Accident Analysis and Prevention* 8:51-66.

Zimring, Franklin E. (1978) Policy experiments in general deterrence: 1970-1975. Pp. 140-186, in Alfred Blumstein, Jacqueline Cohen, and Alfred Nagin, eds., *Deterrence and Incapacitation: Estimating the Effects of Criminal Sanctions on Crime Rates*. Washington, D.C.: National Academy of Sciences.

COMMISSIONED PAPERS

Temperance and Prohibition in America: A Historical Overview

PAUL AARON *and* DAVID MUSTO

INTRODUCTION

In a recent book review about marijuana, Albert Goldman (1979, p. 250) wrote:

> The only controls should be those that are imposed to protect the public from bogus or polluted merchandise. With the dreadful example of Prohibition before us, it seems nearly unthinkable that we should have done it again: taken some basic human craving and perverted it into a vast system of organized crime and social corruption. When will we learn that in a democracy it is for the people to tell the government, not for the government to tell the people, what makes them happy?

This "dreadful example" is now so firmly established that it has become a maxim of popular culture, a paradigm of bad social policy, and a ritual invocation of opponents of a variety of sumptuary laws. The record of the 18th Amendment often has been read by libertarians as a morality tale. Detached and abstracted from their historically specific contexts and presented as a single crusade around which cranks and fanatics have clustered for 150 years, temperance and prohibition have been portrayed

Paul Aaron, who was a consultant to the panel, is a graduate student at the Florence Heller School of Social Work, Brandeis University. David Musto, a member of the panel, is at the Yale Child Study Center, Yale University.

This work was supported in part by Alcohol, Drug Abuse, and Mental Health Administration Grant DA-00037 from the National Institute on Drug Abuse.

as touchstones of bigotry. The lineage of reaction is traced straight from sin-obsessed Puritans, to evangelical extremists and Know-Nothings, to nativists and Klansmen, and most recently to McCarthyites and anti-abortionists.

The record of efforts to restrict drinking is, of course, far too complicated to warrant such axiomatic disparagement. But despite important, recent scholarship, and scientific validation of arguments once ridiculed, claims established by dint of repetition have achieved a kind of incantatory truth and ultimately have been enshrined as pieces of political folk wisdom (Warner and Rossett 1975).[1]

During the 1920s, partisan tracts featured titles like *Prohibition Versus Civilization: Analyzing the Dry Psychosis* and *The Prohibition Mania: A Reply to Professor Irving Fisher and Others* (Darrow and Yarros 1927, Barnes 1932).[2] Repeal institutionalized this propaganda and established an ideological legacy that historians came to inherit long after the battles had ended and the moral climate had cooled. As the antiliquor movement disappeared from the nation's political agenda, it also withered as a subject for research and study, not to reappear again until the early 1960s. Two books, *Prohibition: The Era of Excess* and *Symbolic Cru-*

[1] Warner and Rossett (1975), in their article "The Effects of Drinking on Off-Spring," revive a theory that postrepeal reaction rendered out of vogue. They observed that the moralizing tone of pre-Prohibition temperance writers caused Americans to discount previous work on parental drinking. They go on to say that recent renewed interest in the effect of maternal alcohol on offspring is an example of a common phenomenon—that of an old and unfashionable idea being restored to respectability.

Fetal alcohol syndrome, noted as the result of the gin epidemic in London (1720-1750), caused physicians to appeal to Parliament for control of the liquor industry; spirits were identified as a cause of "weak, feeble and distempered children" (Warner and Rossett 1975, p. 1396). Lyman Beecher, who as early as the 1820s in the United States saw liquor as a race poison, wrote: "The free and universal use of intoxicating liquors for a few centuries cannot fail to bring down our race from the majestic, athletic forms of our Fathers, to the similitude of a despicable and puny race of men" (pp. 1401-1402). In July 1979, the concept of drink-induced mental defects was given lurid endorsement in a television news magazine feature on fetal alcohol syndrome.

[2] Harry Elmer Barnes's book *Prohibition Versus Civilization* (1932) is an especially rich compendium of diatribe and invective. Among the central propositions he presented were:

The sense of the invidious at the root of the prohibitionist sentiment. "It is a common and natural trait," he argued "to hate those who are able to enjoy the good things of life from which we are excluded. The austere Puritans of modern vintage, usually cold, undeveloped, and desiccated personalities, shrink before the joyous intimacy promoted by even mild alcoholic indulgence. But we can hardly hate with good conscience things which we approve, even though we cannot enjoy or possess them. To allow the satisfying sentiment of envy and hatred full bloom we must find that the things denied us are bad and wicked. This saves us from the withering effects of overt and uncompensated envy *sade: Status Politics and the American Temperance Movement*, made important contributions to this recovery.

In *Prohibition: The Era of Excess*, Andrew Sinclair (1962) described the prohibitionist movement as a national St. Vitus's Dance.[3] Employing both Freudian and neo-Marxist categories, he attempted to reveal the "aggressive prurience" behind the masks of religious zeal; he argued that dominant economic interests, anxious to distract the gaze of reformers from the problem of the trusts, helped spread this "rural evangelical virus." Sinclair's portrayal of Prohibition as a florid outburst of a persistent, lurking paranoia backed by big business substituted indictment for objective examination. It represented a sophisticated caricature that drew heavily on the stereotypes of earlier critics.

Joseph Gusfield's (1963) book, *Symbolic Crusade*, constituted a fundamental advance beyond the psychohistorical exposé favored by Sinclair. Gusfield treated efforts to curb drinking not as mass hysteria but rather as a middle-class movement designed to defend lost status. He rejected the view of temperance and prohibition as repositories of a Snopes-like aberration and reoriented the terms of discussion. His analy-

---

and puts us in a frame of mind to go out and take these damnable things from our more fortunate contemporaries" (p. 32).

Prohibition as a hypocritical deceit. "A man may desire to cover up dubious economic transactions, hard bargains with widows on mortgages, cruelty in the domestic circle, sex delinquency, and what not. He finds a stern attitude towards drink a splendid alibi and an effective means of securing approval of the good people in the community" (p. 33).

Barnes also creates a rouges' gallery of "health cranks," "sadistic abnormals," "commercial evangelists," "designing capitalists bent on the realization of 'Fordismus,'" "racists," and "boot-leggers and racketeers" (p. 37).

[3] Sinclair's impugnings in *Prohibition: The Era of Excess* (1962) are often remarkably similar to the tales of horror told by antiprohibitionists in the 1920s: "With a terrible faith in equality," he observes, "the prohibitionists often wanted to suppress in society the sins they found in themselves" (pp. 26–27). He quotes G. K. Chesterton approvingly: "When the Puritan or the modern Christian finds that his right hand offends him he not only cuts it off but sends an executioner with a chopper all down the street, chopping off the hands of all the men, women and children in the town. Then he has a curious feeling of comradeship and of everyone being comfortably together." Sinclair goes on: "It was in this wish to extend their own repression to all society that the drys felt themselves most free from their constant inward struggle. Indeed, they defended their attacks on the personal liberty of other men by stating that they were bringing these men liberty for the first time. . . . Of course, in reality, the drys were trying to bring personal liberty to themselves, by externalizing their anguished struggles against their own weaknesses in their battle to reform the weaknesses of others. The conflict between conscience and lust, between superego and id, was transferred by the drys from their own bodies to the body politic of all America; and, in the ecstasy of that paranoia which Freud saw in all of us, they would have involved the whole earth" (pp. 26–27).

sis established a new standard of inquiry—dispassionate, free from polemical shrillness, and motivated by the desire to explain rather than carp or debunk. Nonetheless, Gusfield's work was not primarily directed toward explicating alcohol control as a thing in itself. "Issues like fluoridation or domestic communism or temperance," he wrote, "may be seen to generate irrational emotions and excessive zeal if we fail to recognize them as symbolic rather than instrumental issues." As an example of what he termed "expressive politics," temperance "operates within an arena in which feelings, emotions, and affect are displaced and where action is for the sake of expression rather than for the sake of influencing or controlling the distribution of valued objects" (Gusfield 1963, pp. 11, 23).

Gusfield's approach provided a store of subtle insights, but its conceptual richness tended to overwhelm other investigative strategies. The explicit, self-identified concerns around which people in the antiliquor movement moblilized, the particular regulatory techniques that were experimented with, and the nature of their impact are all areas that, to a large extent, have lain historically fallow. The emphasis on the "expressive" and on rationalization, projection, and displacement as key analytic tools has had the effect of distracting attention from the actual content of the movement and shifting the level of discourse from the literal to the figurative.

Gusfield's influence is a mark of the power of his formulations. But the struggles that people waged in the past to regulate or proscribe alcohol do not necessarily have to be treated as a nexus of symptoms. Without denying the continued usefulness of Gusfield's concept of expressive politics, it is necessary also to recognize the worth of complementary models of investigation. If, as Room (1974, p.11) suggests, "Our chief aim is to open up the range of frameworks within which the prevention of alcohol problems is discussed," and if accomplishing this requires that we better understand how the governing images evolved around that which we orient our current strategies of remediation, then we must attempt to understand the antiliquor movement, both as a symbolic crusade and as a massive, sustained organizing effort with a highly developed set of tactics and coherent, tangible goals.

Any attempt to discover a "usable past" in the history of American temperance and prohibition requires first that investigators abandon contemptuous reductionism and disenthrall themselves from lurid myths; this process has been largely accomplished, and scholars like Gusfield deserve respect and appreciation for breaking ground. Those, however, who seek to develop improved policy instruments around alcohol use and abuse must be creative scavengers willing to approach prior efforts both as cultural artifacts and as a body of experience capable of yielding valuable clues to the possibilities of regulation today. Through examination of how consumption patterns have changed and the basis for computing the social costs of drinking and through identification of various tactics of control, their original settings, and the reasons for their relative success or failure, the historian can develop a perspective that elucidates the policy choices to be debated.

THE COLONIAL PERIOD

The colonists brought with them from Europe a high regard for alcoholic beverages. Distilled and fermented liquors were considered important and invigorating foods, whose restorative powers were a natural blessing. People in all regions and of all classes drank heavily. Wine and sugar were consumed at breakfast; at 11:00 and 4:00 workers broke for their "bitters"; cider and beer were drunk at lunch and toddies for supper and during the evening.

Drinking was pervasive for a number of reasons. First, alcohol was regarded not primarily as an intoxicant but rather as a healthy, even medicinal substance with distinct curative and preventive properties. The ascribed benefits corresponded to the strength of the drink; "strong waters," that is, distilled liquor, had manifold uses, from killing pain, to fighting fatigue, to soothing indigestion, to warding off fever.

Alcohol was also believed to be conducive to social as well as personal health. It played an essential part in rituals of conviviality and collective activity; barn raisings, huskings, and the mustering of the militia were all occasions that helped associate drink with trust and reciprocity. Hired farm workers were supplied with spirits as part of their pay and generally drank with their employer. Stores left a barrel of whiskey or rum outside the door from which customers could take a dip.

Alcoholic drinks were also popular as a substitute for water. Water was considered dangerous to drink and inhospitable and low class to serve to guests. It was weak and thin; when not impure and filled with sediment, it was disdained as lacking any nutritional value. Beer or wine or "ardent spririts" not only quenched the thirst but were also esteemed for being fortified. They transferred energy and endurance, attributes vital to the heavy manual labor demanded by an agricultural society.

Official policy also endorsed consumption as trade in liquor provided an important source of revenue to the early colonists. Beginning in the 1630s, an *ad valorem* tax was levied on both imported wines and spirits and domestic products. The resulting monies were used to finance a wide range of local and provincial activities, from education in Connecticut, to prison repair in Maryland, to military defense on the frontier (Krout 1925, p. 19).

People drank, too, because alcohol was readily available. Although domestic production of gin, rum, or whiskey did not commence until the latter part of the 17th century, fruit brandies and especially cider were native beverages of daily consumption. Alcoholic drink was a staple that individual farmers created from local stuffs; people wanted it because they thought liquor was good for them and because they connected its production and consumption with traditional forms of civility. To brew ale or press cider were activities that supplied a valuable food, helped domesticate a natural wilderness, and helped restore a sense of continuity with a distant mother land.

Drunkenness was condemned and punished, but only as an abuse of a God-given gift. Drink itself was not looked upon as culpable, any more than food deserved blame for the sin of gluttony. Excess was personal indiscretion. Although Georgia did attempt an initial ban on ardent spririts, the colonial period was otherwise notable for a loosely pragmatic approach to control (Krout 1925, pp. 57-58). But while pragmatism reigned and strategies of regulation were improvised according to the distinctive conditions in each colony, there were basic models that could be found in all regions. Beyond sanctions for drunkenness, which ranged in severity from fines, to whipping, to the stocks, to banishment, conventional mechanisms of control were: (1) limits on the hours that taverns could stay open, on the amount that customers could consume, and on the time that could be spent "tippling"; (2) prohibitions against serving slaves, indentured servants, debtors, or habitual drunkards; (3) laws that proscribed certain activities in conjunction with public drinking (e.g., gambling or loud music were generally forbidden in taverns); and (4) requirements that taverns provide lodging and food, and that retailers sell only for home consumption—not small amounts to be drunk on the premises.

Although acceptable patterns of consumption were thus set forth in law, informal social controls played a much more significant role than legislation. Throughout the colonial period, legislatures delegated to boards of selectmen or county courts the authority to grant tavern licenses. Since the bodies holding this power were composed of the socially prominent, it naturally developed that licenses were issued to men of similar station. This arrangement proceeded less from the wish to maintain a class monopoly on a lucrative trade than from a deep sense of civic obligation with which the clergy and the leading men of property were imbued. As a community institution—a place that provided the amenities of life to travelers as well as a comfortable setting for local recreation—the tavern was a resource whose administration had to be both moral and efficient. The proprietor was expected not only to dispense food, drink, and hospitality, but also to monitor behavior by relying on the deference and respect accorded his social position to keep customers in check. In return for such a responsible oversight, the innkeeper was often granted an exclusive operating right within a particular area; franchise was awarded, or sometimes imposed, for maintaining this service along crucial thoroughfares or adjacent to key bridges.

Tavern owners were often men of rank, as evidenced by the early records of Harvard University, where the names of students, listed by social position rather than alphabetically, showed that the son of an innkeeper preceded that of a clergyman (Krout 1925, p. 44). It was often the case that leading citizens would conclude their public career, having served as town clerk, justice of the peace, or deputy to the General Court, by securing a license to run a public house. Men habituated to moral surveillance could thus continue their scrutiny.

There was always circumvention of rules, however, regulating the flow of liquor. Unauthorized sellers, for example, evaded the prohibition against keeping a tippling house by taking advantage of an ancient right of Englishmen to brew and sell without a license in brush houses at fair times. Thus, in Virginia, "divers loose and disorderly persons set up booths, arbours, and other public places where, not only the looser sort of people resort, get drunk, and commit many irregularities, but servants and Negroes are entertained, and encouraged to purloin their master's goods, for supporting their extravagancies" (Pearson and Hendricks 1967, p. 21). Though slippage existed, the system of control did work. Drunkenness was inveighed against, but it had not become recognized as a serious social problem.

As the 17th century came to a close, this "stable, conservative, well-regulated" system changed (Rorabaugh 1979, p. 29). As large quantities of imported West Indian molasses began to arrive in New England, the domestic distilling trade burgeoned. Soon rum was being manufactured in large enough quantities to supplant French brandy in the triangular slave trade. As hard liquor achieved an increasingly central commercial role, "the public accorded it," Krout wrote "that approbation which attaches to most things indispensible to the world of business" (Krout 1925, p. 50).[4] But while leading citizens amassed fortunes from trading rum for Africans, a glut of alcohol began to erode the structure of class control by which drinking behavior had been regulated. As the price for rum plummeted (in 16 years the cost per gallon was cut almost in half), demand increased and violations of licensing laws became noto-

[4] For another account of the commercial role liquor played in the dealings of fur traders and other merchants with North American Indians, see MacAndrew and Edgerton (1969). These authors also argue that the Indian tribes had no exceptional natural urge toward drunkenness or alcohol consumption.

rious (Rorabaugh 1979, p. 29). As the regulations against the selling of drams by retailers were less frequently observed and as the services that taverns were required to provide shrunk from minimal to nonexistent, the enforcement capacity of local officials was overwhelmed. In Boston, surrendering to pervasive circumvention, officials expediently granted licenses to many of the violators. Operating permits, once awarded only after assessment of the character of the propsective tavernkeeper, were now dispensed pro forma, and their number increased from 72 in 1702 to 155 in 1732 (Rorabaugh 1979, p. 25).

By the middle of the 18th century, management as a moral guardianship and community service gave way to management as a business venture. The innkeeper had lost status; his son fell from rank at Harvard as the occupation as a whole was increasingly dominated by the common folk (Krout 1925, p. 45). John Adams bemoaned the deterioration of control: "I was fired with a zeal," he wrote, "amounting to an enthusiasm, against ardent spirits, the multiplication of taverns, retailers and dram shops and tippling houses. Grieved to the heart to see the number of idlers, thieves, sots and consumptive patients made for the use of physicians in these infamous seminaries, I applied to the court of sessions, procured a committee of inspection and inquiry, reduced the number of licensed houses, etc. But I only acquired the reputation of a hypocrite and an ambitious demagogue by it. The number of licensed houses was soon reinstated, drams, grogs, and setting were not diminished, and remain to this day as deplorable as ever" (Kobler 1973, p. 31). The futility that Adams felt in trying to curb this disorder was shared by other representatives of his class. The breakdown of traditional controls and the social turmoil seen to proceed from it were associated with the increasing commercial exploitation of distilled liquor. Once a largely imported substance whose distribution was an aristocratic monopoly, it had become democratized by the end of the colonial period. Cheap rum from Boston and Providence widened the availability of hard liquor (90 proof, compared with the milder and less potent domestic fruit brandies). People drank more and did so in a context that was less strictly monitored than when taverns had been under the aegis of a proprietary civic elite.

THE DECLINE OF AUTHORITY

The Revolution accelerated the breakdown of class deference and mythologized the public drinking place as a bastion of liberty of the common man. Indirectly, the war also helped to topple the domestic supremacy of rum and replace it with cheaper domestic whiskey. Because trade with the West Indies was disrupted, thereby cutting off the source of rum, a need developed for a substitute hard liquor. Scotch-Irish settlers, arriving in America during this period, brought with them a tradition of pot still whiskey making.

In 1789, the first Kentucky whiskey was made by a Baptist preacher named Elijah Cook; by 1810, the known distillers totaled 2,000 and the annual overall production was more than 2 million gallons (Roueché 1960, p. 42). The rum-producing states attempted to defend themselves against this encroachment. In 1783, Congress voted to help finance the central government by taxing imports; the ratification of this legislation was delayed until 1789 when the New England states, afraid that an excise on molasses would price rum out of the domestic market, succeeded in having whiskey taxed at a level that maintained the preexisting ratios between the two drinks. The farmers of western Pennsylvania resisted this compromise, and their 2-year rebellion, during which tax collectors were tarred and feathered, was crushed only after President Washington (acting after Governor Mifflin refused to call out state militia) occupied the region with 15,000 troops. The imposition of the tax did nothing, however, to stem the decline of rum and its displacement by whisky. Prices for rum had doubled during the 1780s. Annual imports fell from one gallon per capita in 1790 to less than one-half gallon by 1827 to below one-fifth gallon in 1850. The repeal of the whiskey tax in 1802 simply made the position of whiskey even more advantageous (Rorabaugh 1979, p. 68).

The whiskey trade became an indispensible element in the economic expansion westward. H. F. Willkie, writing in 1947, noted: "There were no roads in the new territory and most of the trade was by packhorse. It cost more to transport a barrel of flour . . . than the flour would have sold for on the eastern markets. If the farmer converted the grain into whiskey, a horse, which would carry only four bushels in solid grain, could carry twenty-four bushels in liquid form. Practically every farmer, therefore, made whiskey. So universal was the practice that whiskey was the medium of exchange" (Roueché 1960, pp. 39-40). Albert Gallatin, drafting an appeal to Congress in 1792, wrote: "Distant from a permanent market, and separate from the Eastern coasts by mountains, we have no means of bringing the produce of our lands to sale either in grain or meal. We are therefore distillers by necessity, not choice" (Rorabaugh 1979, p. 54).

Though the estimates of the per-capita consumption vary, it is generally agreed that, beginning at the turn of the 19th century, demand for distilled liquor exploded. In 1972, when the population was 4 million, domestic production was 5.2 million gallons and imports almost 6 million gallons more. Within the next 18 years, the number of distilleries increased 6 times; production tripled. According to the most conservative estimate, per-capita consumption of hard liquor went from 2.5 gallons to almost 5 gallons. Other estimates place the consumption levels much higher, doubling the figure to 10 gallons (Clark 1976, p. 20; Asbury 1950, p. 12). Whatever the most accurate computation, there is consensus that the market for distilled alcohol was inundated. For example, rye, which wholesaled in 1820 for 60 cents a gallon, was 30 cents a gallon within a few years (Rorabaugh 1979, p.68).

The sudden and dramatic increase in production and consumption coincided with a rapid demographic change. Between 1790 and 1830, the population doubled in Massachusetts, tripled in Pennsylvania, and increased five times in New York. In the 20 years following Washington's inauguration, the overall population of the country jumped nearly 100 percent. While only 100,000 people lived in the west in 1790, by 1810 there were 1 million. The population of Philadelphia quadrupled; New York City's population increased 600 percent. Geographic mobility and staggering population increases were accompanied by newly emerging economic relations. Factory towns sprang up, and by the beginning of the 19th century an urban proletariat was evolving (Rothman 1971).

As America became a new society, sloughing off its hierarchical, agricultural, and colonial past, a belligerent pride and enthusiasm developed. Americans believed, Tocqueville observed, "that their whole destiny is in their hands." He went on, however, to comment on the often desperate tone of this optimism. He identified an acute ambivalence at the core of the exaggerated self-confidence and suggested that the sudden disappearance of traditional boundaries left people bereft and disoriented. "The woof of time is every instance broken and the track of generations effaced. Those who went before are soon forgotten; of those who come after, no one has any idea; the interest of man is confined to those in close propinquity to himself" (Lasch 1979, p. 9).

It was during this period of brutally rapid disjuncture that alcohol began to be widely perceived as a serious threat to social order. The experience of discontinuity, which has been called "unparalleled in the world," fragmented networks of deference and respect (Clark 1976, p. 29). Throughout the colonial period, authority was embodied in direct personal relations. The tavern owner was a civic overseer as well as a tradesman. Beyond a fixed pattern of roles, the dimensions of obedience also depended on a belief in man's immutable nature. The colonists did not have to agonize about drunkenness because sin and human weakness were understood to be predictable and internal. Only after deviance grew to be regarded as an inextricable function of environmental disequilibrium did excessive drinking become the target of organized reform.

Sanctions to regulate conduct, operating within an overall context of civic cohesiveness, were intended to shame the offender before the community. The stocks or the wearing of the letter "D" subjected the drunkard to ridicule, and such ceremonies of public humiliation were assumed to have a deterrent power. However, with frenzied economic and geographic mobility, exile became self-imposed. The rootless individual, seeking his fortune, living by his wits, and answerable to no social superior, became celebrated as the national character ideal. The stable, self-policizing community was demolished; the forms of behavioral management that grew out of an inherited concept of reciprocal rights and obligations became obsolete.

One can only speculate on the degree to which a coherent and fixed social order would have been able to absorb, with fewer ill effects, the dramatic rise in the consumption of alcohol. In attempting to account for the shift in the conceptual paradigm of drinking—that is, from an occasional nuisance to a permanent menace—it is impossible to construct a rigid hierarchy of causation or precisely factor out the variables. We do know that the combination of precipitate and bewildering change unmoored people from their sense of place, both social and physical. We do know that there was more drinking of hard liquor in settings that no longer even offered the pretense of other activities. The tavern or inn, where food and lodging provided a milieu that militated against intense drinking, gave way almost exclusively to the grogshop, essentially an early version of the saloon. Drinking became detached from earlier safeguards. And whereas before this process of detachment had provoked attempts to reassert controls, efforts at regulation became increasingly listless and ineffectual. During the first decades of the 1800s, as people drank more and more in places specifically and exclusively designed to cater to consumption of alcohol and as laws governing operating hours or sales to minors were regularly ignored, public drunkenness grew to be defined as a social problem. As in England, when the gin epidemic spread during a period of social and economic transformation, the sharp rise in the amounts of alcohol consumed coupled with the deterioration of drinking behavior reflected the deepening cultural turmoil and impaired the capacity of institutions to relegitimate themselves (Rorabaugh 1979, p. 144).[5]

[5] Rorabaugh (1979, p. 144) notes: "During the early nineteenth century, a sizable number of Americans for the first time began to drink to excess by themselves. The solo binge was a new pattern of drinking in which periods of abstinence were interspersed every week, month, or season with one to three-day periods of solitary inebriation. It was necessary to devise a name for this new pattern of drinking, and during these years the terms 'spree' and 'frolic' came into popular usage."

Lyman Beecher, who in 1812 organized other leading churchmen and established the Connecticut Society for the Promotion of Good Morals, was an avowed conservative determined "to save the state from innovation and democracy. Our institutions, both civil and religious," he wrote, "have outlived that domestic discipline and official vigilance in magistrates which rendered obedience easy and habitual. The laws are now beginning to operate extensively upon necks unaccustomed to the yoke, and when they become irksome to the majority, their execution will become impracticable. To this situation we are already reduced in some districts of the land. Drunkards reel through the streets day after day, and year after year, with entire impunity. The mass is changing. We are becoming a different people" (Krout 1925, p. 86). Clearly, the significance that Beecher and others of his class attached to drunkenness cannot be separated from their anticipation of the downfall of the standing order. Personal insobriety was feared because it was a harbinger of social chaos. Despite Beecher's admitted antidemocratic sentiments, his perception that drunkenness was more common and more overt in its display was shared by a wide spectrum of Americans. The meaning of this phenomenon may have been interpreted differently, but what people observed seemed to be the same thing.

Democrats, concerned about the survival of popular government, were fierce in their opposition to increasing drunkenness. Jefferson bemoaned the rise in the consumption of hard liquor and was particularly concerned that it sapped civic virtue. "The habit of using ardent spirits in public office," he wrote, "has often produced more injury to public service, and more trouble to me, than any other circumstance that has occurred in the internal concerns of the country during my administration. And were I to commence my administration again, with the knowledge that from experience I have acquired, the first question that I would ask with regard to every candidate for office would be, 'Is he addicted to the use of ardent spirits?' " (Kobler 1973, p. 33).

The concept of addiction was borrowed by Jefferson from his friend and advisor Benjamin Rush and became integrated into a theory of individual behavior and social obligation that was essentially optimistic and progressive. Rush, a signer of the Declaration of Independence, a medical pioneer, and an inveterate activist, published in 1785 an enormously influential tract, *An Inquiry into the Effects of Ardent Spirits upon the Human Body and Mind*, of which 200,000 copies were distributed in the first three decades of the 19th century. Rush's tract was intended to change public opinion and overthrow the commonly held faith in the efficacious properties of hard liquor. People didn't need spirits for health and stamina, Rush argued, and were actually poisoning themselves through drink. He laid out an elaborate description of this disease syndrome and emphasized the moral as well as physical decay that alcohol brought on. Spirits progressively deranged the will as they sickened the body. Hard liquor debilitated self-restraint and incited pathological excess. It changed people and induced compulsive behavior by short-circuiting natural mechanisms of self-control.

Rush's analysis of the effects of drinking in terms of addiction had great explanatory appeal to Americans at the turn of the century. Dramatic changes were apparently taking place in patterns of liquor consumption; these changes were especially ominous in their correspondence to rapid social and political irregularities. Rush's model was persuasive not only as a diagnosis. The fatalism implicit in the Puritan view of drunkenness as deliberate, informed self-abuse, an expression of a sinful nature, was replaced, as Levine observes, by Rush's explanation that the alcoholic was a person compelled and controlled, a victim of a substance extraneous to himself (Levine 1978).

This diseased condition of dependence could be cured, according to Rush, only by total abstinence from hard liquor. A variety of treatments were recommended to effect this cure: fright, bleeding, whipping, aversive therapy with emetics. But to restore the drunkard to self-determination, every method had to operate within a context of public support and concern. Rush urged the churches to unite in a campaign of education and political pressure; the number of grogshops must be limited; and the social stigma attached to the sale and consumption of ardent spirits made more harsh. "The loss of 4,000 American citizens, by yellow fever, in a single year," he wrote, "awakened general sympathy and terror, and called forth all the strength and ingenuity of laws, to prevent its occurrence. . . . Why," he asks, "is not the same zeal manifested in protecting our citizens from the more general and consuming ravages of distilled spirits?" (Rush 1814, p. 27).

Rush believed that "spirits are anti-Federal . . . companions of all those vices calculated to dishonor and enslave our country," and that the government, to protect itself as well as the well-being of its citizens, should adopt a public health approach and control the epidemic of addiction. These propositions became the central constructs for the temperance movement that began slowly in the early 1800s and burgeoned 20 years later.

MOBILIZING FOR THE CRUSADE

With the formation of the Union Temperance Society of Moreau and Northumberland, New York, in 1808 by farmers, inspired by Rush's tract, agitation against ardent spirits and the public disorder they spawned gradually increased. This dramatic surge of popularity for temperance, unparalleled in the development of any mass movement, occurred in the 1820s. The American Society of Temperance, created in 1826 by clergymen, spreading the gospel of antidrink through the network of the ministry, inaugurated a crusade of revivalistic fervor. Within 3 years, 100,000 people had pledged to abstain from hard liquor. By 1831, membership had nearly doubled; in 1833, 5,000 chapters were spread across the country, and by 1835, of a total of 13 million citizens, 1.5 million had vowed never to consume ardent spirits again. In 1837, the Eighth Annual Report of the New York City Temperance Society listed 88,076 members, of a total city population of 290,000 (Cherrington 1920, p. 93).

The temperance movement diversified and fragmented as the initial phase of mobilization subsided. Fierce debates about its tactics and ultimate purposes were waged. At the national convention in 1836, radicals pushed for and won a ban on all alcoholic beverages—not just ardent spirits. Rejecting the tradition of Rush and Jefferson, according to which beer and wine were exempted from censure and, in fact, were praised as "temperance drinks," the radicals argued that logical consistency demanded a ban against all intoxicants, even those less concentrated in their "poison" than rum and whiskey. Resistance, however, to the new "long" pledge was initially strong; amid strenuous doctrinal disputes about the character of the wine served at the Last Supper, membership in temperance societies dropped and contributions fell off. (In New York State in 1837, 220,000 people belonged to temperance organizations; 3 years later there was a decrease of 100,000.) Despite the toll that factionalism took, by 1840 temperance had become largely synonymous with teetotalism (Krout 1925, p. 61).

Beyond this debate over the proper extent of abstinence, other struggles took place around the question of political intervention. Although Rush had seen educational and political activity as consistent and mutually reinforcing, the temperance movement had gained its adherents in its first 10 years through an emphasis on moral suasion. But as moral reform began to be increasingly undermined by an organized traffic, legislation came to be regarded as a necessary weapon. Converting the liquor dealers gave way to passing laws that curbed their "mercenary ruthlessness." Some people, it was argued, were simply impervious to example or exhortation and shameless in their pursuit of profit. Social action, not just individual action, was required against this group of moral bandits, just as it was against the commercial slave-traders.

In 1838, known as the "petition year," appeals were made to six state legislatures to restrict the sales of alcoholic beverages. From then until

1855, a majority of the states passed one form or another of regulatory control. Local option was adopted in some states; laws were passed to abolish public drinking by limiting in bulk the amount of the purchase (from 1 to 15 gallons, depending on the state); still other states imposed "high license"—a tax ranging from $100 to $500—on the theory that such expensive taxation would both curb the sale of alcohol by grocers and drive out the small, low-life dive. Actual prohibition of hard liquor was passed in 13 states from 1851 to 1855.

There were endemic problems in the enforcement of all these disparate mechanisms of control. Laws were constantly being experimented with, revised, and repealed; of the 13 states that had prohibition in 1855, only 5 remained dry in 1863.

In 1855, Connecticut passed a prohibition law, only to have it repealed 2 years later. The breakdown of prohibition, according to Governor Dutton, was caused by the sabotage and greed of state attorneys and enforcement officers, "men who made use of the law for the purpose of making money" (Grant 1932, p. 5). Massachusetts's 15-gallon law was attacked as class legislation that left the rich free to consume but penalized the working man. The law was openly flouted, and, in some cases, near riot broke out when grogshop patrons attempted to prevent the owners from being arrested (Krout 1925, p. 271). In Maine, elaborate subterfuges were concocted to evade the prohibition law passed in 1851. A pickle or wedge of cheese would be sold at an inflated price accompanied by a free drink—the law, of course, banned only the *sale* of alcohol.

Despite the persistent difficulties in enforcement, and varying degrees of control imposed by the states, the first wave of the temperance movement (1825 to 1855) did accomplish dramatic reductions in the level of consumption of hard liquor. Although beer drinking increased sharply after 1850 (Coors, Pabst, and Anheuser established breweries to supply recently arrived immigrants; the first lager brewery and beer hall opened in Philadelphia in 1840), consumption of whiskey and rum decreased by at least half between 1820 and 1850 (Clark 1976, p. 47).

It is difficult to assess the extent to which this decrease can be attributed to temperance organizing. Changes in styles of consumption having little to do with the antiliquor movement were significant. But it nonetheless seems likely that temperance agitation did accelerate this process. The social legitimacy of spirits was undermined as a new status became attached to self-control.

Temperance had a broad-based support. While the urban poor, increasing numbers of whom were immigrants, never responded to antiliquor appeals and the rich were only marginally involved in the crusade, the middle class—the skilled mechanics and tradesmen—represented the solid force. Many of the groups that belonged to the movement were unaffiliated with the church. As fraternal organizations, they were dedicated to mutual aid. The Washingtonians, self-styled ex-drunkards who recruited other drunkards, the Independent Order of the Rechabites of North America, the Sons of Temperance, and the Independent Order of Good Templars all developed outside traditional ecclesiastical control. In fact, their unconventional approach to proselytizing and their disinterest in religion often caused evangelical temperance exponents to view these groups suspiciously.[6]

The enormous popularity of this secular wing (in 1850, the Sons of Temperance had 230,000 members) derived primarily from their emphasis on creating a cohesive structure of reciprocal obligation. Membership allowed participation in an active social life free from alcohol and a sense of brotherhood and collective identity. Family integrity was given special prominence. Auxiliaries for wives were formed; children were moblilized into "cold-water brigades," complete with uniforms and marching bands. Regalia, rituals, and group outings, along with various group insurance programs, were all designed to maintain family intactness within a context of organizational support.

One of the most persistent themes in the temperance movement, cutting across factional lines, was the need to aid the family in its struggle with the saloon. The perception of the family under siege was constantly expressed. Family breakdown was blamed for pauperism, crime, and dependency; the blandishments of the grogshop were regarded as fundamentally subversive to the security of the home. Beginning in the 1820s, prison officials regularly made inventories of the causative factors that were at the root of the crime. Of 173 biographies compiled at Auburn Prison, two-thirds were interpreted to show that a malfunction in family control provoked crime and that, in most of the cases, intem-

[6] Krout (1925) quotes from an exchange between two rival journals in 1842. The editor of *The Essex Washingtonian* had suggested that if the Washingtonians prevailed, then the church-based temperance movement would be "blown to the moon." *The Journal of the American Temperance Movement* responded: "In the late extraordinary reformation of drunkards, a subject of thankfulness, however transient, but more especially when resulting as it has in thousands of cases of permanent sobriety, a deep sympathy was felt for this unfortunate class—prejudiced against religion . . . by long absence from the Sabbath's influence and by subjection to the vile and debasing principles of the bar-room and dram-shop. To induce them to sign the pledge, these prejudices, it was supposed, must be consulted. The wharf, the markethouse, the public hall, rather than the church, must be the place of meeting. . . . In the prevalence of pure Washingtonianism, the reformation of drunkards and the relief of the miserable, every philanthropist, every patriot, and Christian ought to rejoice, but when the unprincipled improve a temperance meeting to revile the Clergy, the Churches or the Magistracy, they should receive the withering rebuke of every virtuous citizen" (pp. 203–204).

perance was the underlying reason for the disordered home life (Rothman 1971, p. 65). The New Prison Association proclaimed that intemperance was "the giant whose mighty arm prostrates the greatest numbers, involving them in sin and shame and crime and ruin." Behind it, "never let it be forgotten, lies the want of early parental restraint and instruction" (Rothman 1971, p. 74). A vicious cycle was thus identified: drinking made parents irresponsible and slack; children growing up without discipline were apt to inherit both a craving for alcohol and a defiant attitude toward authority; as adult offenders, they would pass along to their children the same disabilities.

As the community became less and less coherent and caring, as contacts between employer and workers became discordant and estranged, as the capacity of the church to serve as a mediating force eroded, the family was invested with a redemptive and protective sacredness. Mass migrations, whether impelled by desire, ambition, or economic duress, were disorienting. A recently arrived family on the frontier or in a burgeoning city was forced to turn to itself. The support system that once had helped sustain family life was either primitive or nonexistent. As strangers in new surroundings, the family had to serve as its own refuge from isolation. Given the fragile nature of the requisite self-reliance, anything that might weaken this familial membership became a jeopardy to survival. It was precisely such a threat that alcohol, and the public setting in which it was drunk, constituted.[7]

[7] The connection has often been observed between loss, change, and the anxiety, depression, and nostalgia they provoke and the use of alcohol. Clark (1976) comments on the "guilt-ridden but sentimental reverence" for a mythological past that is reflected in the songs of the early 1800s: "Home Sweet Home," "Old Oaken Bucket." The names of whiskies also conjure up wistful remembrance: "Old Grand-Dad," "Southern Comfort" (p. 29).

The founder of Alcoholics Anonymous recalls drinking at a tavern in the 1920s. His account is infused with a sense of loss and longing: "The warm, friendly smell of wet sawdust, stilled beer and whiskey . . . the feeling of being at home, the feeling for the men. In later years, I would think of that and nothing else. I wanted it again." Jack London also wrote about drinking in a saloon as an experience laden with emotion: "A newsboy on the street, a sailor, a miner, a wanderer in far-off lands, always where men come together to exchange ideas, to laugh and boast and dare, to relax, to forget the dull toil of tiresome nights and days, always they came together over alcohol. The saloon was the place of congregation. Men gathered to it as primitive men gathered about the fire at the mouth of the cave" (London 1913, pp. 6–7).

The maternal, caretaking ambience of the tavern and saloon to which this testimony alludes supports the thesis that a high consumption level of alcohol is correlated with punishment or deprivation of dependency needs. Bacon et al. (1965, p. 43) found that "frequent drunkenness or high consumption, or both, tend to occur in cultures where needs for dependence are deprived or punished, both in childhood and in adult life, and where a high degree of responsible, independent, and achieving behavior are required."

Beyond surpassing the home in decorativeness, if not physical comfort, the public drinking place was able to offer recreation and companionship. And alcohol soothed and consoled. In sum, the saloon and the substance it sold supplanted the most precious functions of domesticity.

Throughout the voluminous literary propaganda that temperance organizations printed, throughout the public readings of tracts and the conversion testimony of the Washingtonians (both of which constituted the popular moral theater), the high moment of melodrama was the scene of drink-induced domestic violence. Scenes of children being abandoned to poverty and shame by their drunken father or of wives being brutally beaten were constantly depicted. Justin Edwards's classic tract, *Temperance Manual*, presents a litany of evils attendant on drink. There is gruesome medical evidence, complete with descriptions of swollen and discolored organs and of bodies exploding when alcoholic breath gets too near a candle; the "poisoning of the seed and a legacy of death and disease" are likewise luridly detailed. But according to Edwards, the most powerful proof of the threat that drinking posed is the insanity and savage familial violence it provoked. Edwards lists episode after episode of murder and mayhem: "A father took a little child by his legs and dashed his head against the house, and then, with a bootjack, beat out his brains. Once that man was a respectable merchant, in good standing, but he drank alcohol" (Edwards 1847, p. 37). The temperance plays, *The Drunkard, or the Fallen Saved*, *One Cup More, or the Doom of the Drunkard*, similarly evoked the terror of husbands out of control, subject to fits of violent rage or total neglect, visiting on their families broken bones or economic dispossession.

Tocqueville observed that the narrowed range of community trust and cooperation had "confined the interest of man to those in close propinquity to himself." This "propinquity" could become murderously claustrophobic. But amidst a period of pervasive social disorder, the family was looked to as a bastion of self-restraint, the bulwark of rational affection. The family was expected to be the primary structure of attachment through which social purposes were embodied.

Alcohol subverted this structure. Justin Edwards described it as "the great deceiver and mocker," creating "new, artificial, unnecessary and dangerous appetite. . . . It cries for ever 'Give, give,' and never has enough. Hence the reason why the incautious youth, or the sober man who had unhappily formed this appetite, went on step by step with increasing velocity, to the drunkard's grave. Not a man on earth can form this appetite without increasing his danger of dying a drunkard" (Edwards 1847, pp. 28-29). "The grog-shop," General James Appleton wrote, "decoys men from themselves and from their self-control."

It has been suggested that this preoccupation with alcohol as a moral snare and with the saloons as an arch-rival to the family became a fetish used to ward off a more comprehensive and disturbing diagnosis of social ills. But while the temperance movement, during its first wave, did identify liquor as the root cause of pauperism, crime, and family disintegration, its adherents were also convinced that defective social structures had eroded traditional collective defenses against drunkenness and had condoned and even encouraged the omnipresence of corrupting influences. Walter Channing, who lectured widely in the 1830s on intemperance and pauperism, accused the rich of materialism and selfishness and of a betrayal of the partriarchal trust that as a class they had once assumed toward the poor. Class division had excluded the lower orders from civilizing associations and abandoned them to vice and drink. "By our mode of life—our house—our dress—our equipage: in short, by what is strictly internal to us, men detach themselves from their neighbors—withdraw from the human family . . . in its even recognized relationship of brotherhood" (Rothman 1971, p. 174).[8] Joseph Tuckerman, minister and temperance activist in Boston during this same period, also indicted the community for callousness and neglect. "Has not society a large share of responsibility for the evil of intemperance?" he asked. Did not the citizen involved in the liquor traffic "minister to the utter ruin of his fellow beings?"

The millions of Americans who were part of the first temperance wave joined the movement for reasons other than deference to the stewardship of men like Channing and Tuckerman. As Gusfield suggests, drunkenness was more than a metaphor encapsulating and accounting for a nexus of disorders, abstinence more than a gesture of "moral athleticism." For those who became members of temperance groups—particularly the fraternal organizations that operated as mutual aid societies—such structures of affiliation helped to restore a sense of belonging. Amid economic and demographic dislocation that affected large numbers of the

[8] Durkheim (1972) was one of the most profound and most deeply pessimistic diagnosticians of "the malady of infiniteness which torments our age. . . . For men to see before him boundless, free, and open space, he must have lost sight of the moral barrier which under normal conditions would cut off his view. He no longer feels those moral forces that restrain him and limit his horizon. . . . The notion of the infinite, then, appears only at those times when moral discipline has lost its ascendency over wants; it is a sign of the attrition that occurs during periods when the moral system which has prevailed for centuries is shaken" (p. 174).

The individual could not cure himself of his morbid insatiability; "since the individual has no way of limiting [his passions], this must necessarily be accomplished by some force outside him. A regulative force must play the same role for moral needs which the organism plays for physical needs" (p. 176).

poor, those on the edges of society clutched at whatever they could to enhance dignity and self-possession.

By 1855, the approach of Civil War had fragmented and regionalized the temperance movement (a prohibition movement now in all but name) and shifted the focus of moral energy. Northern temperance societies were militantly antislavery and had inundated subscribers to southern temperance journals with abolitionist literature. The war further split apart a movement whose unity had already disintegrated. But in 1869, when the national Prohibition Party was formed from the remnants of the temperance forces, its supporters proudly saw themselves as the direct descendants of the abolitionists. A national war of liberation was called for in the keynote speech. Just as the government had taken a moral stand to end slavery, so too should it now unshackle the drunkard from the manacles that the dramshop had slipped on him. Although its platform was built around the destruction of the saloon—"that manufacturer of madmen and murderers . . . and millions of rum-ruined and unutterably wretched homes"—the party from the very beginning aligned itself with a range of other causes. During its first national campaign in 1872, not only did it endorse the direct election of United States senators, but it also was in the vanguard of advocacy for complete and unrestricted suffrage for women and full, adequately protected voting rights for blacks. A meager turnout of supporters—5,600 votes in all—simply encouraged the party to become even more broadly based in its appeal for support.

New styles of organizing and agitation were employed. For the first time, women were recruited systematically into the antiliquor movement and given a leading role in the struggle. Beginning in 1873 in Ohio, crusades were mobilized to shut down the saloon. Hundreds of women demonstrated, kneeling in the streets before the barroom, singing and praying. In one famous episode of direct action, the women of Cincinnati laid siege for 2 weeks to a particular saloon, keeping a round-the-clock vigil and even rigging up a locomotive headlight to expose what was taking place behind the swinging doors.

Such strenuous and imaginative efforts coincided with a severe economic depression and an accelerating process of monopolization within the brewing industry. These factors were crucial in accounting for the drop in production in the United States of malt beverages to 5.5 million gallons between 1873 and 1875 as well as the disappearance of 750 breweries. But the Women's Crusade, "the whirlwind of the Lord," also made important contributions to this temporary decline (Asbury 1950, p. 85).

As with earlier forms of temperance activity, the crusade provided—this time to women—a sense of collective purpose and solidarity. As one of the crusaders wrote: "The infectious enthusiasm of these meetings, the fervor of the prayers, the frankness of the relation of experience, and the magnetism that pervaded all, wrought me up to such a state of physical and mental exaltation that all other places and things were dull and unsatisfactory to me. I began going twice a week, but soon got so interested that I went every day, and then twice a day in the evenings. I tried to stay home to retrieve my neglected household, but when the hour for the morning prayer meeting came around, I found the attraction irresistible. The Crusade was a daily dissipation from which it seemed impossible to tear myself. In the intervals at home, I felt as I fancy the drinker does at the breaking down of a long spree" (Kobler 1973, p. 129).

This experience of moral comradeship, which provided an outlet for the social needs of isolated women in the context of service to the family, led to the creation of the Womens' Christian Temperance Union (WCTU) in 1874. Under the leadership of Frances Willard, the WCTU became one of the national forces in the fight for prohibition. Raising the banner of "home protection," local chapters were formed across the country. In 1880, Francis Willard gave her support to the Prohibition Party, which, to commemorate the alliance, changed its name temporarily to the Prohibition Home Protection Party.

Beyond being engaged in work around prohibition, the WCTU was also deeply concerned about a range of issues: womens' rights, prostitution, prison reform, the struggles of labor, kindergartens, and smoking. Within the WCTU 45 separate departments carried on special campaigns and won victories ranging from the installation of female guards in womens' prisons to an increase in the age of consent to 18 (it had been age 10 in 20 different states). Of particular signifiance was the department of scientific temperance instruction. Under its aegis, an elaborate curriculum was developed that school systems all across the country soon made mandatory. Temperance education repeated the tales of earlier propaganda: "The majority of beer drinkers die of dropsy. . . . When alcohol passes down the throat it burns off the skin leaving it bare and burning. . . . Alcohol clogs the brain and turns the liver quickly from yellow to green to black" (Furnas 1965, p. 285). But though these clichés had been the stock in trade of temperance literature for nearly 50 years, they gained new credibility and exposure. When prohibition won its later electoral victories, voters had been systematically inculcated with this lurid pseudoscience.

Frances Willard's intelligence and energy as well as the espousal by the WCTU of a variety of progressive causes vitalized the Prohibition Party. The coalition she helped engineer increased its share of the vote in the election of 1884 to 151,809, a dramatic surge from the 10,000 votes recorded 4 years earlier. The party also gained great prestige by cutting away votes from the Republicans and swinging the election to the Democrats. In 1888, the prohibitionist forces increased their total to 250,000, and in 1892, to 270,000.

The period from 1880 to 1890 marks the second great prohibition wave. More state legislatures voted on prohibition and submitted to the people the question in the form of state constitutional amendments than at any other time. But though the issue was fiercely contested in three-quarters of all the states, only six enacted prohibition laws by 1890.

In some regions, particularly in the South, policies evolved that attempted to strike a balance between statewide prohibition and completely free-license systems. Following the example established by the Gothenberg Plan, by which corporate bodies of local government were allowed to control the liquor trade, a number of different towns and counties in North Carolina, Georgia, Alabama, and Virginia began to operate local dispensaries. The aim of these municipally run distribution networks was to further the cause of temperance, presumably by removing the profit motive from alcohol sales, and to generate revenues for the public welfare. The compatability of these twin purposes was given its most extensive test in South Carolina; there, in 1893, "Pitchfork" Ben Tillman, assuming the governorship after an election in which the majority of voters supported statewide prohibition, struck a compromise by creating a state board of control to administer and regulate sales. A state dispensary commissioner, appointed by the governor and approved by the board of control, was the chief officer. By law, he was required to be a man of "sound moral character" as well as "an abstainer from intoxicants" (Grant 1932, p. 7).

Under the system that Tillman set up, all production of alcohol was banned within South Carolina. Bulk orders were placed by the board and commissioner with manufacturers outside the state; these orders arrived at the dispensary at the capital; there, repackaging was conducted, along with shipment to county or city retail dispensaries. Operated by a county board, these outlets sold liquor only during limited hours; the purchaser had to submit a signed request, and anyone of intemperate habits or bad social reputation was theoretically to be refused.

This system, on paper, had appeal to a wide spectrum of opinion on the liquor question. Prohibitionists and antiprohibitionists were initially united, and the first year of the experiment seemed to bear out the promise of this immediate approach. In the year before the new law went into effect, there were 613 licensed saloons in the state; once these were shut down, the 146 dispensaries constituted the only points of legal distribution (Grant 1932, p. 8). Besides reducing the number of outlets, the state-run system produced significant revenues; the profits amounted to one-half million dollars a year, roughly a third of the total raised from all state sources. Despite an auspicious beginning, the state-operated scheme of regulation soon proved to have major flaws. The chief problem that emerged was pervasive malfeasance in the management and enforcement apparatus. The state board of control and the county board whose members it approved became notorious for their venality. As an investigating committee of the legislature reported in 1906, "officials of the dispensaries 'have become shameless in their abuse of power, insatiate in their greed, and perfidious in the discharge of their duties' " (Grant 1932, p. 10). State agents were imprisoned for various conspiracies to receive bounties and rebates from distilleries; local dispensaries regularly violated the letter of the law, which required written application, and the enforcement officers, called "spies," were as suspect of corruption as those whose conduct they were supposed to monitor.

The failure of the dispensary system in South Carolina and its eventual abolition in 1907 represented a victory for prohibitionists. The lesson educed was that half measures could not work, especially when they implicated the state in a dirty business. Dry representatives felt confirmed in their belief that government regulation was not only ineffective, but counterproductive; once the state sought to rehabilitate the liquor trade by becoming an accomplice, the principles of temperance were irrevocably compromised. Efforts to control were viewed as misguided and expedient; prohibition, it was concluded, was the only consistent response to a traffic intrinsically dedicated to excess and disorder.

If attempts at control like the one experimented with in South Carolina were regarded as futile and misconceived, prohibitionists saw in federal regulation an even more flagrant example of law serving the greedy interests of the liquor traffic rather than curbing them. In 1890, the Supreme Court, in its Original Package Case, reversed a ruling that had been in effect since 1847, and then held that a dry state was powerless to bar a liquor dealer from importing alcohol and then reselling it in its original package. This ruling essentially annulled state prohibition by holding that interstate commerce and the sanctity of trade took precedence over dry laws.

Congress responded to the Supreme Court decision by passing the Wilson Act in the same year. This act affirmed the right of a dry state to impose its law on liquor arriving from outside its boundaries. The Supreme Court, in turn, defined the word "arrival" to mean that state laws applied only to shipments that actually "arrived at its destination upon direct delivery to the consignee" (Grant 1932, p. 6). This loophole permitted easy circumvention. Through the use of traveling salesmen, often going door-to-door, and through advertising and other forms of solicitation, liquor dealers in wet states took orders from citizens in dry states and shipped in liquor COD. If warehouses could not be set up in dry states, then stocks could still be widely distributed through express and freight companies acting as informal agents.

A demonic pattern was detected behind compromise measures like those enacted in South Carolina and in the apparent unwillingness of the federal judiciary to allow dry states to enforce the law. Prohibitionists blamed the organized, conspiratorial might of the liquor industry for whatever failures dry forces suffered. These assertions have usually been treated by historians as paranoid tirades; the corruption and plot-filled intrigue against which people reacted is more than apocryphal.

In 1862, the alcohol beverage industry had begun to organize to defend its interests; by 1880, the Personal Liberty League of the United States was ensconced in Washington and state capitals to prevent passage of any measures that might jeopardize sales. Efforts were especially active in cities. Of 24 aldermen elected in New York City in 1890, 11 were saloonkeepers; in 1884, 633 of 1,002 Republican and Democratic conventions and primaries were held in saloons (Asbury 1950, p. 107).

Large amounts of money were raised to support antiprohibitionist candidates; contributions were crucial to the successful operations of urban machines. The connection between political corruption and the traffic in liquor was well established. The funds raised by the industry through subscriptions from the 900-member Brewers' Association, from distillers, and from the ancillary businesses—hotels, grain dealers, coopers—were the war chests used to bribe politicians and buy votes (Timberlake 1963, pp. 110-115).[9]

[9] "For the privilege of breaking the law, saloons delivered to the politicians both money and votes. Money, of course, was needed to finance elections and to satisfy the politicians' more personal needs. The method of collecting this graft and the amount saloons were required to pay varied from place to place. In New York City, for example, a saloon paid $5 a month for Sunday openings, $25 a month for harboring prostitutes, and $25 a month for permitting gambling. These fees were collected by a Tammany wardman who later divided the money with those higher up. The patrolman got his graft from the local retail liquor dealers association, supported by a monthly contribution of about $6.50 from each saloon" (Timberlake 1963, p. 112).

"An extreme example of the power of the saloon in politics was furnished by Louisville in 1905. In an election that year, the local city machine defeated a reform movement only by dint of faithful work on the part of its saloon allies. On election day nearly a hundred bartenders and saloonkeepers were qualified as election officials, and at least 4,500 fraudulent votes were cast, of which 4,000 were registered as residing in the upper rooms of saloons. Voting places in ten precincts were moved, nine of them to the rear of some saloon, and in each of these ten precincts the voters were found to have cast their ballots in alphabetical order. With the help of police, dozens of reformers serving as election watchers were thrown out of the polling places, and some were knocked down, clubbed, and beaten" (Timberlake 1963, p. 113).

Brewers owned or had controlling interests in nearly 80 percent of all saloons. With the development of refrigerated railroad cars, and the crown bottle cap, fierce, unrestrained expansion became endemic. The big brewers of St. Louis and Milwaukee began establishing local district offices and made loans at easy terms to saloons for licenses, fixtures, and stock. Once such deals had been struck, the saloonkeeper was driven to sell as much as possible. The saloon became increasingly a "putrid festering spot" because survival in a market flooded with drinking places encouraged, or even required, violation of existing laws. At the turn of the century, saloons in Washington state stayed open 24 hours a day, 7 days a week, in complete defiance of regulations (Clark 1976, p. 58).

The liquor industry was also crude and belligerently aggressive in its self-promotion. There was little effort made to soften its image or mollify public concerns. At a convention of dealers in Ohio in 1874, the question of serving minors was raised. A delegate argued: "Men who drink liquor, like others, will die, and if there is no new appetite created, our counters will be empty as well as our coffers. . . . The open field for the creation of appetite is among the boys, and I make the suggestion, gentlemen, that nickels expended in treats to the boys now will return dollars to your tills after the appetite has been formed" (Timberlake 1963, p. 102).

The liquor industry was a perfect archetype of unbridled, plutocratic self-interest, and the prohibitionist campaign against it represented a critique of both the economic and moral order. The traffic threatened America along several fronts: it debased electoral politics by corrupting voters; it allowed conglomerates to gain unfair advantage in the marketplace and mock the principle of equal competitive opportunity; and it spread environmental blight. "Why make the brewer or the distiller impound their tailings? They draw the young men of the country into their places of business, they crush them, they extract from them all that is precious—and they throw their tailings out upon society. They make society pay for the insane, the pauper and the criminal" (Binkley 1930, p. 24). The liquor industry was perceived as quintessentially parasitic; the descriptive images of the prohibitionists were loaded with metaphors of cholera. Alcohol infiltrated the drunkard's body and the social body, breaking down self-control, and eventually resulting in a befouled and excruciating death.

The prohibitionists who worked to eliminate this public health hazard largely felt themselves to be innoculated personally from the disease. The biographies of 641 leaders in the antiliquor movement between 1890 and 1913 show a group of solid, middle-class men and women. Most were born in the Northeast, and more came from urban backgrounds than the national average. Four out of five belonged to evangelical denominations, primarily Methodist and Presbyterian. Of those who fought in the Civil War, three out of four were on the Union side; previous party affiliation was overwhelmingly Republian (Blocker 1976, p. 8).

The prohibition movement in the late 19th century involved an attempt by "the decent classes" to create a morally coherent national culture. The solid division that had appalled Channing 50 years before had grown more gaping and antagonistic; massive immigration, the gross materialism of the new rich, and the rising violence of labor struggles all were signs of the times. A great war had to be fought to reestablish unity, to reconcile the opposing forces, and to stabilize a social order that seemed to be coming apart. Restraint—through inculcating self-restraint and imposing social control—had to be institutionalized. From the same basic concern for social and moral cohesiveness, earlier temperance advocates had also invoked the right of governmental intervention. But by the last decade of the century, the detachment of classes from one another, which Channing had seen ominously evolving, appeared now as a commonly accepted fact of life. Given these fixed, pervasive polarities of class, it became incumbent on the state to act systematically as mediator and, when necessary, even to impose moral and civil equilibrium.

The mandate that the prohibitionists encouraged government to accept was designed to curb extremist behavior. Rich and poor, the exploited and the exploiter were locked in a *folie à deux*. The alcohol addict and the millionaire brewer were each impulse-ridden. The gilded capitalist and the beer-swilling slum dweller were both dedicated to consumption. By contrast, a prohibitionist writer defined his fellow activists and himself: "We are the whole better class: the working, paying, thriving, home-loving masses, whose lives are lived between and distinct from the idle on the one hand, and the vicious on the other. . . . We represent," he said, "teachers and professional men, clerks, skilled mechanics, and railroad men. Drunkenness and tippling belong now to the very rich, the reckless, the ignorant, the vicious, and the very poor" (Blocker 1976, p. 16).

In this survey, there are clearly tensions between politically conservative and progressive impulses. Censure of those above and below provided no clear theoretical basis on which to plan strategy. As self-styled protectors of the social outcasts, the prohibitionists went beyond simple paternalism. At the same time, there was deep anxiety about alliances with other political groups that shared a commitment to social justice but were not equally dedicated to the drive against alcohol. During the 1890s, a struggle was fought within the Prohibition Party; its outcome would decide the ultimate future of the entire antiliquor movement.

"NARROW" VERSUS "BROAD-GAUGE"

Frances Willard led the faction that urged coalition building. The Populists, the Farmers' Alliance, and the Union Party all were identified as potential allies. In 1890, the beginnings of the kind of coalition that Willard recommended were established on the state level, with Populist and Prohibition candidates running as a slate. But despite Willard's attempts to build on this collaboration, it never evolved beyond this preliminary stage. The Populists themselves finally decided that their need for a farmer-labor alliance precluded a prohibitionist stand that might seriously alienate the urban, immigrant voter. While paying lip service to prohibition in their plank of 1892 ("Resolved: that the saloon is the great enemy of reform and the chief fountain of corruption in our politics. We denounce its pernicious influence upon our country and demand its suppression."), the Populists refused to endorse legal controls (Blocker 1976, p. 54).

The Populists' rejection of fusion intensified rather than dampened the debate about the search for common ground. Momentum began to develop for an unequivocal, single-issue approach. "To succeed," wrote Thomas Carskadon, a Methodist farmer, "we must bring to us that steady, Christian, patriotic element now constituting the best element of the old parties, and the populist party, too." He went on to denounce broad-gaugers in the Prohibition Party who sought to attract "money-loving, unsanctified brother farmers." While dwindling in influence, there continued to be those who argued for fusion: "Ours is a political party," declared one promiment fusionist, "and not a church. If we do not desire votes, and take measures to secure them, we had better go out of politics and organize a hard-shell temperance society at once." Another fusionist argument made clear the crux of the policy dispute; prohibition of liquor, E. J. Wheeler declared, was critical only because the power of the traffickers was used to block "the channels of legislation" through which other reforms had to pass. "To put an end to gin-mill government is the first, but by no means the last, thing required in order to reach the satisfactory solution to our industrial problems"—a solution that necessitated, he said, government ownership of monopolies, women suffrage, and more (Blocker 1976, pp. 72-73).

Though in 1893 the party national paper, *The Voice*, could still comment on Berkman's attempt to shoot Henry Frick by noting that it was becoming increasingly hard "to reconcile the teachings of Christ, who said, 'bear ye another's burden,' with the competitive system of today that says 'get what you can, keep what you get and devil take the hindmost,' " (Blocker 1976, p. 88); the interparty debate was drawing to a close by 1896, with the single-issue approach becoming more dominant.

The formal split between the fusionists—committed to coalition politics, hostile toward established economic power, and often disaffected from the churches of America—and the narrow-gaugers—more suspicious of the dispossessed than the possessors, deeply involved in organized religion, and fearful of becoming entangled in lower-class, civil issues such as currency or the wage system—took place at the party convention in Pittsburgh. "There is a wide difference between us [and the Populists]," one narrow-gauger remarked. What these masses need is a strong moral party to rise and abolish these sources of crime and poverty, take away those irresistible temptations, and turn the price of labor into legitimate channels of trade that will bring comfort and contentment into the home. Labor organizations ought to aid us in this, and, instead of constantly demanding higher wages, should systematically teach their members how to get the most real good out of it. 'By industry and economy one becomes rich' is old but always and eternally true. All the ages have been oppressive to those who 'have wasted their substance with riotous living' " (Blocker 1976, p. 103). Other antifusionists recoiled at the idea of a partnership with disreputable elements. "We had better die a natural death, and leave an honorable record behind us, than be choked to death in an attempt to swallow socialism, communism, and anarchism in one gulp. Look at history, and point to an epoch in which the ragamuffin elements of society have won a battle. You will find ambuscades and bloody streets. You will not find a case in which civilization has been advanced. When you have brought into the Prohibition Party Debs and Altgeld and Coxey, and Blood-to-the-Bridle Waite, and all their tattered and dirty followers, and have washed from them the stains of the gutters . . . you shall have a sight to behold," another majority peroration warned. Concluding, the speaker said: "And while you are drawing these men in, you are proclaiming to the world that Prohibition is an [association] with all those which speak of disorganization and revolution. . . . There is a way to build up a party in America, and this is to build on American principles" (pp. 103-104). Such polemics were felt necessary to rebut what still were powerful, if minority, opinions. Fusionists countered by arguing that a single-issue focus doomed prohibition to political irrelevance: "Those who would debauch our people with liquor, double their debts with a dishonest dollar, enslave the toiler and lay out the heaviest burdens of taxation on the poor, are united in support of these indescribable wrongs, while reformers are fighting in factions. . . . He who thinks we can get the voters of the country this year to drop all other issues and settle the liquor question, is too visionary to deserve, and certainly will not get, the respectful attention of the people" (p. 108). A broad-gauged representative drew the line: "We are not here as a Methodist camp meeting, a Salvation Army or a church. We are holding a political convention. For 25 years the Prohibition Party has been run by narrow-gauge managers. Tons of prohibition literature has been sent out, but not a leaf on any other issue. There are 20,000,000 church-members in the country, 4,000,000 church voters, and yet we can only poll 270,000 votes. It is time we quit slobbering over the church" (p. 112). The broad-gaugers urged a platform that included: (1) government issue of money, and free silver; (2) land grants only to actual settlers; (3) government ownership of railroads and utilities; (4) income tax and free trade; (5) abolition of convict labor; (6) women's suffrage; and (7) one day's rest in seven.

The platform was rejected in favor of a call to the party to take on a missionary role: "Let us for once open up every church in the land and unshackle the voice of the preacher on this monstrous crime. . . . Give the church a chance. . . . Let us enable the banker of New York and the silver-miner of Colorado to stand shoulder to shoulder, and I assure you that every church that I can touch will ring out death to the rum traffic" (Blocker 1976, p. 113).

In response, the minority faction bolted and formed the National Party. The Prohibition Party, left now entirely in the hands of staunchly one-dimensional crusaders, suffered a dismal defeat at the polls, receiving less than half the votes it received 4 years earlier. This humiliation merely confirmed that the party had strayed too far from the paths of righteousness and demonstrated the need for an unqualified identification with the church. Toward this end, a depoliticizing of the party was undertaken. By 1900, former traces of sympathy for the underdog were all but effaced. In its place were only smug self-congratulation and moral primping; the party's self-image was as the "large and enthusiastic gathering of intelligent, sober, clean and prosperous citizens, willing to sacrifice year after year for the good of others." But such protests of social conscience had grown wholly ritualistic; fear and loathing of the "rumsoaked rabble" had tipped the balance against joining a movement for social change in which sober citizens might be sullied by the blood or dirt of the misbegotten.

By 1905, the Prohibition Party had surrendered leadership of the movement to the Anti-Saloon League. The league was militantly single-issue and denounced partisan politics and expediency. Instead, the league, billing itself as "the church in action," began to implement a strategy based on supporting major party candidates solely on their willingness to back antiliquor legislation. By its very name, it suggested a sober moderation of purpose—not to build a broad-based, mass movement, or even to prevent people from drinking, but rather to attack the seat of boss government—the symbol of decadence.

Even while the league intended from the very beginning to move toward national prohibition, it saw this ultimate victory accomplished only as the result of a series of local successes. The country would be dried up piecemeal. The popular will would be appealed to and mobilized along a wide continuum to oppose institutionalized debauchery. An elaborate hierarchical structure was created: beginning at the level of leagues sponsored by individual congregations and proceeding to town, country, and state chapters, the policy called for dry votes to be bartered as a solid block with political candidates in return for a commitment to support a designated regulatory approach. The league pushed for local option and then moved on to state prohibition. With hundreds of full-time, professional organizers funded through subscriptions from local churches and donations from national corporate leaders and with a publishing house that, during its first 3 years, printed the equivalent of 250 million book pages, the league brought bureaucratic rationality to the antiliquor movement. Its leaders, from H. H. Russell to Wayne Wheeler, acknowledged that their techniques derived from business practices, particularly those of the liquor industry itself. In its opportunism and pragmatism, in its single-issue approach, in its delivery of electoral payoffs to its friends, and in its ability to destroy its enemies, the league became a machine.

Politicians were intimidated. As one legislator said to a league official: "While I am no more of a Christian than I was last year, while I drink just as much as I did before, you have demonstrated to me that the boasted power of the saloon in politics is a myth. . . . I shall stand with you . . . if you give me your support" (Binkley 1930, p. 29). But the great successes that the prohibitionist forces achieved cannot be attributed simply to the ruthless pressure politics and lobbying skill of the league. Most of the dry victories came about through referenda, not legislative amendments. In 1906, only 3 states had prohibition; by 1913, there were 9, with campaigns under way in all the others. By 1916, there were 23 dry states, and in 17 of these states the measure was approved by the direct vote of the people.

The last great wave of prohibition sentiment that led directly to the adoption of the 18th Amendment represents far more than simply a public relations coup. No matter how organizationally adept and politically manipulative the league was, it could only engineer, not manufacture, the juggernaut. Just as references to "Wheelerism" in no way account for majority support for prohibition, so too is analysis insufficient that explains the triumph of the antiliquor movement as either the result of war hysteria or rural animus toward urban ascendency. Both were contributing factors, but not decisive ones. Of 25 referenda victories between 1914 and 1918, 16 preceded America's entry into the war. In 1914, the House voted for national prohibition by a 197 to 190 majority. Blocker, in examining referenda in rural states (defined as those states with more than half of the population living in areas of less than 2,500 people), concluded that class was a more significant correlate than ruralism for approval of prohibition (Blocker 1976, p. 240).

PROHIBITION

Admittedly, the article that the Anti-Saloon League drew up for Congress in 1917, which was ratified in January 1919 and took effect 1 year later, defined prohibition in far stricter ways than had most state laws. Only 13 states (with only one-seventh of the total population) had a bone-dry ban before the 18th Amendment. In the other 23 states with prohibition, numerous provisions allowed citizens to import a specified monthly amount or even to ferment wine. So many exceptions were included that between the years 1906 and 1917, the per-capita consumption of legal hard liquor increased from 1.47 to 1.60 gallons (Kobler 1973, p. 216). But even though the 18th Amendment went beyond abolishing the saloon—the goal that had provided the basis for unity for the antiliquor movement—and imposed a degree of abstinence that was unfamiliar to residents of most dry territories, the elimination of these loopholes was accepted as a more thorough purifying. For almost 40 years national prohibition had been associated with deliverance; for most Americans, its precise form was best left to league experts to elaborate.

In fact, the 18th Amendment was not nearly so rigorous and uncompromising as the league made it out to be. Its three brief sections prohibited, after 1 year, "the manufacture, sale or transportation of intoxicating liquors," gave states "concurrent power" to enforce the article, and made enactment of the amendment contingent on state ratification. Significantly, the amendment avoided proscribing the purchase or consumption of liquor and provided a grace period during which stocks could be put away by those who had the desire or the money. And by outlawing manufacture and sales only, it countenanced through omission patronage of the bootlegger. In addition, the substitution of the words "intoxicating liquors" for "alcoholic beverages" was more than semantic choice. Endless debates would ensue about whether the source of abusive behavior was inherent in a particular drink or in the particular psychology of the drinker. In addition, failure to define the nature of the "concurrent powers" of the states and the federal government gave rise to a dispute that lasted as long as Prohibition.

The Volstead Act, which detailed enforcement of the 18th Amendment, consisted of 72 sections full of complicated and equivocal codes. Designed with the help of the league, it was intended to integrate the best features of various state prohibition laws. However, far from crystallizing the collective experiences of Prohibition, the Volstead Act was a bewildering mélange. Cross-references, contradictions, and modifications of ordinary criminal procedures created enormous administrative and legal problems.

Under the act—which was intended to "prohibit intoxicating beverages," regulate "the manufacture, production, use and sale of high-proof spirits for other than beverage purposes, . . . insure an ample supply of alcohol, and promote its use in scientific research and in the development of fuel, dye and other lawful purposes"—the Treasury Department was charged with overseeing a vast system of permits, preventing illegal diversions, and arresting bootleggers. In 1920, little more than $2 million was allocated by Congress to put 1,500 agents in the field; as a group they were untrained and underpaid (the average salary was $1,500). Furthermore, they were all political appointees, the league having lobbied Congress to exempt the bureau from Civil Service so that men of "strong prohibitionist principle" could serve.

This use of the spoils system filled the Prohibition Bureau with men whose incompetence and venality discredited the enforcement apparatus with juries, courts, and the public. In 1921, 100 agents in New York were fired at one time for malfeasance; the same year, a federal grand jury for the southern district of New York declared that "almost without exception the agents are not men of the type of intelligence and character qualified to be charged with this difficult duty and federal law" (Kobler 1973, p. 247). But even had the bureau been staffed with more honest and able men, the tasks that they were assigned and the laws governing their performance made successful enforcement nearly impossible.

Agents of the bureau were responsible for tracking down the illegal production of alcohol; but the Volstead Act reinforced the federal statute relating to search and seizure by adding a clause making issuance of any warrant dependent on proof that the liquor was for sale. No matter how much liquor a person had at home, no matter how it was obtained or what use was intended, agents had to have positive evidence that a commercial transaction was involved. What such a requirement actually did was to permit home manufacture, both for personal use and as a cottage industry organized as part of large criminal networks. Small stills, primitive but effective, were set up by bootleggers in apartments, with the liquor being collected on a regular basis. More common were beermaking and winemaking for genuine home consumption. Malt and hops shops proliferated; big food chains openly advertised ingredients and apparatus—hoses, gauges, capping machines. Section 29 of the act, which permitted fruit juices to be made and consumed for personal use, led to a boom for the California grape industry. During the first 5 years of Prohibition, the acreage of vineyards increased 700 percent; the grapes were marketed as concentrate in "blocks of port," "blocks of Rhine Wine," and so forth and came with a warning: "After dissolving the brick in a gallon of water, do not place the liquid in the jug away in the cupboard for twenty days, because then it would turn into wine" (Binkley 1930, p. 108). Mayor LaGuardia of New York sent out instructions on winemaking to his constituents along with a bit of free legal advice: "The beverage may be called wine or beer, but must not be labeled as such. . . . The question of the intoxicating character of the beverage is not determined by any fixed or arbitrary code. . . . The average homemade wine may be considered nonintoxicating within the meaning of the law. It cannot be given to strangers; even though the beverage is nonintoxicating, it loses its legal character if sold" (Binkley 1930, p. 25). Such charades not only made it innocent and amusing to circumvent Prohibition but also revealed a pervasive ambivalence at the root of the 18th Amendment and its enforcement statutes. Even while possession of liquor illegally obtained was unlawful, the act of drinking was maintained as privileged. This anomaly suggests first that prohibitionists understood the politically feasible limits of regulating individual behavior. The league and other defenders of the 18th Amendment regarded as a slur any suggestion that they supported sumptuary laws and were deeply embarrassed and resentful of the popular concern of the dry snooper and killjoy on the prowl to mind someone else's business. There was, among prohibitionists, a pervasive if implicit recognition of the practical inviolability of private conduct.

But the failure of prohibition laws to intrude directly on the mores of the consumer was not primarily a tactical decision made on the basis of calculating how deeply public policy should be allowed to invade private experience. Once the destruction of the 170,000 saloons had been achieved, and the systematic spread of addiction stopped, it was believed that the appetite for drink would wither away without the artificial stimulation of an organized traffic. The taste for alcohol was a false need, implanted by moral brigands. Ever since Benjamin Rush elaborated the concept of drink as disease, the consumer was simultaneously exempted from fundamental blame and diminished in the degree of autonomy. For being freed from censure and sin, the drinker paid the price of compulsory wardship. In the tradition of temperance and prohibition, the person who submitted to alcohol was the quintessential victim. The environment needed to be decontaminated; there was no reliable way to immunize people or even to identify those particularly at risk. Instead, the body politic had to be kept free from infection. Once the liquor industry was removed, then health would be restored. "It is part of the philosophy of Prohibition that the final triumph of the cause, the definitive solution of the liquor question, requires that there should be an unsullied generation which would regard drinking as a moral perversion and the purveyor as a felon" (Binkley 1930, p. 25).

According to prohibitionist doctrine, Americans had once been pure. A nefarious trade had robbed people of their reason and corrupted domestic and social integrity. The 18th Amendment represented a millennial triumph inaugurating personal self-restraint and national solidarity. To police compliance was contradictory. It made no sense to enforce a cure that would naturally take effect.

But even with the saloon effaced, there continued to be both a demand for alcohol and a traffic to meet it. The Volstead Act had built in wholly inadequate safeguards against the reproduction of the trade.

Legal alcohol was among the largest source of illegal liquor. The total production of distilled spirits rose from 187 million gallons in 1912 to 203 million gallons in 1926. Most of this increase, of course, can be attributed to the greatly expanded use of industrial alcohol. As a basic chemical, it was required in the manufacture of cosmetics, leather goods, dyes, and synthetic textiles (one rayon plant used 2 million gallons of denatured alcohol annually). The single most significant factor in the expanded production of denatured alcohol was the spectacular burgeoning of the auto industry. In 1919, there were 7.5 million passenger cars and trucks registered in the United States; in 1926, there were 22 million. And whereas before, 90 percent were open and therefore not suitable for cold-weather driving, by 1926, 70 percent were closed, suggesting the possibility of operation in cold weather. Antifreeze was thus a sudden necessity, and three-quarters of the total output of completely denatured alcohol went for that purpose. A considerable portion of the enormous increase in production undoubtedly was diverted to bootleggers. Emory Buckner, the United States Attorney for the southern district of New York, estimated that in 1925, 60 million gallons were sold as beverage alcohol (Dobyns 1940, p. 283). Since it is likely that this congressional testimony was intended to secure additional appropriations for enforcement, the claims of such mammoth diversions are suspect. However, even more modest figures suggest systematic fraud.

The technique of diversion most often employed took advantage of the complicated chain of subsidiary denaturing plants allowed by the Volstead Act. Large distillers produced denatured alcohol but also sold alcohol in its natural state to smaller, independent firms for them to process. These firms could also pass along pure alcohol to other, even smaller plants. During this series of transactions, alcohol could be illegally diverted at any point. There was no way to monitor sales and verify them as legitimate. Distillers might sell 10,000 gallons to a dummy perfume manufacturer or chemical plant in receipt of a legal permit; the alcohol would then be sent to bootleggers. Some would be reserved to make whatever item the company was expected to produce, and this commodity (cologne, for example), with records faked to show a volume large enough to account for the diversion, would then be shipped to a wholesaler known as a cover house. The bill for the inflated amount of cologne would be paid and the deal completed. Such a ruse was almost foolproof. Inspection was woefully inadequate; prohibition agents were prevented from investigating beyond the point of the initial sale of alcohol to the manufacturer; and the number of legitimate companies requesting permits was increasing as a consequence of rapid industrial development.

The reluctance of the Volstead Act to monitor more stringently how businesses used their allocations of alcohol stemmed not simply from an unwillingness to expose the inherent problems of enforcement. Rather, it betrayed Prohibition's own sense of almost paralytic ideological bewilderment.

For the duration of the Volstead Act, there was a pervasive anxiety not only about interfering with personal behavior, but also about intruding on the marketplace. Just as individual consumption was protected as a right, so too was private enterprise sanctified. Under the direction of the league, the antiliquor movement had aligned itself more and more closely with the cult of efficiency and had come to regard corporate leaders as benefactors rather than as plutocrats. Two years before the passage of the 18th Amendment, a dry activist had written: "Every psychological trend in modern life is today sweeping towards Prohibition. . . . The cry of efficiency is on everyone's lips. It has come up out of the added pressure on our modern life, which has driven us to a scientific attitude towards life, a determined effort to find out just how much human force there is in every man or woman, and to get it all out, and not let any of it go to waste, to stop the leak at any cost" (Banks 1917, p. 48). But once this leak had been stopped through the law, then further governmental tinkering might risk damage to the delicate gears of the machinery of production. The individual, unfettered from seductive influences, would return to a natural state of sobriety;

likewise, prohibitionists assumed business, left to its own devices, would be a strong ally, grateful for the profits extracted from a now abstemious labor force.

Throughout much of the history of temperance and prohibition, a tension can be traced between a public health perspective—one that emphasized the need for collective action on behalf of those disabled by drunkenness, stressed the accountability of a powerful, profiteering business in the creation of disease and human misery, and accepted the premise that genuine democracy could not exist so long as pervasive, systematic polluting of the social environment was allowed—and an essentially antagonistic free enterprise perspective—one that emphasized discipline, efficiency, and self-restraint.

Only for a brief period during the 1890s did at least some elements within the movement accept the implications of the call to liquidate an entire industry. But while a broad anticorporate agenda was rejected and though the politics of prohibition grew increasingly racist and demagogic as the league took command, the central contradiction could not be evaded. There was no convincing way to reconcile the confiscation of private property with a faith in the sacred rationality of the market. Prohibitionists invoked their descendency from abolition because such a heritage served to validate the justice of the cause as well as restrict the meaning of reform to a millennial moral triumph. The precedent of abolition had established that an evil traffic could be destroyed and owners dispossessed of their tainted wealth, all without threatening the legitimate property rights of other economic sectors.

There were, however, fundamental differences between commerce in beer and commerce in human bodies. Despite tirades against liquor as the Simon Legree of the soul, Prohibition could never successfully equate literal physical slavery and slavery to alcohol. No matter how aggressive and seductive the marketing of liquor might be, drunkenness still involved self-induced surrender of independence. Prohibitionists did argue that the power of alcohol to form moral lesions discounted the possibility of free choice, but this largely metaphorical line of reasoning was specious since its extensions could not easily be controlled. Although temperance and prohibition were always imbued with nostalgia for a bucolic and mythic past in which behavior was managed by a patriarchal guardianship, the concept of government as moral trustee did not belong exclusively to those who were looking backward. Socialists could just as easily appropriate it and demand that government intervene to protect the individual against the predatory attacks of a whole range of industrial buccaneers. Once the principle was established that the conduct of citizens was vulnerable to a pathogenic milieu, then regulating that milieu became legitimate.

As the business community recognized even before the passage of the 18th Amendment, the limits of this regulation were difficult to confine. *The Commercial and Financial Chronicle* declared "that it is a singular component of reform that nothing is so important as the task at hand, whether it be the manufacture and sale of intoxicants, or the eight hour day, or daylight saving, or the removal of signboards from vacant lots. . . . The moral of it all is that we cannot preserve either our liberties, our institutions, or our peculiar form of government, if we are to let self-appointed guardians of the public weal seek the cover of general law for the purpose of obtaining their self-satisfying ends. This prohibition measure and mandate is but one of these ends. It is . . . a theory of the proper social life. In precisely the same manner theorists are seeking to control individual life in commerce" (Timberlake 1963, p. 78).

In the enforcement of Prohibition there was always an underlying sensitivity to the boundaries of regulation. Even in the approach to its most notorious enemy—the brewery industry—prohibitionists proceeded with great caution.

The authors of the Volstead Act refused to adopt statutes, which were included in some state laws, that banned entirely the production of malt liquors. Instead, a clause was included that permitted breweries to manufacture beer as long as it was dealcoholized to 0.5 percent. This allowance of "near-beer" created further opportunity for evasion; agents were unable to monitor which part of the overall production wound up as legal beverage and which part as real beer was drawn off and secretly piped into vats for bootleggers (Dobyns 1940, p. 279).

The exemptions allowed under the Volstead Act seriously compounded the problems of enforcement. Not only were the energies of the inadequate and poorly trained bureau hopelessly divided between a myriad of potential points of violation, but more important, the special allowances contaminated the whole body of antiliquor law. Official disinterest in evasion encouraged noncompliance; an obvious lack of consistency in the way that availability was controlled made people question whether public policy actually intended to achieve a dry America.

Gusfield, in attempting to account for the gaping loopholes of the Volstead Act, has proposed that Prohibition never seriously aimed to curb drinking. The 18th Amendment was, he said, a ceremonial victory for the middle class, "a dramatic event of deviant designation." The legislation itself, rather than its application, "affirmed one cultural standard of conventionality, and derogated another." As a "symbolic gesture," it thus made no difference if Prohibition was enforced. "Laws may be honored in their breach as much as in their performance," he declared (Filstead 1972, p. 70).

Gusfield clearly makes an important distinction between law as a "designating ritual," and law as purposive in instrumental terms. But the hesitation and apparent arbitrariness that compromised enforcement need not only be accounted for as part of "a patterned evasion of norms." There were tremendous difficulties in preventing production and distribution and serious political constraints imposed on eliminating loopholes. It must also be borne in mind that for all the corrupt, inept, and discretionary ways that that law was written and enforced, it did reduce the amount that Americans drank.

There is now little dispute about the fact that the annual per-capita consumption level declined as the result of Prohibition. Because of the long war of statistics fought between wet and dry forces, data, no matter how useful, have often been assumed to be little more than disguised polemics. There was, of course, a surfeit of spurious evidence churned out by both sides. But there is a body of credible information suggesting that the 18th Amendment had a substantial impact on drinking patterns.

Those who opposed Prohibition argued that Americans were drinking less during the decades leading up to the 18th Amendment, and that a severely restrictive policy was both unnecessary and counterproductive. "All observant Americans more than fifty years of age," W. H. Stayton wrote, "had, up to 1920, noticed a marked national change in the direction of sobriety. With the coming of national prohibition, that tendency was reversed" (Stayton 1923, p. 34). Statistics, however, reveal an entirely different picture of per-capita consumption. Between 1900 and 1904, consumption was 1.36 proof gallons of spirits and 16.94 gallons of beer annually. Between 1910 and 1914, the annual figures had risen to 1.46 proof gallons and 20.38 gallons (U.S. Department of Commerce 1923, p. 397).

In the 1932 issue of *The Annals of the American Academy of Political and Social Science* that recapped the experience of national prohibition, Clark Warburton presented data indicating a dramatic decline in consumption during the early years of Prohibition and a leveling off as the 1920s ended. In a persuasive summary of research on this question, Norman H. Clark concluded:

> [Warburton's] synthesis was conjectural and his projections were admittedly rough but they were refined in 1948 by the century's most prominent and indefatigable researcher in alcohol studies, the late E.M. Jellinek. Even more recently, Joseph Gusfield has re-examined both these studies and concluded that "Prohibition was effective in sharply reducing the rate of alcohol consumption in the United States. We may set the outer limit of that at about 50 percent and the inner limit at about one-third less alcohol consumed by the total population than had been the case . . . [before Prohibition] in the United States (Clark 1976, pp. 146-147).

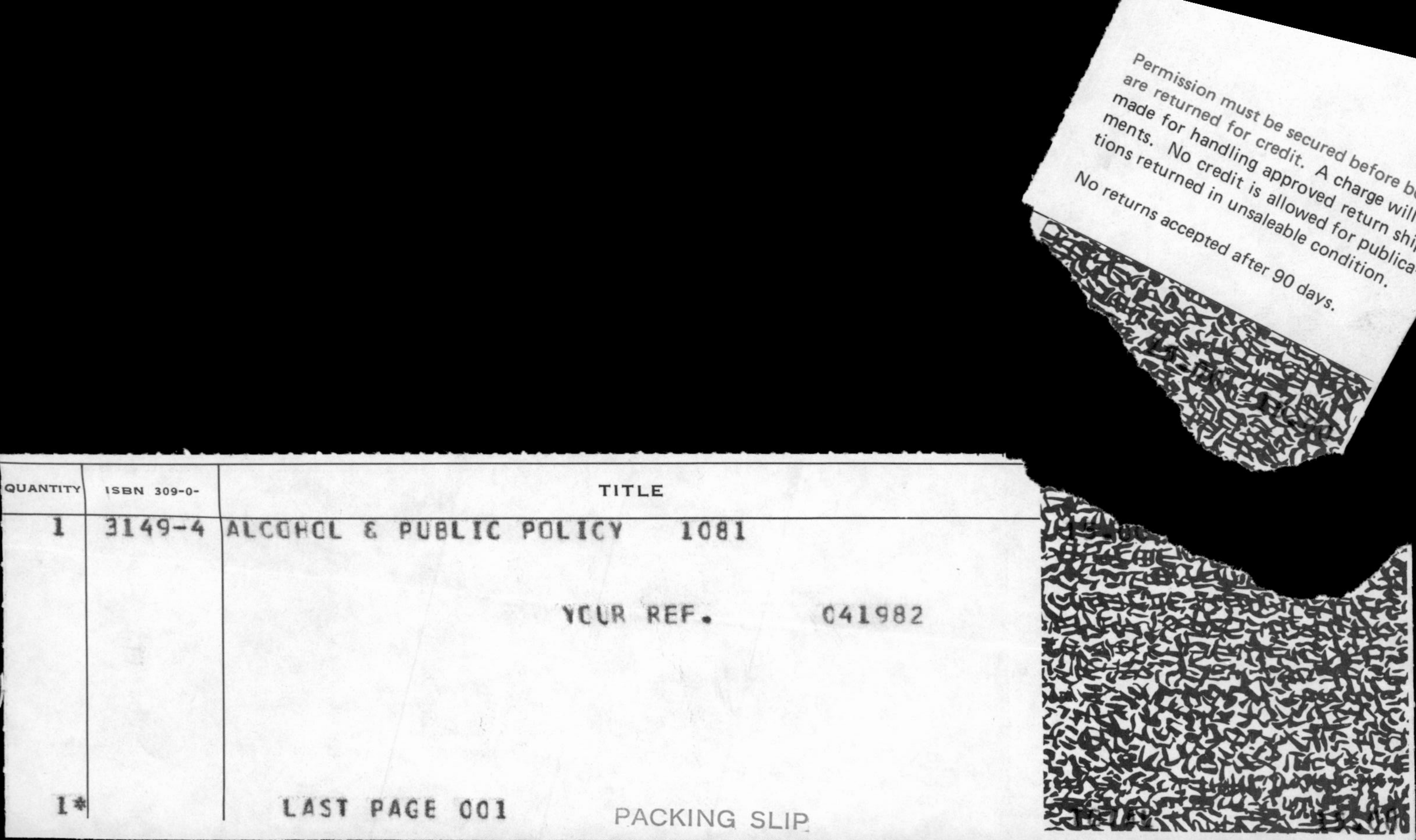

QUANTITY	ISBN 309-0-	TITLE
1	3149-4	ALCOHOL & PUBLIC POLICY 1081
		YOUR REF. C41982
1*		LAST PAGE 001

PACKING SLIP

Permission must be secured before be... are returned for credit. A charge will... made for handling approved return ship... ments. No credit is allowed for publica... tions returned in unsaleable condition.

No returns accepted after 90 days.

Various data tend to confirm his judgment. Death rates from cirrhosis were 29.5 per 100,000 in 1911 for men, and 10.7 in 1929; admissions to state mental hospital for disease classified as alcoholic psychosis fell from 10.1 in 1919, to 3.7 in 1922, rising to 4.7 by 1928 (Emerson 1932, p. 59). In two predominately wet states, the decline in alcoholic psychosis was even more dramatic. In New York, it fell from 11.5 in 1910, to 3.0 in 1920, to 6.5 in 1931, and in Massachusetts, from 14.6 in 1910, to 6.4 in 1922, to 7.7 in 1929 (Emerson 1932, pp. 59-60).

National records of arrest for drunkenness and disorderly conduct declined 50 percent between 1916 and 1922 (Feldman 1927, p. 367). Reports of welfare agencies from around the country overwhelmingly indicated a dramatic decrease among client population of alcohol-related family problems (Feldman 1927, p. 136).

Although the 18th Amendment, even supporters admitted, was so full of compromise and disparity as to be "amphibious" rather than dry, it still managed to fulfill some of the fundamental expectations about Prohibition. The principal benefits accrued to the group most vulnerable in the past to the devastating effects of drink. Observers, during Prohibition and since, have been unanimous in concluding that the greatest decreases in consumption occurred in the working class, with estimates of a 50-percent decrease (Sinclair 1962, p. 249).[10] In large measure, intoxicants priced themselves out of the market. In 1928, when the average family earned $2,600 annually, a quart of beer cost 80 cents, 6 times more expensive than 12 years earlier; gin was $5.90 a quart, 5 times more expensive; and whiskey was $7.00, 4 times more expensive (Clark 1976, p. 54).

Defenders of Prohibition claimed improvements in the physical health and sobriety of the labor force as a vindication of their cause. Workers were more fit for production and more likely to avoid sinking into costly deviance or dependency. Spending money, once squandered at the saloon, was now used to purchase cars and refrigerators. Thus, the 18th Amendment deserved credit for helping to stabilize the behavior of the laboring class and pave the way for prosperity.

[10] In testimony before the Wickersham Commission, Whiting Williams concluded: "Very much of the misconception with respect to the liquor problem comes from the fact that most of the people who are writing and talking about the prohibition problem most actively are people who, in the nature of things, have never had contact with the liquor problem in its earlier pre-prohibition days, and who, therefore, unduly are impressed with the changes with respect to drinking that they see on their own level; their own level, however, representing an extremely small proportion of the population."

"The great mass, who, I think, are enormously more involved in the whole problem, of course, in the nature of things, are not articulate and are not writing in the newspapers" (Sinclair 1962, p. 249).

Given the dutiful service that the prohibitionists saw themselves delivering to industrial coherence, it came as an incomprehensible betrayal when major elements of the business community abandoned the dry alliance. This defection helped obscure the accomplishments of the 18th Amendment, shift the focus away from the public health benefits and toward the costs paid in terms of personal freedom and increased class conflict, and spawned a campaign to nullify the law through noncompliance.

The Association Against the Prohibition Amendment (AAPA) represented more than a front for the alcohol industry or a plot hatched by the du Ponts. Both of these interpretations are superficially plausible; breweries and distillers vigorously supported the organization of this earliest and most powerful prorepeal group. Pierre du Pont took over the AAPA in 1926 and solicited other millionaires on the basis that an end to Prohibition would save them huge amounts in corporate and personal income taxes. But it is a mistake to regard the AAPA as simply a crude tool wielded by venal interests.

The AAPA and its various subsidiaries (The Crusaders for Young Men and the WONPR, the Women's Organization for National Prohibition Reform) attacked Prohibition as "wrong in principle, has been equally disastrous in consequences in the hypocrisy, the corruption, the tragic loss of life and the appalling increase of crime which have attended the abortive attempt to enforce it; in the checking of the steady growth of temperance which had preceded it; in the shocking effect it has upon the youth of the nation; in the impairment of constitutional guarantees of individual rights; in the weakening of the sense of solidarity between the citizens and the government which is the only sure basis of a country's strength" (Dobyns 1940, p. 107).

Some of the claims contained in this litany are legends that can be quickly dismissed. Before the 18th Amendment, people were drinking more, not less. There is no convincing evidence that Prohibition brought on a crime wave; homicide had its highest rate of increase between 1900 and 1910; organized rackets, while expanding to take advantage of new opportunities, had been firmly established in urban areas before the 18th Amendment (Feldman 1927, p. 249). According to census statistics, the rate of death from wood or denatured alcohol remained almost constant from its peak year in 1920 of 369 (Feldman 1927, p. 401).

Mystifications constantly appeared in newspapers and magazines as fact. The image became imprinted of Prohibition as a public health hazard; the AAPA, with 800 prominent journalists and cartoonists working as a special committee, popularized the myth of "the tommygun and the poisoned cup" (Dobyns 1940, p. 107).

Even more important to the AAPA in its campaign to undo the 18th Amendment were threats of class strife. Al Capone, while a convenient symbol of the failure of Prohibition, was less terrifying than the spectre of Bolshevism. Advocates of repeal stole the thunder from the dry forces. Prohibitionists argued that "the most characteristic feature of the world today is the breakdown of authority . . ." (Fisk 1923, p. 8). Respect for the law and the Constitution were bulwarks against "hysterical unrest and anarchy." The AAPA conjured up the same menance but accused "Volsteadism" of being the Reds' unwitting dupe.

W. H. Stayton, head of the AAPA, spoke directly to the business community and criticized it for being "singularly blind" to the real content of the 18th Amendment. "Surely that class of our citizens who should be most concerned to fight against laws confiscating private property is the employer class. . . . A few years ago a barrel of whiskey was private property; objectionable property, if you will, but property none-the-less. . . . Then came forward people saying, 'We do not approve of that kind of property; we think it works harm to the people.' " Stayton then pitched his warning: "But there are many people in this country who do not approve of accumulated or inherited fortunes, believing them to be harmful to the people; indeed, some of those among us do not approve of any kind of private ownership. When the time comes that these classes demand confiscations to suit their beliefs, the employers will be in no position to turn to the working man for help in sustaining property rights; for the poor man may well reply, 'No, it was *you* who made this precedent, and you made it for no good purpose, but with the intent to rob me of my hours of relaxation, so that you might get more work and more profit out of me' " (Stayton 1923, p. 32).

It was this stock in trade nightmare of the Anti-Saloon League that the AAPA appropriated as its own. While the AAPA did borrow and reconceptualize other classic prohibitionist images of dread (the saloon as "the Siamese twin of syphilis" became the hip flask, "the insidious moral danger to which maidens are exposed"), class struggle remained the dominant theme of jeopardy.

The defenders of Prohibition were both enraged and baffled by the betrayal of corporate America. A dry advocate declared that the failure of Prohibition "thus far lies with two elements of our population: the immigrants established in congested city colonies, and a fraction of the élite, to whom the use of alcoholic drinks has become not so much an inveterate habit as the sign and symbol of a luxurious 'kultur' " (Woods 1923, p. 123). A Methodist minister was more mystified than angry by the apparently self-destructive behavior of the rich: "It is our observation that the laboring man and the poor are not the lawbreakers, but it lies more largely with the rich and the well to do, who seem to think it smart. . . . Their smartness . . . is the rankest stupidity, for as a class they would suffer most should the lawless get control and break up all the law" (Tomkins 1923, p. 19).

In their deep resentment and bewilderment, backers of Prohibition did not understand that the very prosperity for which the 18th Amendment took credit had gradually begun to create the need for a new labor force. The worker as exclusively a producer was being transformed into worker as consumer. Fearing national deterioration, prohibitionists argued that "this is an age when we need not more personal liberty, not more pampering and self-indulgence and influences contributory to ease and comfort, but less self-indulgence, less pampering, and more courage to face life's struggle" (Fisk 1923, p. 8). But important elements of the ruling class had to be incited rather than suppressed; a huge advertising industry began to develop in the mid-1920s to manufacture new needs.[11] Workers would achieve equality in the marketplace, and their position there had to be respected and enhanced. Freedom of consumption required personal sovereignty; forward-looking companies understood this to be a tenet of economic growth and industrial peace. "Stopping the leaks"—the aim that prohibitionists had for so long advocated—was discarded in favor of a strategic policy of encouraging, or even creating, leaks.

This evolution in the concepts of corporate management took place organically; the ideological shift occurred not as part of a grand conspiratorial design, but rather as a gradual and intuitive reassessment. The worker needed to "blow off steam," to be vented. As Samuel Gompers, as much a speaker for the employer class as a labor leader, observed: "Prohibition is not a matter of right or wrong. It is not a question of whether we approve of drinking or not. It is a habit, and when you invade a man's habit, you unsettle him. You find that the man who has heretofore been contented to labor as he had been laboring, becomes restive and discontented. . . . Harmful as vodka was, it enabled the Russian peasant to find surcease from dull monotony. Without it,

[11] H. A. Overstreet, one of the pioneers of the advertising industry and an early master of epistemological self-righteousness, helped make popular the view that being "ill-liberal" was tantamount to insanity. "A man," he wrote, "may be angrily against racial equality, public housing, the TVA, financial and technical aid to backward countries, organized labor, and the preaching of social rather than salvational religion. . . . Such people may appear 'normal' in the sense they may be able to hold a job and otherwise maintain their status as members of society; but they are, as we now recognize, well along the road to mental illness" (Van den Haag 1975, p. 123).

he found only trouble and torment and the desire to tear down what he could not rebuild" (Stout 1921, p. 135).[12]

It was largely from the corporate-dominated movement opposing Prohibition that the concept of alcohol control emerged. As Levine suggests, the phrase was rarely used before the passage of the 18th Amendment; during the 1920s, however, it came to represent a pivotal idea in the struggle for repeal (Levine 1979, p. 3). Basically, advocates of control took a position that they defined as a just mean between dangerous polarities. Permitting saloon power to run amok had given rise to futile attempts at imposing moral fetters. A policy of control avoided this dual extremism. Instead of efforts to repress consumption through a total curb on availability, alcohol control emphasized effective regulation. Whereas Prohibition required that government assume the role of punitive moral arbiter, "control" conjured up the image of the state as manager. As part of the critique developed by the AAPA, the argument was made that bureaucratic rationality should underlie any state intervention.

The experiences of foreign countries were seen as especially instructive. The direct involvement of the state in Canadian, Swedish, and English systems of alcohol sales proved, according to proponents of control, that the best safeguards to public order were technical mechanisms rather than sumptuary laws. The example of Carlisle was among the most frequently cited. Here, a board of control bought out the entire alcohol industry in a 500-square-mile area of Scotland and managed the business themselves. The experiment, a wartime measure initiated in 1916, was undertaken in order to reduce problems of drunkenness that had increased sharply because of the doubling of the adult male population brought about by sudden industrial concentration. The city had the highest proportion of criminal convictions to population in all of England. The board proceeded to exert control through restrictive regulation and constructive arrangements. The number of licensed premises was reduced from 203 to 69; the practice of selling a mixture of beer and spirits was prohibited; Sunday closings were enforced, as were bans on sales to young persons. On the constructive side, the board encouraged the sale of food and nonintoxicants. As Lady Astor, an important publicist for the experiment, wrote: "There was a deliberate attempt to

[12] In *The Cup of Fury*, Upton Sinclair (1956) takes the opposite tack; "The Communists use liquor as a sort of Geiger-counter, probing for the weaknesses of men and women. They have used it to gain recruits, they have used it to steal a nation's most guarded secrets" (p. 123).

conduct a refreshment trade, and food taverns were opened, where a properly cooked meal was offered at a reasonable price" (Astor 1923, p. 274).

The results of direct government involvement were impressive; drunkenness declined rapidly, and the system earned the support of all sectors of the community. In the United States, those who believed that Prohibition constituted a perversion of federal power, pointed to Carlisle as a case in which the state had demonstrated its capacity for creative control. Those who designed public policies surrounding alcohol had to be free from narrow partisan prejudice; according to the AAPA and other adherents of the control model, neither the liquor industry nor the sectarian zealots of the Anti-Saloon League were competent to maintain social stability. Government action to regulate the distribution of liquor was promoted as the antidote to the poisonous viruses of fanaticism. The state, through direct supervision of sales, could both structure patterns of consumption and appropriate and defuse a social issue that had generated conflict and division for nearly 100 years.

The paradigm of control became a powerful theoretical alternative to Prohibition because it refined the principle of government responsibility rather than rejecting it. Control posited a fundamental substitution; instead of moral stewardship, government would provide administrative expertise. Under bureaucratic auspices, a new rationale would evolve. To overcome what John Koren, a leader in the control movement, called "the moral chaos, a morass" resulting from efforts to legislate abstinence, government had first to accept the fact that drinking would go on (Koren 1923, p. 52). However, by exercising its legitimate authority as defender of the public order, government was also entitled to place checks on drinking. George Catlin, author of the book *Liquor Control*, wrote that the curbs on consumption "on the one hand [must be] sufficiently lenient to be consistent with public opinion and susceptible of competent enforcement by the police without drawback to their more important duties, and, on the other hand, sufficiently vigorous to remove the drinking of intoxicants from the list of significant threats to social order" (Catlin 1932, p. 183). Catlin stressed the importance of particular techniques of regulation being geared to local conditions. Federal law should sustain the authority of the states. Beyond this, Catlin urged that "spirits be compulsorily diluted, discriminately taxed, and available only during certain hours" and recommended the establishment of chartered companies that would have a monopoly control of liquor within their areas (Catlin 1932, p. 184). Although advocates of control were in favor of experimenting with a variety of different forms, there was wide agreement on certain basic principles.

The guiding assumptions for Catlin—that spirits posed a greater threat than beer and wine and therefore required a more restrictive tax and that the liquor trade must not be run for private profit—were tenets of the control position.

In discussing the alcohol control movement it would be a mistake to exaggerate the degree of its influence in changing public opinion. As a theoretical concept, management of consumption was an approach set apart equally from Prohibition and laissez-faire; its stance was too scientific and too dispassionate to gain a following. While the AAPA incorporated in its membership those who made serious attempts to devise regulatory policies as alternatives to Prohibition, the movement for repeal mobilized its support primarily through the use of polemics and propaganda. Despite the earnest and objective inquiries of men like Catlin and Koren, such policy analysis had little impact on the immediate struggle. The battle lines were split between wet and dry, and ideological third parties were anathema. It must be understood, too, as the decade of the 1920s came to an end, that the forces favoring the maintenance of Prohibition were more than holding their own.

The gradual displacement of one model of individual and social behavior (virtue inherent in austerity and self-restraint) by another (virtue inherent in moderate consumption and easy-going compliance) did not constitute a sudden ideological transformation that swept away support for the 18th Amendment. For all the heavily funded organizing of the AAPA and for all the scandals involving both agents of the bureau and Anti-Saloon League members themselves (these latter exemplars of rectitude became implicated in various shady transactions), Prohibition was still strong in 1928. In the election of that year, dry political candidates swept the field. Hoover was an overwhelming victor; 80 of 96 senators, 328 of 424 House members, and 43 of 48 governors elected were backers of the 18th Amendment. The Wickersham Committee, appointed in 1928 to investigate the 18th Amendment, came out with a report that, while including minority opinions, nonetheless endorsed Prohibition and urged stronger enforcement. Even William Randolph Hearst, one of the principal opponents of Prohibition, as late as 1929, regarded repeal as out of the question. The prize-winning essay in the national contest sponsored by his papers proposed redefining beer as nonintoxicating; a more direct attack on the 18th Amendment was seen as political adventurism (Hearst 1929).

Prohibition was part of the Constitution and thus protected by an aura of the immutable. Breaking the law defied but did not overthrow official legitimacy. "Repeal of the Eighteenth Amendment is pure nonsense—thirteen states with a population less than half of New York state can prevent repeal until Halley's Comet comes in," Clarence Darrow had observed (Dobyns 1940, p. 389). But beyond this practical consideration, repeal involved an element of the blasphemous.[13]

Economic collapse created the possibility for extreme and unprecedented measures.[14] By 1930, the AAPA-backed referenda drives had produced victory for repudiation in nine states. The primary argument now made for repeal was no longer the demoralizing effects of Prohibition on civil liberties or class harmony; the end of the 18th Amendment was presented as the key to economic salvation. In the pamphlet *Prohibition and the Deficit*, the AAPA declared that "by the end of 1931 annual liquor tax collections since 1920, if national prohibition had not intervened, should have totalled practically eleven billion dollars. This money might have been used (if all other sources of revenue had been availed of) to reduce our 1931 indebtedness from $16,801,000,000 to $5,801,000,000, If we continue our estimate, by the end of 1933 we should have a balanced budget and a public debt of $7,306,000,000 instead of $20,341,000,000" (Dobyns 1940, p. 377). Alcohol was now linked to a patriotic cause; just as Lincoln had turned to the liquor trade

[13] Felix Frankfurter declared that "if the process by which this Amendment came into the Constitution is open to question, one can hardly dare contemplate the moral justification of some of the other amendments, or of the Constitution itself. Whether we like it or not, the 18th Amendment *is*!" (Frankfurter 1923, p. 193).

[14] Historians have debated the extent to which the Great Depression and New Deal constitute a fundamental break or turning point in the American experience. Degler has written that "as the Civil War represented a watershed in American thought, so the depression and its New Deal marked the crossing of a divide from which, it would seem, there could be no turning back." Louis Hacker uses the description of "the Third American Revolution," and Hofstader argues that a "drastic new departure in the history of American reformism" was set in motion after 1929. On the other hand, Richard Kirkendall cautions against exaggerating the extent to which the changes produced during this period should be construed as a "divide," and emphasizes instead the important continuity with the past that the decade of the 1930s still maintained (Braeman et al. 1964, pp. 146, 148, *et passim*).

But whether one holds with Kirkendall that the Great Depression simply accelerated "the rise of a collectivistic or organizational type of capitalism evident since the third quarter of the 19th century," or whether one is persuaded that some more fundamental cleavage took place, it seems clear that economic collapse did require people to take stock of a whole range of traditional beliefs and values. In the New Deal's pragmatic approach to reconstruction, and in the corresponding advent of the "guarantor state," the bulwarks of Prohibition disintegrated. The ideal of individual abstinence came to be perceived as wrongheaded and even cruel, an artifact of a discredited ideological system. As Gusfield puts it, "The Great Depression dissolved the magic power of the old symbols. . . ." (Braeman et al. 1968, p. 305); the sources for public disorder and misery so obviously transcended personal indulgence that the struggle against drink took on an almost antedeluvian irrelevance.

in 1862 to finance the war effort, so too did Roosevelt campaign in 1932 on the promise to repeal the 18th Amendment and revive an industry that could provide both jobs and tax revenue. Nine days after his inauguration, Roosevelt sent before Congress a piece of legislation modifying the Volstead Act and legalizing the sale of beer. During the summer of 1933, the administration stumped for repeal; with James Farley in charge of the effort, and the AAPA and WONPR providing organizational support, repeal was promoted as a key element of a recovery program.

In a series of state elections to select delegates to conventions, the degree of shift in popular sentiment was measured. Michigan, which had a plurality of 207,000 for the 18th Amendment in 1919, voted for the 21st Amendment by 547,000; overall, repeal triumphed 3 to 1 (15 million to 4 million). The necessary 35 states ratified the amendment by December 1933, and what Roosevelt described as the "damnable affliction of Prohibition" came to an end (Blocker 1976, p.242).

EPILOGUE

While retreating from prohibition enforcement, the federal government retained responsibility to regulate the legitimate production of distilled spirits, wine, and beer and to prevent the illegal production of these products. These functions have been consolidated since 1972 in the Treasury's Bureau of Alcohol, Tobacco, and Firearms (BATF). After repeal, most questions regarding alcohol devolved to the states. Seven continued with prohibition, though 5 of these declared beer to be nonintoxicating; 12 states decided to permit liquor, but only for home consumption; 29 states allowed liquor by the glass. Legislators vowed ritualistically to prevent the return of the saloon and exclude the liquor traffic from political influence.

The Alcohol Beverage Control (ABC) laws that the states did adopt were designed in part to curb the most notorious abuses of the pre-Prohibition era. Restrictions were imposed on hours and days of sales in an effort to diminish the bar's seduction of the breadwinner from his domestic obligations. Sunday closings were observed; liquor could no longer be sold on election days: the "tied-house," blamed for inciting extreme forms of consumption behavior, was banned. Visibility requirements were instituted; in some states they mandated that bars be open to public inspection, in others they kept the spectacle of the drinking act safely hidden from the eyes of children or decent citizens.

These laws were full of conceptual confusion. On one hand, they embodied a ceremonial deference to those Americans sensitive to the morally fraught nature of alcohol use. But on the other, it was precisely this problem-oriented attitude toward alcohol that repeal had undermined. The issues of public sensibility or public morality were inextricably linked to a discredited past; temperance, which once had denoted moderate use, now conjured up the image of prohibitionist fanaticism. The role of the state was to oversee orderly distribution rather than to curtail availability.

A wariness of moral intervention was only one element in the position that the states assumed. Economics was another major factor. The exigencies of the fiscal crisis forced many commonwealths to turn to the beverage industry for help. Between 1933 and 1935, 15 states adopted monopoly systems; these states were broke and in order to stock their chains of stores had to buy on credit from the distillers. In Ohio in 1935 the Department of Liquor Control boasted that "without one cent of capital, the Department faced the problem of purchasing on credit a sufficient amount of liquor to supply Ohioans with safe, palatable, and legal liquor" (Harrison and Laine 1936, p. 119). This dependency created a pattern whereby revenue rather than social control became the guiding concern; indebted to the industry and desperate to generate funds to help finance local government, states found themselves in the position of stimulating demand and participating in what only a few years before was still widely considered "the nefarious trade."

The alliance between state government and the liquor industry produced revenues that were often earmarked for special purposes; hospitals, schools, drought relief, and mothers' aid all received funds that served to heighten enthusiasm for sales. A trade magazine underscored the industry's own promotion of these benefits (Dobyns 1940, p. 418).

> A little child is playing happily in the streets of a big city. With all the strength of a twelve-year-old, he throws the ball against the side of a building. It bounces off his hand on the rebound. Quickly the youth runs after the ball into the middle of the street. Brakes screech wildly. One anguished scream rends the air. Johnny lies unconscious beneath the wheels of a big truck, his two legs broken.
>
> Were it not for alcoholic beverages, Johnny might go through life a helpless cripple. Thanks to the revenue derived from liquor taxes, however, the state has been able to build and maintain a large hospital just for cases like this.
>
> And this is only one of the many splendid causes to which liquor revenue is put. Publicity has been often given in the past to the so-called evils of liquor while the sale has been, and is, attacked vigorously by varying numbers of drys. Small stress, on the other hand, is given to the enormous benefits derived from liquor taxes.

However, the postrepeal rehabilitation of the liquor industry stemmed from more than such putative public service. Not only did the manufacturers of alcohol subsidize indirectly the general welfare, but they also produced a commodity that had become decontaminated. The experience of the 18th Amendment endowed drinking with a new prestige, both social and moral. Consumption was exhibited as a badge of tolerance and civility. Conversely, efforts to regulate availability grew tainted; as Roosevelt had said in his repeal proclamation, the proper interventionist role of the government was limited to "educating every citizen towards temperance." Responsible individual behavior could be encouraged and even taught, but not imposed or coerced. After Prohibition, the problems associated with alcohol were seen more and more as ones of personal choice or personal disability.

Today the emphasis on individual accountability and the distaste for vigorous governmental action appear more firmly enshrined than ever. Self-care and a corresponding antagonism toward a beneficient "Big Brother" have become tenets of a popular critique of the welfare state. Institutionalized altruism is increasingly perceived as counterproductive, the cause rather than the cure for the ills that afflict us. Overregulation is now a code phrase conjuring up an elaborate set of inept and self-righteous rules, impossible to enforce and corrosive in the sweep of their mandate. Citizens, it follows, are best protected by being left to their own devices. Of course, a basic education for living should be provided. ("Many of the same decision making mechanisms involved in deciding how to use alcohol will be involved in deciding how to drive a car, how to handle finances, when to get married, and how to plan for a future life," Chafetz writes [1974].) But once such skills have been imparted, then their application depends on free, individual choice.

Health has become conceptualized as a duty: "One has an obligation," Leon Kass writes, "to preserve one's own good health. The theory of a right to health flies in the face of good sense, serves to undermine personal responsibility, and in addition, places obligation where it cannot help but be unfulfillable." In this same context, John Knowles groups "sloth, gluttony, alcoholic intemperance, reckless driving, sexual frenzy and smoking" together as at-risk behaviors that people select as part of a personal life-style (Crawford 1978, pp.14-16).

But while powerful, this concept that individuals are held accountable for themselves is still not unchallenged. An important countervailing theoretical perspective has emerged. The environmental movement, the antismoking campaign, the protests against atomic power, and the oil companies all have a collective view of hazards. For example, a growing contention is that pollution is so pervasive that individuals cannot avoid being exposed to its hazards.

The political configurations of this alliance are unclear. However, a concern for the social and physical ecology has already led some groups to develop sophisticated lobbying techniques and mobilize successful and massive campaigns of public education.

Prohibition will certainly never return. Above and beyond the mechanical problems of enforcement, it failed originally because it created no stabilizing vested interests. No reform movement can survive unless it is rooted in new institutions. But while extreme forms of controlling consumption of alcohol are utterly lacking in feasibility, there is the chance that state policy may once again assume a more interventionist role. The boundaries between personal and governmental responsibility constantly shift. Although a mass movement to curb drinking will never reemerge, one can conceive that new, extensive regulation of the liquor industry might be integrated into a paradigm of environmental safeguards and corporate responsibility.

Antialcohol organizing reached its pinnacle of influence during historical periods in which agitation against the plundering of the social and physical landscape was most intense. Efforts to curb drinking emerged from broad reformist sentiment. The relative obscurity today of any alcohol control movement may be deceptive. The conditions are present for a revival of widespread interest in the problems of alcohol. No one can predict if such a resurgence of popular concern will, in fact, develop, but the record of the past suggests that movements once thought safely interred do not always remain in their graves.

A brief review of the shifting attitudes toward cigarette smoking demonstrates how quickly a substance once thought innocuous or even beneficial can be redefined as dangerous and deviant. (The material presented here is a synopsis of Nuehring and Markle 1974, and Markle and Troyer 1979.) Cigarettes were brought back to America in the 1850s by tourists returning from England. Smoking was initially tolerated and even endorsed; rations of cigarettes provided to soliders during the Civil War were regarded as crucial to their well-being. By the 1870s, however, cigar manufacturers, wary of competition, attempted to discredit cigarette smoking. Lurid accusations were made: cigarettes were laced with opium and the paper bleached with arsenic; the content was derived from garbage and packaged by Chinese lepers. Transformed into a vice, cigarette smoking began to be taken up by the temperance movement; pledges against smoking and drinking were often made together by churchgoers. Cigarettes and drinking were attacked as an evil partnership threatening to undermine physical and moral health. Young people were considered especially vulnerable, and delinquency and school failure were often traced directly to indulgence in these dirty and debilitating habits. So widespread was the association between smoking and antisocial behavior that 14 states passed prohibition laws against cigarettes during the period 1895 to 1914.

For many of the same reasons that the 18th Amendment was repealed and drinking returned to respectability, smoking also underwent a rehabilitation in public opinion. By 1927, the bans had all been overturned. By the 1930s, cigarettes were grouped with alcohol as aids to economic revival through their provision of important tax revenues. In addition, smoking along with drinking became raised to the status of the normative. Nonsmokers, just like teetotalers, were suspect as antisocial eccentrics.

By the late 1950s, the discovery of the link between cancer and cigarettes began to erode the legitimacy of smoking. But the identification of these risks was incorporated into an assimilative rather than coercive approach. Consumers were to be alerted; once sufficient information was provided, then presumably those smokers, or potential smokers, would abandon or avoid self-destructive behavior. These educative principles underlay congressional hearings held in 1957 on the hazards of smoking, the antismoking curricula adopted by public schools (Florida passed a law in 1965 requiring students be taught "the adverse health effects of cigarette smoking"), and the Federal Communications Commission's decision in 1967 to have warning labels attached to all packages.

Such initial efforts, though targeted at the consumer as opposed to the product, nonetheless aroused the deep concerns of a well-organized and powerful industry. Tobacco is the fifth largest cash crop for the entire United States and is probably the best investment of all farm products. The manufacturers have a lobbying arm, the Tobacco Institute, which Senator Edward Kennedy called the most effective in all of Washington. Tobacco has significant alliances with other sectors of the economy (for example, the industry contributed $400 million to advertisers in 1977 alone) as well as major links to the federal government. Some $1.3 billion were added to the balance of payments through foreign sales in 1977, and $2.3 billion were paid in excise taxes during the same year.

All of these resources, substantial in terms of both money and political capital, were mobilized to defend the position of the industry. Cigarette manufacturers made strenuous efforts to revamp their image. Companies donated funds to a whole range of worthy civic causes, from bookmobiles in poor areas to the production of documentary films about American Indians. Contributions to cancer and heart disease drives were also made; the Tobacco Institute proclaimed that the industry as a whole was committed to research into tobacco-related health issues, but emphasized that "the answer to this unsolved problem cannot be sidestepped merely because an apparent statistical association has spotlighted a convenient though probably innocent suspect" (Nuehring and Markle 1974, p. 523). Going beyond disclaimers, the industry voluntarily imposed on itself new regulations that restricted appeals to youth.

Despite these defensive maneuvers thrown up by the industry, the antismoking movement, far from being placated, has grown increasingly militant and prohibitionist since the middle 1960s. Changes in rhetoric reflect important shifts in the movement's operating premises. As Markle and Troyer (1979) observe, attacks on smoking have taken on a distinctly coercive quality. Smokers, who before were appealed to as unenlightened, are now regarded as noxious. Their habit, once considered a piece of personal behavior that should be voluntarily shed for the good of the smoker, has been redefined as an invasion of the rights of the nonsmoker that must be aggressively resisted. Gaining strength from a general upsurge in public concern for the environment, the antismoking movement has declared that freedom of choice, i.e., whether to smoke or not, is as spurious a privilege as that invoked by industrialists dumping pollutants into a river. From this sense that the common good necessarily takes precedence over perverse private satisfaction, stringent laws curtailing smoking have begun to be proposed. Federal regulatory agencies have taken positions in support of the movement, and initiatives on the state level have been widespread. In California, a proposition that would have relegated smoking to the confines of private homes was beaten back after the tobacco industry spent $5.6 million to defeat it.

Although a number of parallels between the history of antismoking and antidrinking agitation present themselves, it is not the intention here to enumerate them or to suggest a pattern of rigid correspondence. What must be recognized, however, is that cigarettes once appeared as inviolate to such public discrediting as alcohol now seems to be. The tobacco industry was respectable and politically well connected; smoking was so well accepted in American life that opposition was tantamount to faddism or bigotry. The disintegration of this apparently solid structure of legitimacy resulted from a convergence of forces; growing awareness of the hazards of smoking, rising concern about environmental contaminants, the emergence of a cultural style whereby individual purity (the backpacker, the jogger, the natural-food eater) is defined as the feasible span of self-determination.

The same constellation of elements may not coalesce in precisely the same way to form a revived antiliquor movement. We must be aware, nonetheless, that cycles of organized opposition to smoking and drinking have coincided, and that in the past, many of the same impulses inspired both. The times today are volatile, and one can easily imagine that "moral athleticism"—the term Gusfield applied to the temperance movement—could once again have a broad appeal.

REFERENCES

Asbury, H. (1950) *The Great Illusion: An Informal History of Prohibition.* New York: Doubleday.

Astor, Lady (1923) The English law relating to the sale of intoxicating liquors. *Annals of the American Academy of Political and Social Science* 109(Sept.):265-278.

Bacon, M. K., Barry, H., and Child, I. L. (1965) A cross cultural study of drinking: II. Relations to other features of culture. *Quarterly Journal of Studies on Alcohol*, Supplement 3:29–48.

Banks, A. L. (1917) *Ammunition for the Final Drive on Booze.* New York: Funk and Wagnall.

Barnes, H. E. (1932) *Prohibition Versus Civilization: Analyzing the Dry Psychosis.* New York: Viking.

Binkley, R. C. (1930) *Responsible Drinking: A Discreet Inquiry and a Modest Proposal.* New York: Vanguard.

Blocker, J. S. (1976) *Retreat from Reform.* Westport, Conn.: Greenwood Press.

Braeman, J., Bremner, R., and Walters, E., eds. (1964) *Change and Continuity in Twentieth-Century America.* Columbus: Ohio State University Press.

Braeman, J., Bremner, R., and Brody, D., eds. (1968) *Change and Continuity in Twentieth-Century America: The 1920's.* Columbus: Ohio State University Press.

Catlin, E. G. (1932) Alternatives to prohibition. *Annals of the American Academy of Political and Social Science* 163(Sept.):181-187.

Chafetz, M. E. (1974) The prevention of alcoholism in the United States utilizing cultural and education forces. *Preventive Medicine* 3(Mar.):5-10.

Cherrington, E. H. (1920) *Evolution of Prohibition in the United States of America.* Westerville, Ohio: American Issue Press.

Clark, N. H. (1976) *Deliver Us From Evil: An Interpretation of American Prohibition.* New York: Norton.

Crawford, R. (1978) You are dangerous to your health. *Social Policy* 8(4):10-20.

Darrow, C., and Yarros, V. S. (1927) *The Prohibition Mania: A Reply to Professor Irving Fisher and Others.* New York: Boni and Liveright.

Dobyns, F. (1940) *The Amazing Story of Repeal: An Expose of the Power of Propaganda.* Chicago: Willett, Clark & Co.

Durkheim, E. (1972) *Selected Writings.* A. Giddens, ed. and trans. Cambridge: Cambridge University Press.

Edwards, J. (1847) *Temperance Manual.* New York: American Tract Society.

Emerson, H. (1932) Prohibition and mortality and morbidity. *Annals of the American Academy of Political and Social Science* 163(Sept.):53-60.

Feldman, H. (1927) *Prohibition: Its Economic and Industrial Aspects.* New York: Appleton.

Filstead, W. J., ed. (1972) *An Introduction to Deviance: Readings in the Process of Making Deviants.* Chicago: Markham.

Fisk, E. L. (1923) Relationship of alcohol to society and citizenship. *Annals of the American Academy of Political and Social Science* 109(Sept.):1–14.

Frankfurter, F. (1923) A national policy for enforcement of Prohibition. *Annals of the American Academy of Political and Social Science* 109(Sept.):193–195.

Furnas, J. C. (1965) *The Life and Times of the Late Demon Rum*. New York: Putnam.

Goldman, A. (1979) *Grass Roots: Marijuana in America Today*. New York: Harper & Row.

Grant, E. A. (1932) The liquor traffic before the Eighteenth Amendment. *Annals of the American Academy of Political and Social Science* 163(Sept.):1-9.

Gusfield, J. (1963) *Symbolic Crusade: Status Politics and the American Temperance Movement*. Urbana: University of Illinois Press.

Harrison, L. V., and Laine, E. (1936) *After Repeal: A Study of Liquor Control Administration*. New York: Harper.

Hearst Temperance Contest Committee (1929) *Temperance—or Prohibition?* New York: J. J. Little and Ives Co.

Kobler, J. (1973) *Ardent Spirits: The Rise and Fall of Prohibition*. New York: Putnam.

Koren, J. (1923) Inherent frailties of Prohibition. *Annals of the American Academy of Political and Social Science* 109(Sept.):52-61.

Krout, J. A. (1925) *The Origins of Prohibition*. New York: Knopf.

Lasch, G. (1979) *The Culture of Narcissism*. New York: Norton.

Levine, H. (1978) Discovery of addiction: Changing conceptions of habitual drunkenness in America. *Journal of Studies on Alcohol* 39(1):143-174.

Levine, H. (1979) The inventors of alcohol control: Alternatives to prohibition from the Committee of Fifty to the Rockefeller Commission. Draft manuscript, Social Research Group, School of Public Health, University of California, Berkeley.

London, J. (1913) *John Barleycorn*. New York: Century.

MacAndrew, C., and Edgerton, R. B. (1969) *Drunken Comportment: A Social Explanation*. Chicago: Aldine.

Markle, G., and Troyer, R. (1979) Smoke gets in your eyes: Cigarette smoking as deviant behavior. *Social Problems* 26(5):612–623.

Nuehring, E., and Markle, G. (1974) Nicotine and norms. *Social Problems* 21(2):513-526.

Pearson, C. C., and Hendricks, J. E. (1967) *Liquor and Anti-Liquor in Virigina, 1916-1919*. Durham, N.C.: Duke University Press.

Room, R. (1974) Governing images and the prevention of alcohol problems. *Preventive Medicine* 3(1):11-23.

Rorabaugh, W. J. (1979) *The Alcoholic Republic: An American Tradition*. New York: Oxford University Press.

Rothman, D. (1971) *The Discovery of the Asylum: Social Order and Disorder in the New Republic*. Boston: Little, Brown.

Roueché, B. (1960) *The Neutral Spirit: A Portrait of Alcohol*. Boston: Little, Brown.

Rush, B. (1814) *An Inquiry into the Effects of Ardent Spirits upon the Human Body and Mind*, 8th ed., with additions. Brookfield, Mass.: B. Merriam.

Sinclair, A. (1962) *Prohibition: The Era of Excess*. Boston: Little, Brown.

Sinclair, U. (1956) *The Cup of Fury*. Great Neck, N.Y.: Channel Press.

Stayton, W. H. (1923) Our experiment in national Prohibition. What progress has it made? *Annals of the American Academy of Political and Social Science* 109(Sept.):26-38.

Stout, C. T. (1921) *The Eighteenth Amendment*. New York: Mitchell and Kennerly.

Timberlake, J. H. (1963) *Prohibition and the Progressive Movement, 1900-1920*. Cambridge, Mass.: Harvard University Press.

Tomkins, F. W. (1923) Prohibition. *Annals of the American Academy of Political and Social Science* 109(Sept.):15-25.

U.S. Department of Commerce. (1923) *Statistical Abstract of the United States 1922*. Washington, D.C.: U.S. Department of Commerce.

Van den Haag, E. (1975) *Punishing Criminals: Concerning a Very Old and Painful Question*. New York: Basic Books.

Warburton, C. (1932) Prohibition and economic welfare. *Annals of the American Academy of Political and Social Science* 163(Sept.):89–97.

Warner, R. H., and Rossett, H. L. (1975) The effects of drinking on offspring: An historical survey of the American and British literature. *Journal of Studies on Alcohol* 36(11):1395–1420.

Woods, R. A. (1923) Notes about Prohibition from the background. *Annals of the American Academy of Political and Social Science* 109(Sept.):121-132.

Alcohol Use and Consequences

DEAN R. GERSTEIN

INTRODUCTION

In a perceptive and scholarly analysis, Harry G. Levine (1978) has identified four distinct parts to the American perspective on alcohol use, each of which was initiated in, and was characteristic of, a distinct historical period, although none has disappeared from American consciousness, politics, or scientific discourse. The first component, largely dominating the 150-year colonial era of American history, focused on sentimental or customary attachment to drinking and the moral qualities of drunkenness. For most Americans in this period, drinking and drunkenness aroused as much interest as eating and obesity—which is to say, while there was plenty of both, they were widely accepted as normal. Those few who did choose to stand against drink did so on religious grounds: habitual drunkenness was sinful, a dissipation of the moral energies that colonial society expected to be devoted to God's work on earth. Yet the fault was not laid to alcohol itself, but to the defective moral character of those few who besotted themselves entirely too often. The problem was not addiction, but an overweening love of intoxication.

During the expansionary period of American history, approximately the century and a half from the Revolutionary War to the first decades of the current century, the notion took hold that alcohol was addicting

Dean R. Gerstein, study director of the panel, is senior research associate of the Committee on Substance Abuse and Habitual Behavior, National Research Council.

and that this addiction was capable of corrupting the mind and body. With this concept, alcohol itself became the focus of concern. By 1850 the temperance movement had set out to remove the destructive substance, and the vertically monopolized industries that promoted its use, from the country. The movement held that while some drinkers might escape unharmed, even the most moderate of them flirted with danger at the rim of every cup.

This view of alcohol and its purveyors as a public menace was carried into force by the 18th Amendment and the Volstead Act. But despite the broad public majorities voting in favor of prohibition, the older customary attachments to drinking proved to be deeply rooted. The consequent unwillingness of most jurisdictions to adopt Draconian enforcement measures, or (in the days of Harding, Coolidge, and Hoover) to commit more than a bare minimum of public funds to such activities, ensured that illegal marketeers, buoyed by the willingness of drinkers to pay three to four times the prewar going rates, developed a strong black market in booze.

Prohibition was swept off the books in the first months of the new Roosevelt Administration. In the ensuing period, two rather different perspectives have grown up and coexisted. The disease view of alcoholism has shifted from the earlier focus on alcohol as an invasive agent to an approach based on a scientific model of chronic disease. Alcoholics have come to be viewed as a population with a special vulnerability to alcohol, while all other people can be exposed to it at virtually no risk to themselves. In contrast to the colonial view that alcohol is physically and morally innocuous but that some morally defective individuals take to perpetual drunkenness as a sign of their dissipation, this view holds that while alcohol is innocuous for most, a minority—fine people in all other respects—cannot touch it without succumbing to the addictive disease process, for which there is no cure except total abstinence. This is the professional viewpoint of the "alcoholism movement" and its organizational focus, Alcoholics Anonymous. As an ideological foundation for widespread public support of alcoholism treatment, this definition of the situation has proven to be quite effective.

The fourth perspective, which has roots in the repeal movement, places less emphasis on the qualities of alcohol or individuals who drink it than on the social arrangements and social consequences that surround the practice of drinking. This perspective is most closely identified with social and behavioral scientists, originally at Columbia University around 1930 and more recently at the Addiction Research Foundation in Toronto, the Finnish Foundation for Alcohol Studies in Helsinki, and the Social Research Group at the University of California at Berkeley.

The principal ideas in this public health perspective are that:

- the rates of alcohol-related problems in society are the focus of concern;
- these rates are contingent on the social arrangements involving both drinking and the activities that drinkers may engage in; and
- it is eminently rational to try to minimize the incidence of alcohol-related problems by managing these arrangements properly rather than only treating their results.

The following discussions both criticize and draw on certain tenets of this public health view. I first discuss the measurement of drinking per se and its distribution through societies, which has been of particular importance for the public health perspective. Then I discuss evidence on the relation of drinking to a series of biosocial functions and dysfunctions that are thought to be related to alcohol use. The concern throughout is with issues of measurement and attribution in the empirical analysis of alcohol consumption and associated behavior.

ALCOHOL USE: INTOXICATION, DRINKING PATTERNS, AND TOTAL CONSUMPTION

The relation between alcohol use and consequences is seldom simple, direct, or universal, despite many of our common sense conceptions about the matter. The central scientific issue for policy research on alcohol is how different patterns, levels, or circumstances of alcohol use are translated into different types, ranges, and degrees of consequences.

First, imagine alcohol use alone, stripped of any considerations of place, person, companions, or consequences. How can an individual's alcohol use be measured or described in an efficient, useful way? Use of alcoholic beverages is generally conceived to occur on "drinking occasions" or "drinking events," much the way we conceive of eating as occurring at meals or occasional snacks. There are three distinct ways to think about such occasions. First, we might inquire into the amount consumed on an occasion, to try to determine just how drunk, how intoxicated, the drinker became. Second, we can ask about the frequency and distribution, the pattern, of such occasions—especially, how a sustained pattern of intoxication impinges on everyday life. Third, we can ask about the total amount drunk across a number of such occasions. Each of these three ways of thinking about drinking emphasizes a different aspect of alcohol use—and ultimately, a different order of associated consequences.

1. Interest in the amount drunk on a given occasion focuses concern on the degree of intoxication or "drunkenness" achieved. Intoxication is not a simple consequence of the amount of alcohol consumed, since body weight, spacing of drinks, and metabolic rate (which may reflect previous experience with alcohol) intervene between the amount drunk and the biobehavioral states involved in being drunk. The most common technical measure of intoxication, blood alcohol content (BAC),[1] represents the volume of alcohol circulating in the body at a given time, once these intervening variables have had their effect. Since the human liver detoxifies alcohol at a substantial rate, a very important feature of intoxication is its temporary, transient nature. BAC tends to rise during a drinking occasion and then decline toward zero not long after drinking ceases.

The decisions that individuals make about where to drink and how to behave while intoxicated have great bearing on the effects that accrue while being drunk. Most of the behavioral and subjective states that we find easy to ascribe to drunkenness—sociability or sadness, daring or tranquillity, aggression or passivity—do not spring directly from the bottle but find expression due to the social circumstances and personalities in which drunkenness is brought into play. Even the effects most consistently tied to intoxication—clumsiness and befuddlement—are only consequential if the drunk tries to engage in complex performances or enters a dangerous environment.

2. The second way of characterizing drinking involves its frequency and distribution in time—most importantly, the frequency and distribution of intoxication. The question is how this pattern of intoxication fits—or misfits—other patterns of responsibility and role performance; to what degree it is inappropriate, unwanted, or disruptive. All of these features—impropriety, undesirability, or disruption—have to do with the social and physical setting of intoxication. What the question of pattern emphasizes is not how drunk someone might become on occasion, but how repeated occasions of drunkenness impinge on everyday life.

3. The last measure, total consumption, involves the accumulation

[1] BAC level (expressed as a decimal percentage of a gram per liter, e.g., 0.03) correlates reliably, in dose-dependent fashion, with degradations in psychomotor coordination, especially of such complex performances as precise reasoning, eye-hand coordination, and balance while in motion. At a BAC of 0.05, which for most people requires more than two drinks within an hour, performance begins to consistently degrade. At 0.08–0.10, most jurisdictions consider a motor vehicle operator "legally intoxicated" and unable to drive safely. At 0.15–0.30, consciousness may be lost, while a BAC above 0.30 is considered potentially lethal (Haberman and Baden 1978).

of all the alcohol that an individual has drunk over a specified period of time. Total consumption usually means during a year or a number of years, although shorter periods are often used to estimate (extrapolate) these accumulations and to make the figures more easily understood. Thus, total consumption is often reported as an average amount of pure alcohol (e.g., 80 grams, 100 milliliters, or 3 ounces) per day. This is known as a volume index: it implies that this average is sustained over long periods and might be reported as 10 gallons (38 liters) per year. This way of looking at alcohol use tends to emphasize the way in which sustained exposure to volumes of alcohol may be related to various risks of morbidity and mortality.

A great deal of the evidence used in the recent scientific analysis of drinking by partisans of the public health perspective hinges on the appropriate use and interpretation of total alcohol consumption statistics in local, regional and national populations (and samples). The major argument made in the manifesto of this perspective (Bruun et al. 1975) is that: ". . . changes in the overall consumption of alcoholic beverages have a bearing on the health of the people in any society. Alcohol control measures can be used to limit consumption: thus, control of alcohol availability becomes a public health issue" (p. 90).

The "overall consumption of alcoholic beverages . . . in any society" is a key term in this argument, for overall or total consumption figures have been regarded as good measures of the central tendency of alcohol use by whole populations. This particular line has been linked to a larger argument that the general consumption level of any alcohol-using population, which reflects the societal "climate of acceptance" of drinking, determines the prevalence of heavy alcohol users and the rate of alcohol-related morbidity, mortality, and related costs such as lost production values and extra medical care.

It therefore is important to closely examine the measurement of aggregate or overall consumption, to see how sound a basis such statistics can be for scientific or policy judgments.

METHODS OF MEASUREMENT

The most common form of alcohol statistic used is unquestionably the annual per-capita consumption attributed to the drinking age population (typically interpreted as older than 14) of a nation or state. Derivation of the annual per-capita figure (technically a "period ratio") requires enumerations or estimates of two quantities. First, the numerator: the total quantity of pure or "absolute" ethyl alcohol contained in beverages drunk by the population during the year; Second, the denominator: the average number of individuals meeting the specified age criteria living in the area during the year.

Estimates of the denominator are usually derived from census enumerations and extrapolations. These data generally exclude institutionalized populations and are known to disproportionately underestimate various segments of the population by unknown amounts. Inner-city residents, migrant laborers, and illegal aliens are most often cited as undercounted groups in the United States.

The numerator is usually compiled from figures supplied by producers or wholesalers of alcoholic beverages; specifically, the "tax paid withdrawal" of beverages from wholesale stocks available at the beginning or produced during the course of the year, along with their reported alcohol content. In the United States and most industrialized countries, the regulation of manufacturers by government agencies ensures a fairly uniform content of alcohol in respective classes of such beverages; but lapses have occurred, for example, in notorious French and Italian wine frauds. In the liberal and fairly inexpensive mass markets of recent years, it seems safe to assume that the quantity of consumed alcohol that has *not* been produced and counted under taxation and licensing control is probably small (Mäkelä 1978, Gavin-Jobsen Associates 1978). This includes illicit beverage production for sale or home consumption, laboratory and commercial ethanol diverted to beverage use, trace alcohol in soft drinks and other consumables, and drink produced legally but diverted from taxed inventories. The comparability of U.S. data in this regard with data from other countries is probably satisfactory for recent years in industrialized states.

Taxed withdrawal from local wholesalers' stocks, however, is not a direct measure of consumption by local residents. Consider the rank-ordering of states by per-capita sales in Table 1. The three locations with the greatest apparent consumption—Nevada, New Hampshire, and the District of Columbia—all have small local populations relative to large tourist trades. While all three would be very likely to rank in the upper end of the list without the tourism factor, it is probable that about half of their apparent per-capita consumption can be attributed to visitors not counted in the census of local residents. (This may also account to some degree for the high rate in Vermont due to its ski resorts.) In addition, alcoholic beverages are cheaper in New Hampshire and the District of Columbia than in surrounding areas and the border-crossing "liquor run" is a well-known local custom (Rooney and Butt 1978).

As these considerations suggest, gross estimated per-capita consumption figures are subject to error, the more so as migration, transiency,

TABLE 1 Apparent Consumption of Absolute Alcohol, Population Aged 14 Years and Older

State	U.S. Gallons Per Capita	State	U.S. Gallons Per Capita	State	U.S. Gallons Per Capita
Nevada	7.05	Delaware	3.04	South Dakota	2.49
District of Columbia	5.72	New Mexico	3.03	Virginia	2.41
New Hampshire	5.66	Illinois	2.96	Pennsylvania	2.38
Alaska	3.73	Oregon	2.90	Missouri	2.37
Vermont	3.70	New York	2.86	Iowa	2.30
California	3.57	Michigan	2.84	Ohio	2.28
Colorado	3.54	Texas	2.82	Mississippi	2.24
Wyoming	3.51	North Dakota	2.79	North Carolina	2.22
Wisconsin	3.42	New Jersey	2.79	Indiana	2.18
Arizona	3.41	Minnesota	2.78	Oklahoma	2.10
Florida	3.40	Connecticut	2.76	Alabama	2.03
Hawaii	3.28	Louisiana	2.75	Tennessee	2.02
Montana	3.20	Maine	2.74	Kansas	1.96
Rhode Island	3.17	Idaho	2.69	Kentucky	1.92
Maryland	3.15	South Carolina	2.68	West Virginia	1.89
Massachusetts	3.15	Nebraska	2.64	Arkansas	1.85
Washington	3.12	Georgia	2.55	Utah	1.77

Source: Adapted from Hyman et al. (1980, p. 4.)

or census errors affect the accuracy of the denominators (people count) and as untaxed production/distribution affects the numerators (gallon count). As a result, both random errors and systematic biases may enter the calculations. As a rule of thumb, in comparisons of a single jurisdiction across time where the systematic bias is not likely to change rapidly, year-to-year estimates using invariant procedures are likely to have confidence margins of 1–5 percent. In comparisons of different areas, error margins on the order of 5–50 percent may be assumed, depending on knowledge of such factors as recordkeeping efficiency, extent of migration and tourism, tax evasion, beverage quality control, and the like.

USES OF PER-CAPITA CONSUMPTION

At this point we an put aside the matter of accuracy of estimates and inquire what it is that per-capita period consumption averages should tell us about alcohol use in populations. First, we should observe that per-capita consumption is an arithmetic mean. Usually a mean is computed from data directly about individuals. For example, we measure the height of everyone in a group, sum the measurements, then divide by the number of individuals. Mean alcohol consumption using tax-paid withdrawals is arrived at quite differently. Instead of observing and measuring the consumption of individuals, we measure produced alcohol at one step in its distribution for sale, assume that all alcohol destined for drinking passes through this step, assume that little is thrown away, assume that retail and private stocks beyond this step do not increase or decrease much relative to the total production and flow of alcohol in a year, and, finally, assume that we have aimed our population count at all and only the people who might possibly drink the alcohol being measured. While none of these assumptions seems to unduly threaten the validity of derived means, except as noted above, all serve to indicate that per-capita consumption is an index that has been constructed to represent an implied set of behaviors.

What, then, do we know about the relation between this index and the set of consumption behaviors? Most of our knowledge about this derives from two sources: surveys of individual drinking practices and comparison of per-capita consumption with public health statistics about diseases strongly related to very heavy drinking. Before discussing this, however, a brief statistical note is in order.

We are accustomed to thinking about the mean, and assigning statistical meaning to it, in terms of the familiar bell-shaped normal distribution. If we think consumption is normally distributed, then the mean lies at the peak of the symmetrical bell-shaped curve that graphs the frequency distribution of individuals across alcohol consumption rates. In a normal distribution, more people consume the mean amount than any other amount (the mean is therefore also the mode); half drink more than the mean and half less (it is therefore equal to the median); and, depending on the degree of dispersion (indexed by the variance or standard deviation) of the curve, the majority of people consume an amount that is fairly close to the mean figure, with fewer people drinking at a given rate the farther it is from the mean.

The mathematics that underlie the bell-shaped curve lead us to expect that an approximate normal distribution will generally occur when there is a large number of independent causes, each of which contributes a small fraction of the total variance in a dependent variable. Thus, if individual rates of annual alcohol use were normally distributed, one would reason that the sufficient causes for single drinking events (person-drink occasions) were many in number and that variations in the occurrence of these causes were largely independent or uncorrelated with each other.

Investigation of the actual distribution of consumption rates has occurred largely in Europe and Canada, spurred by the work of Ledermann

(1956). Ledermann wanted to develop a new basis for estimating the prevalence of harmful effects of alcohol use. He developed a rather bold, theoretically based hypothesis: that frequency distributions of consumption in a population are always logarithmic normal distributions (the normal frequency function, not of the consumption rate, but of its logarithm). The resulting curves are skewed to the right so that the median and mode lie below the mean (see Figure 1). Ledermann fit these data to various small samples of drinkers. Subsequent studies (deLint and Schmidt 1968, Mäkelä 1971, Skog 1971) have confirmed that lognormal curves may be roughly fit to a variety of survey data on consumption involving other, larger populations.

Bruun et al. (1975) went a step further, arguing that "*differences as to dispersion between populations with similar levels of consumption are quite small*" (p. 32, authors' emphasis). These authors go on to treat the dispersion across levels as practically invariant.

There are two important aspects to this argument. First, the lognormal function involves only two variables,[2] the mean and the standard deviation. If dispersion, and thus standard deviation, can be treated as invariant, then we have a "one-parameter" distribution: knowledge of the mean figure alone enables us to know the exact proportions below, within, or above any given consumption level or range. In order to calculate the rate of alcoholism, then, given the mean consumption, one need only define a consumption rate beyond which a diagnosis of alcoholism has a known likelihood: a "hazardous level of consumption" (Schmidt and Popham 1975-1976; Popham and Schmidt 1978). The term "hazardous consumption" can then stand as a quantitative proxy for the prevalence of clinical "alcoholism."

Under the lognormal/one-parameter (also called the single-distribution) model, the relationship between mean consumption and the prevalence of hazardous consumption is well defined, and the mean, appropriately transformed, becomes an index for the rate of alcoholism. Since the transformation is exponential (parabolic), an increase (or decline) in mean consumption translates into a more pronounced increase (or decline) in alcoholism. Hence, changes in mean consumption may provide quite dramatic evidence in building a public health perspective.

I have stated that there are two important aspects to the argument. The second aspect has to do with interpretation of the mean not only as an index useful for estimating the size of a high-risk subgroup, but as an indicator of the alcohol-using behavior of the population as a

[2] More exactly, two degrees of freedom, which are captured by two independent parametric variables or parameters (metric attributes of a whole series or population or measurements).

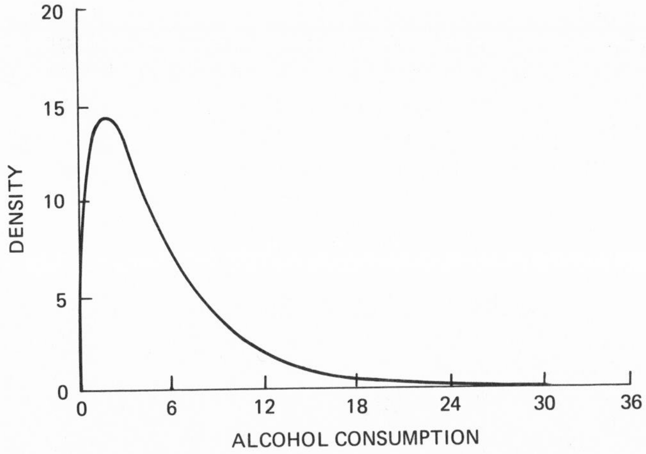

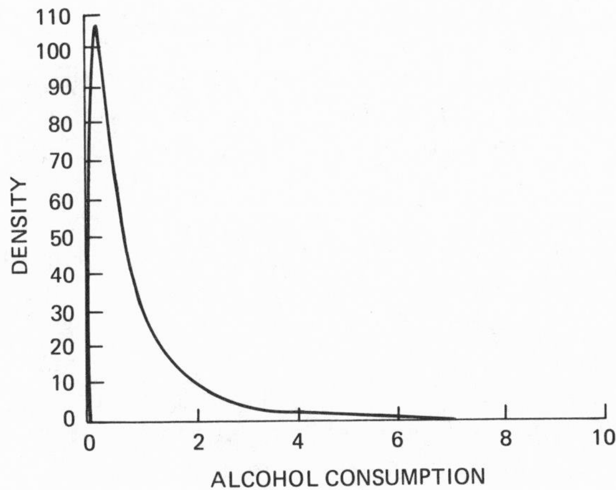

FIGURE 1 Exemplary logarithmic normal curves.

whole. If dispersion is truly invariant, then any change in the value of the mean implies change in the number at *all* consumption levels—not only at the highest ones. The impression is given that a *whole population changes its alcohol use together*. This impression is reinforced by one of the mathematical assumptions that underlie the lognormal function; namely, that the many causal impulses involved in generating empirical lognormal distributions are not additive but multiplicative. In addition to these considerations is the latent habit of thinking that the mean, as central tendency, directly describes the approximate behavior of a large proportion of the population.

Actually this *modal* property accrues only if the mean value is close to the mode and if the dispersion of the curve is relatively small. In virtually every empirical consumption survey analyzed in the available literative (Skog 1971, 1979; Guttorp and Song 1977, 1979), one or the other of these conditions does not appear to be met. But this habit is strong, especially in the presence of limited data. Thus, Bruun et al. (1975) are led to say:

> The reasons for this invariance are unknown; little research has been done on this subject [The apparent stability in dispersion'] and given our present state of knowledge no plausible explanation can yet be offered, other than that the level of consumption of each individual may be presumed to reflect his social milieu. One might well say, to paraphrase Euripides, "Tell me the company you keep, and I'll tell you how much you drink" (p. 34).

There are three difficulties with this way of thinking about consumption data. First, the invariance of the lognormal curve is not so apparent. Skog (1979), on whose analyses Bruun et al. relied for this finding (Skog being one of the alii), has subsequently argued that the gamma distribution is a better fit to most available survey data than the lognormal and that "there seems to exist an inverse relationship between per capita consumption and the dispersion of the distribution" (p. 3; but see Guttorp and Song 1977, 1979).

Second, adding to the uncertainty regarding the universal adequacy of the single-distribution model, if the consumption surveys that underlie the one-parameter model are used to estimate the mean statistics, their results diverge sharply from the figures derived by the production or sales/census method. Survey-based estimates of mean consumption have been found to be only one-third to one-half as large as those computed for the same populations by the production method (Houthakker and Taylor 1970, Mäkelä 1971, National Institute on Alcohol Abuse and Alcoholism [NIAAA] 1978).

Room (1971) reported that "our best survey questions uncover about two-thirds of the total expected consumption." By way of comparison, Warner (1978) estimated that in tobacco-use surveys, people report smoking about 75 percent of the cigarettes that manufacturers report selling—a figure that was 90 percent a decade ago.

Third, the theoretical interpretation is not compelling. Let us suppose that the statistical evidence favoring the single-distribution model was strong and exact, rather than tentative and approximate. Still, this evidence is all correlational. Rather than arguing that the overall company of users spreads its influence widely among members, pushing them into (or pulling them back across the edge of) hazardous consumption, one could just as plausibly argue that the small fraction of relatively dedicated drinkers "sets the pace" for the rest. It does this not only by boisterous example, but also by supplying the bulk of the economic ballast for the alcoholic beverage industry. This price-conscious high consumption base provides a foundation for economies of scale in manufacture, for aggressive marketing, and for antitax lobbying—all of which keep prices down for high-end and low-end consumers alike. Thus, it is just as plausible to think that more heavy drinking leads to overall consumption increases, as the other way around. Without further investigation, which has not yet occurred, the mean index, even under strong statistical interpretation, does not yield definite causal accounts of the sort that have been attributed.

Fortunately, the mean statistic, even as derived from production/census reports, need not be supported by invariant distributional properties or strong theoretical interpretations in order to be of value. Even if the observed distributions can be fitted with a wide range of curves, the evidence is persuasive that these curves are all similar in shape: unimodal and skewed strongly to the right. If we were to reduce the overall U.S. consumption curve to a representative sample of 10 drinking-age adults, their annual consumption of absolute ethanol would not be very different from the following rough approximation: 3 nondrinkers, 3 drinking a gallon among them, and the others drinking 1.5, 3, 6, and 15 gallons, respectively (extrapolated from Cahalan et al. 1969). The total is 26.5 gallons or 2.65 gallons per capita, roughly the national figure for the past decade. The point to be noticed is that one drinker (10 percent of the population) consumes 57 percent of the total; two drinkers (20 percent) consume 78 percent of the total. If we examine changes in the mean statistic across time, we will largely be tracking the drinking behavior of this fraction of the population, the 20 percent, 10 percent, or less of heaviest drinkers—whatever changes may be occurring (or not occurring) among the remainder (see also Room 1978). As a statistical observation, we can simply note that given a skewed distribution, the mean is fairly responsive to the behavior of the long tail of the curve.[3] This was the original point of Lederman's (1956) investigations, and without further regard to specific aspects of the distribution function, it justifies the conclusion of Bruun et al. (1975, p. 45): "A substantial

[3] For lognormal and gamma curves that have been fit to available consumption data, the median or the mode (rather than the mean) would give much better estimates of how the bulk of the population is behaving, since neither responds so much to the tail on the right of the curve. The proportion abstaining is a good longitudinal proxy for the mode in most populations. Of course, it tells us very little about heavy consumption, as Bales (1944, 1946) observed.

increase in mean consumption is very likely to be accompanied by an increased prevalence of heavy users."

On the other hand, this by no means justifies such unqualified attributions as "a fall in the average level of consumption . . . will lead to a fall in the number of heavy drinkers" (Hetzel 1978, p. 84). It is at least as plausible to say that the reverse is true, as is implied by Skog's theoretical demonstration of the long-term effect of a large shift in consumption preference by a small fraction of the members of a hypothetical social drinking network (1980).

To sum up the methodological implications thus far: the mean alcohol consumption statistic is a reasonable gross index of the amount of potentially hazardous consumption in a society. It is most valuable for this purpose when used as a time series or longitudinal index. To see this in operation, see Figure 2, which gives a 150-year record of this index in the United States.

The long trend of per-capita consumption in the United States can be divided into four distinct periods: pre-1850, 1850-1914, 1915-1945, and 1946 to the present. Prior to 1850 the trend was downward from the high level (6-7 gallons annually per adult) that characterized colonial and revolutionary America. From 1850 to 1914, the level hovered in the area of 2 gallons, but began to rise appreciably at the turn of the century, reaching its highest level just prior to U.S. entry into World War I. During the third period (the time of the world wars), state and federal prohibition laws reduced per-capita consumption to about half or slightly more of the pre-World War I level, from which it recovered to about two-thirds that level by the end of World War II. After 1945 the rate hovered about the 2-gallon level for some 15 years, then rose during the 1960s back to the pre-World War I level of roughly 2.7 gallons per capita.

For comparative purposes, Table 2 supplies a two-point time series for the postwar era for a number of industrial (mostly European) countries. As indicated above, caution is necessary regarding the accuracy of comparison of these consumption statistics between different countries, but there are two indisputable general observations to be made. First, the United States is unremarkable in its total consumption during this period relative to other similar national societies. It is neither especially dry nor especially wet. Moreover, the postwar increase in U.S. consumption is also unremarkable. With little exception, increases on this order or greater have been common throughout the industrialized world and, for that matter, virtually everywhere else for which statistics have been compiled (Sulkunen 1976, Moser 1979). In both the static and comparative perspectives, the United States is in the middle of the pack formed by the array of mean consumption figures.

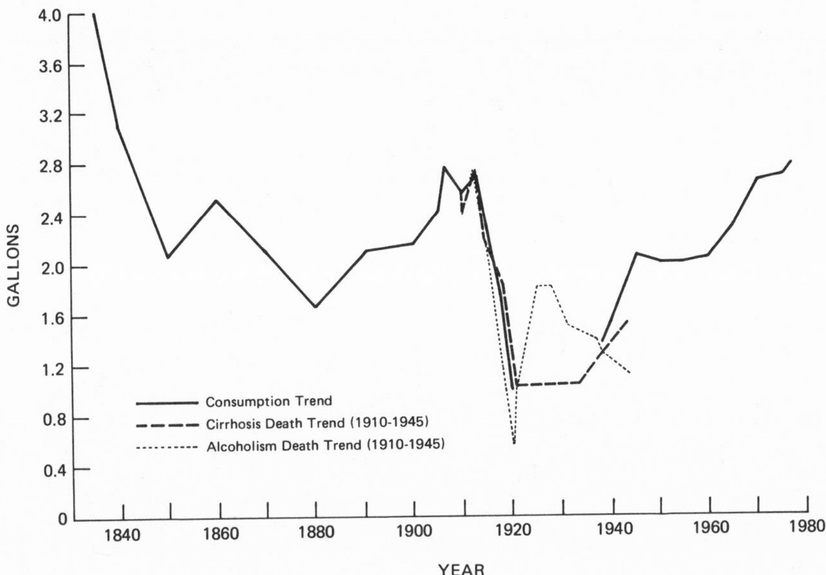

FIGURE 2 Time trend of U.S. consumption of absolute alcohol, per-capita drinking age population, 1830-1977.

Note: Data points are less dense and less reliable prior to 1900. All estimates adjusted to reflect population 15 years and older.

Sources: Consumption statistics from Keller and Gurioli (1976); alcoholism death trend from Jellinek (1947-1948, p. 40); cirrhosis death trend from Bureau of the Census (1975, p. 58). In estimating Prohibition-era consumption, Jellinek favored use of cirrhosis-based estimators; Warburton (1932) principally used alcoholism death rates and production estimates.

DISAGGREGATED CONSUMPTION STATISTICS

I have devoted considerable attention to the average absolute alcohol use measure, first because it has proven to be of strategic importance for one major set of alcohol-related problems (liver disease), and second because its use in *Alcohol Control Policies in Public Health Perspective* (Brunn et al. 1975) and related literature has generated so much interest and controversy, both scientific and political.

There are, however, compelling arguments for dividing total consumption statistics into finer grain. Currently, four such disaggregating strategies have been employed, singly and/or in combination, in the literature. The most easily accessible of these is the division of aggregate statistics by type of beverage. Pekke Sulkunen (1976) has forcefully argued for the consideration of the "use values" involved in alcohol

TABLE 2 Changes in Apparent Consumption of Absolute Alcohol in 20 Countries in Gallons Per Capita of the Population Aged 15 Years and Over

Country	Year	Total Absolute Alcohol	Year	Total Absolute Alcohol	Percentage of Change
France	1955	6.8	1972	5.9	− 13
West Germany	1959	2.2	1974	3.9	+ 80
Belgium	1958	2.2	1973	3.8	+ 72
Switzerland	1956–1960	3.0	1971–1973	3.7	+ 23
Italy	1958	2.8	1973	3.6	+ 28
Australia	1957–1958	2.4	1972–1973	3.5	+ 44
New Zealand	1957	2.5	1972	3.3	+ 31
Denmark	1958	1.4	1973	2.9	+112
Canada	1958	1.9	1974	2.8	+ 51
Netherlands	1958	1.8	1974	2.8	+ 53
USA	1960	2.1	1975	2.8	+ 34
United Kingdom	1959	1.6	1974	2.7	+ 71
USSR	1957	1.3	1972	2.3	+ 83
Ireland	1959	1.1	1973	2.2	+100
Poland	1959	1.7	1974	2.1	+ 22
Finland	1957	0.8	1973	2.0	+157
Sweden	1958	1.3	1973	1.8	+ 43
Norway	1959	0.9	1974	1.5	+ 68
Iceland	1956	0.7	1973	1.3	+ 84
Israel	1959	0.10	1974	0.9	− 12

Note: Arranged in order of consumption in most recent year available as of 1976. Comparisons are for the available data approximating 15 years prior for each country. See text on problems of comparisons between countries. Even a 1-year difference in dates could substantially alter percentage changes. Certain European countries, not listed due to missing early data points, appear in Table 3.
Source: Adapted from Keller and Gurioli (1976).

consumption. At the most basic level, he identifies three sets of values: intoxication, nutrition, and conviviality. Very roughly, these tend to be differentiated in parallel with the three main types of Western alcoholic beverages: distilled spirits are used as an intoxicating drug, wine as the liquid part of a meal, and beer as an accompaniment to sociable relaxation. These usages are only rough guides, of course, and are historically associated with particular ethnic-national practices. Sulkunen points out that in the postwar period, there has been a mild trend toward "homogenization" of drinking cultures worldwide, with one region's drinking preferences and practices being imported and superimposed on the practices native to other regions. However, traditional wine drinkers, for example, in adopting the use of distilled spirits, tend to use them in ways more consonant with their traditional use of wine; e.g., to use distilled spirits in the form of liqueurs and brandies not far removed from mealtimes.

If we follow Sulkunen's arguments and, as in Table 3, identify countries by their predominant (according to absolute alcohol volume) beverage, we see that the highest-ranked countries (by mean per-capita total alcohol) are wine-drinking, the lowest-ranked consume mainly spirits, and the heavy beer-drinking countries are mainly found in between. There is also, among the European countries, a regional concentration

TABLE 3 Apparent Consumption of Absolute Alcohol in Gallons Per Capita of the Population Aged 15 Years and Older, in 25 Countries, by Beverage Type

Country	Year of Latest Data	Distilled Spirits	Wine	Beer	Total
Portugal	1974	0.46	5.27*	0.54	6.37
France	1972	0.80	4.07*	0.99	5.87
Italy	1973	0.66	3.79*	0.24	4.69
Switzerland	1971–1973	0.75	1.70*	1.39	3.85
Spain	1971	1.07	2.18*	0.53	3.78
West Germany	1974	0.91	0.82	2.02*	3.75
Austria	1972	0.84	1.25	1.63*	3.72
Belguim	1973	0.61	0.63	2.15*	3.39
Australia	1972–1973	0.47	0.45	2.40*	3.32
Hungary	1972	1.00	1.54*	0.69	3.23
New Zealand	1972	0.49	0.35	2.34*	3.18
Czechoslovakia	1973	0.92	0.57	1.49*	2.98
Canada	1974	1.01	0.32	1.51*	2.84
Denmark	1973	0.53	0.35	1.71*	2.79
USA	1976	1.14	0.34	1.31*	2.78
United Kingdom	1974	0.53	0.28	1.94*	2.76
Netherlands	1974	0.97	0.44	1.34*	2.75
Ireland	1975	0.78	0.14	1.56*	2.47
USSR	1972	1.22*	0.91	0.21	2.34
Poland	1974	1.43*	0.29	0.49	2.21
Finland	1976	1.07*	0.24	0.82	2.14
Sweden	1973	0.87*	0.30	0.67	1.84
Norway	1974	0.63	0.15	0.69*	1.47
Iceland	1973	0.94*	0.10	0.12	1.16
Israel	1974	0.48*	0.18	0.20	0.86

*Indicates predominant beverage type.
Source: Adapted from NIAAA (1978).

TABLE 4 Apparent Consumption of Absolute Alcohol, in U.S. Gallons Per Capita of the Population Aged 14 and Older and the Percentage Contribution of Each Class of Beverage to the Total, in U.S. Geographical Regions[a]

	Spirits		Wine		Beer		Total
	Absolute Alcohol	% Contribution	Absolute Alcohol	% Contribution	Absolute Alcohol	% Contribution	Absolute Alcohol
New England	1.41	44	0.45	14	1.35	42	3.21
(minus New Hampshire)	(1.28)	(42)	(0.44)	(15)	(1.30)	(43)	(3.02)
Middle Atlantic	1.06	40	0.33	12	1.24	47	2.63
East North Central	1.03	40	0.13	5	1.39	54	2.55
West North Central	0.93	38	0.19	8	1.32	54	2.44
South Atlantic	1.25	44	0.29	10	1.28	45	2.82
(minus District of Columbia)	(1.21)	(44)	(0.27)	(10)	(1.27)	(46)	(2.75)
East South Central	0.85	42	0.10	5	1.07	53	2.02
West South Central	0.91	35	0.22	8	1.49	57	2.62
Mountain	1.35	40	0.42	12	1.61	48	3.38
(minus Nevada)	(1.17)	(38)	(0.38)	(12)	(1.56)	(50)	(3.11)
Pacific	1.33	38	0.71	20	1.41	41	3.45
TOTAL	1.12	40	0.36	12	1.34	48	2.82

[One U.S. gallon = 3.785 liters = 0.833 imperial gallon.]

[a] The standard regions of the U.S. Census Bureau.

Source: Hyman et al. (1980, p. 7). Reprinted by permission of Journal of Studies on Alcohol, Inc.

of wine countries in the southwest, where the grape vine grows plentifully. The dispersion and range down the table are greatest for wine: a number of countries are separated by geometric ratios in excess of 25 and as high as 50, much more than the differences for beer (10x) and spirits (3x). We do know that ratios of consumption between beer and spirits have been known to shift abruptly in European countries, such as in Denmark, when in 1917 the beverage taxation rates were quite drastically modified (Wilkinson 1970).

Regional distributions are also evident in the United States (Table 4), although they are much milder. Wine varies by about 5x, spirits by 2x, and beer by no more than 1.5x. Although the limited comparability of regional indices is, as indicated above, an inherent caution against generalization, it seems very safe to say that the states are, as a whole, quite uniform in their prevalence of heavy beer drinking and less so in regard to spirits. (Note the high rate in New England versus the low rate in South-Central.) The greatest regional difference lies between the wine preferences of the Pacific states in contrast to the Midwest—although these wine-growing coastal states unquestionably lie well below the prevalence of heavy consumption found in the European wine countries.

We may also examine changes in the consumption of different beverages across time. A 150-year trend for the three main classes of beverage types in the United States is displayed in Figure 3. The overall configurations are quite different. (Note that these graphs are scaled in gallons of beverage rather than gallons of absolute alcohol.)

Wine remained at fairly low stable levels throughout the 19th century and into the 20th century. The World War I mobilization, along with state (then national) prohibitions, succeeded in suppressing the apparent mini-boom about 1910, but home winemaking became popular during the 1920s. Heavy wine drinking recovered rapidly from the Great Depression, and average wine consumption was the most vigorous beverage index during the consumption boom of the late 1960s.

Beer, by contrast, only began to be drunk in the United States about 1850, but its heavy use grew remarkably (the index increasing by 10-fold) in the next 65 years. The drastic effects of Prohibition were not shaken off for a full generation, and the mean consumption of beer today is less than it was in 1915. Most notably, the mode of drinking beer in the prewar period—in draught from tavern barrels—has been very largely replaced by canned and bottled beer (Rooney and Butt 1978).

Finally, consumption of heavy spirits dropped steadily during the 19th century, finally leveling off about 1880. The wartime state prohibitions and the Volstead Act depressed this index for a few years, but it rebounded during the mid-1920s to about 30 percent higher than its Victorian level. The Great Depression appears to have strongly affected the index, although the switch in data base from indirect measures of illicit production (1930) to tax records of legal distillation (1934) exaggerate this. The index was somewhat below the 1880-1915 level after World War II, then rose by 1970 to its recent plateau—still below the levels of heavy spirits drinking estimated during Prohibition.

In brief we find that consumption trends for the three alcoholic beverage types diverged widely during the 19th century and through about 1935. Wine use (nutritional) was stable (until a growing popularity after 1900), intoxication (spirits) fell to 25 percent or less of Jacksonian levels, and heavy beer drinking in the convivial atmosphere of the tavern grew steadily until Prohibition effectively ended it. Since 1935, which is to say since the reinstatement of federal control over production and state control over distribution, the consumption trends of all three types have been quite similar. The largest difference is in the more rapid growth of wine drinking, which now contributes about 13 percent of all absolute alcohol consumed in the United States.

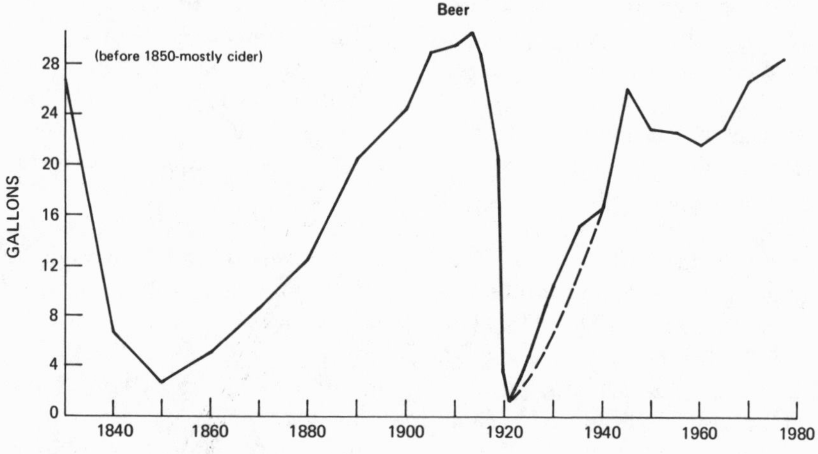

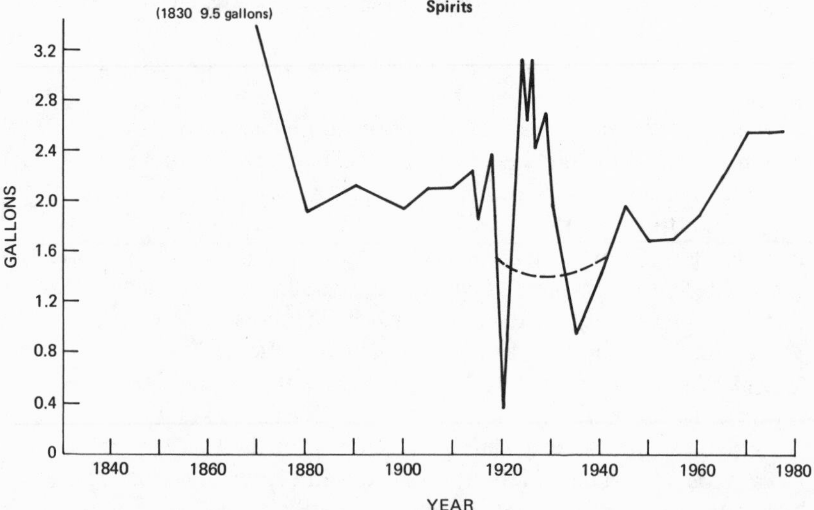

FIGURE 3 Time trend of U.S. consumption of types of alcoholic beverages, 1830-1977.

Note: The solid lines for the years 1920-1935 are consistent with Warburton's (1932) estimates, while the dotted lines represent Jellinek's (1947-1948) views. Other data are derived from Keller and Gurioli (1976) and Gavin-Jobsen Associates (1978). See also notes to Figure 2.

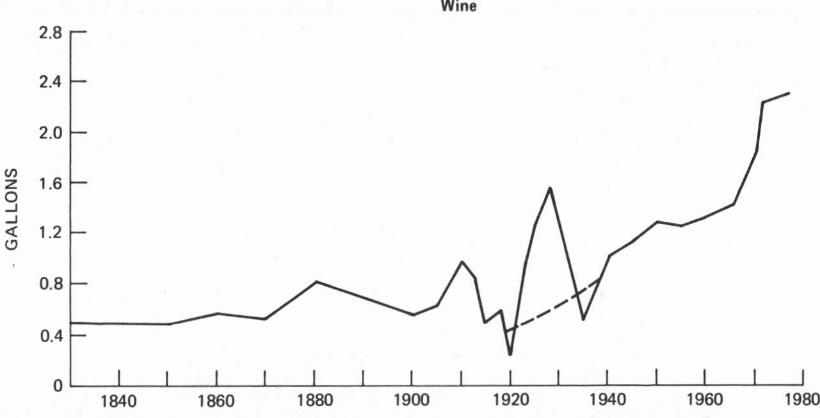

FIGURE 3 (*continued*)

DRINKING PATTERNS IN SURVEY DATA

The total consumption approach draws on sales data for its mean statistic and on sample surveys of drinking for its single-distribution argument. A rather different use of sample survey data is made by those who are interested in the frequency and distribution of intoxication, what has come to be called "drinking practices" or "patterns."

Although the study of drinking patterns has been carried on largely by questionnaire survey research, a few smaller studies have used "daily drinking diaries," which have recently been compared with survey methodology and have demonstrated relatively minor deviations in summary results (Gerstel et al. 1975, Harford 1979). Since large-scale survey research is basically a post-World War II phenomenon, detailed quantitative data about drinking patterns are virtually confined to this period.

One approach that has been widely assimilated is to divide drinkers into categories such as abstainer, infrequent, light, moderate, and heavy drinkers, taking into account the frequency of drinking occasions, the usual quantity consumed, and the frequency and quantity consumed during maximum drinking episodes. This is the quantity-frequency-variability or Q-F-V approach (Mulford 1964, Knupfer 1966, Cahalan et al. 1969, Cahalan and Room 1974). Mapped against the skewed distribution of annual per-capita alcohol consumption, each of these discrete subtypes of drinker would generate its own characteristic curve, "underneath" the aggregate total consumption curve, which the summation of the five would yield. Many of the surveyed individuals whose drinking pattern might qualify them as light or moderate drinkers (small amounts every day; seldom noticeably intoxicated) have a total annual consumption greater than individuals whose pattern classifies them as heavy drinkers (twice a month "blind drunk"). The typological approach identified here is especially sensitive to such occasions of extreme drunkenness and thus to "binge drinking," a type that has haunted the alcohol literature, both empirically and conceptually, since Jellinek (1960) originally formulated this as a distinct type. The typological method clearly assumes that particular consumption characteristics, beyond period consumption per se, dramatically escalate any risks associated with alcohol.

According to successive surveys, the proportion of reported abstainers (no alcohol drunk in the past year) among U.S. adults was about 45 percent in the middle and late 1950s, but declined to about one-third by 1970; the figure has been unchanged since that time. However, there is considerable difference between adult men and women. About 27 percent of men, but 42 percent of women, are now abstainers.

It is conventional to distinguish heavy from moderate drinkers. The criteria for drawing this line vary from study to study. In the major national survey carried out in the 1960s, the designation "heavy" was used for a conglomerate of patterns; it included those who had five to six drinks twice a month, a drink or two nearly every day, five to six drinks "once in a while," and a drink three times a day; and it encompassed patterns combining or exceeding these frequencies. By these lights, 1 in 5 men, and 1 in 10 women were heavy drinkers in 1965. Annual surveys initiated in the 1970s (see Johnson et al. 1977) using a simpler definition of "heavier drinking" (an average of 2 or more drinks per day) yielded virtually identical proportions. These figures too have shown no significant change since 1970.

In a comparative perspective, the United States is remarkable among Western industrial nations for its proportion of adult abstainers, which is higher in the United States than in Canada or in any nation in western Europe (these countries range from 3 percent to 30 percent abstainers; [NIAAA 1978]). There is considerable regional variation in the United States, however. The proportion is lowest in the Mid-Atlantic states (17 percent) and highest in the Bible Belt states of the South (65 percent). It is not possible to make broad international comparisons for proportions of "heavy drinkers" using such survey data due to incompatible or nonexistent data for the different countries.

It is important to note that even where drinking patterns for a population appear to be stable across a number of years, there is continual flux in the drinking patterns of many individuals within the population.

At the extreme, many clinically diagnosed alcoholics alternate between periods of heavy consumption and periods of abstention. There is also a tendency for drinking to decline markedly beyond the age of 50, and there are often significant differences in drinking rates between age and sex cohorts as well as within the same cohort at different points in time. Skog (1979) has argued that these different drinking subpopulations are the primary explanation for the departure of consumption curves from Ledermann's predictions.

USE AND CONSEQUENCES: THE PROBLEM OF ATTRIBUTION

It is one thing to know how much or how often people drink; it is another to pinpoint the effects of that drinking. For virtually every possible consequence of alcohol use that we may be interested in, alcohol is neither a necessary nor a sufficient cause, but rather one in a series of factors that may combine in various ways to yield effects. The presence of specifiable other factors, some permanent but others subject to manipulation, is just as important as drinking per se to the production of alcohol-related effects.

For example, sustained heavy alcohol use is demonstrably associated with the pathological and potentially fatal liver condition called cirrhosis. The data on this connection are as strong as those linking heavy smoking with lung cancer. But the mechanism by which alcohol use causes this gradual accumulation of scar tissue in the liver is still speculative. It is known that cirrhosis can occur in people who have never drunk alcohol. It is known that various nutritional deficiencies or imbalances can not only cause cirrhosis directly, but also can markedly change the vulnerability of liver tissue to alcohol-related cirrhosis. Good nutrition probably cannot prevent alcoholic cirrhosis; but bad nutrition can certainly hasten it along.

Another example is automobile accidents. Drunken driving is notoriously a precursor of traffic fatalities. But at least half of all traffic fatalities in the United States evidence no alcohol "involvement" at all. Perhaps 1 in 500 to 2,000 of legally drunk-driving episodes result in arrest—and there are far more arrests than fatalities. Moreover, an individual can hardly cause a traffic fatality, no matter how drunk, in the absence of a lethal vehicle. This may seem trivial in a nation of 100 million automobiles, but it is not trivial if changes in gasoline costs drastically depress people's use of cars or change the way autos are built and driven. In short, certain of the socially most important consequences of alcohol use are rare, may occur with no alcohol present, and/or require the presence of additional factors in order to happen at all.

These two examples involved the negative effects of alcohol on physical well-being. In looking at psychological and social well-being, the difficulties do not decrease. A number of studies have found some common properties among the personalities of diagnosed alcoholics: depression, anxiety, low frustration tolerance, feelings of powerlessness, etc. But virtually the same properties have been deemed to cause alcoholism. If depression is both a cause and an effect of heavy drinking, how can one decide what part of the alcoholic's depression, or of suicides that follow from depressive episodes, is due to drinking?

Similarly, attempting to pinpoint the exact part played by drinking in problematic social behavior is exceedingly difficult, compounded as it is by interaction with others during numerous situations over time. Statistical methods to separate alcohol consumption from other causes do offer some help in disentangling multiple causes, but since the problems are generally long-term ones, in which both the drinker and "significant others" build up a history of cumulative perceptions and judgments, such attempts to isolate and assign weight to one factor may have little relationship to reality.

All of these problems in attribution apply to the positive effects of alcohol use as well. They are somewhat aggravated by the disinclination of researchers to investigate positive effects, which means that less data are available for analysis. For example, the best-known study of economic effects of alcohol is on the costs of alcohol abuse and alcoholism and does not attempt to estimate effects that may yield economic positives. Mortality studies ask how many deaths alcohol may have been at least partly responsible for and do not concern themselves with the lives alcohol may preserve, even though studies on heart disease have indicated certain advantages possibly associated with moderate drinking.

There is no avoiding the difficulties in attributing the effects of alcohol use; however, it is necessary to have some scheme of accounts for these effects if we are to consider alcohol-related policy. If a central principle of any policy is to preserve and promote good effects while minimizing bad ones, then it is difficult to evaluate any policy without identifying the effects of relevance and observing how they change over time. Since the consequences of alcohol use are manifold, it is important in the first place to identify the ones of greatest social importance, to assess the relative importance of alcohol in generating them as best we can, and then to be systematic and comprehensive in thinking about how these effects can be modified.

A more exact feel for the difficulties of attribution may be achieved by examining recent responsible attempts to estimate the effects of drinking in the United States. Three of these were prepared in the 1970s under the auspices of the National Institute on Alcohol Abuse and Alcoholism (NIAAA) and were prominently featured in the Third Special Report to the U.S. Congress on Alcohol and Health. Each examines a different dimension or measure of effects: mortality (deaths), economic costs (dollars), and psychobehavioral problems (symptoms).

Mortality

Table 5, reproduced in *Alcohol and Health 3* (NIAAA 1978), estimates that the number of deaths related to alcohol use in the United States in 1975 was between 61,000 and 95,000. The associated text notes that other mortality studies have respectively estimated 140,000; 185,690; and "as high as 205,000 deaths per year, which was 11 percent of the total 1.9 million deaths in 1975." The largest figure has received the widest publicity.

A comprehensive study of deaths among clinical alcoholics (Schmidt and Popham 1980, also see Polich et al. 1980) suggests how the totals in Table 5 could have been inflated. For 12,000 former alcoholism patients in Ontario, Canada, who were followed up at an average of 8.5

TABLE 5 Estimated Deaths Related to Alcohol in the United States, 1975

Cause of Death	Number of Deaths, 1975	Percent Related to Alcohol	Estimated Number Related to Alcohol
Alcohol as a direct cause			
Alcoholism	4,897	100	4,897
Alcoholic psychosis	356	100	356
Cirrhosis	31,623	41–95	12,965–30,042
TOTAL	36,876		18,218–35,295
Alcohol as an indirect cause			
Accidents			
Motor vehicle	45,853	30–50	13,756–22,926
Falls	14,896	44.4	6,614
Fires	6,071	25.9	1,572
Other[a]	33,026	11.1	3,666
Homicides	21,310	49–70	10,442–14,917
Suicides	27,063	25–37	6,766–10,013
TOTAL	148,219	29–40	42,816–59,708
OVERALL TOTAL	185,095		61,034–95,003

[a] Includes all accidents not listed above, but excludes accidents incurred in medical and surgical procedures.

Sources: Day (1977) and National Center for Health Statistics (NCHS) (1975b).

TABLE 6 Percentage of Excess Deaths in a Clinical Alcoholic Population in Canada

Heart disease	(20)*
Accidents	(20)
Cirrhosis	(16)
Suicide	(12)
Alcoholism	(9)
Cancers of the mouth, throat, esophagus, windpipe	(9)*
Pneumonia/bronchitis	(6)*
Alimentary ailments	(5)*
Other causes (includes homicides)	(3)
TOTAL	100

* Not counted in Table 5.

Source: Adapted from Schmidt and Popham (1980).

years, 1,062 *excess* deaths (compared with age-standardized population norms) were attributed to the causes listed in Table 6.

The authors of this study caution that alcoholics smoke cigarettes at considerably higher rates than the comparison population and, hence, that excess heart disease and cancer deaths are associated to an unclear degree with tobacco rather than alcohol exposure. They also warn that nutritional and other life-style differences intervene in the relation between drinking alcohol and dying.

Still, 40 percent of the deaths in Table 5 were from causes not directly considered in Table 6; hence, an absolute maximum figure in the range of 150,000 for deaths "related to alcohol" might be acceptable, presuming that all untreated alcoholics have the same death risks as clinical patients. But what does this mean? If we wish to determine how alcohol affects the overall death rate in the United States, i.e., to discover its net causal impact, we would have to estimate both deaths resulting from and deaths prevented by alcohol use. (There is some evidence, for example, that moderate drinking is correlated with decreased risk of death, particularly death involving ischemic heart disease.) However, being related to or correlated with does not amount to causal proof. In this sense, the 150,000 figure is simply a pool within which whatever deaths might be *caused* by alcohol—and thus *prevented* by minimizing its use or misuse—are to be found. The number will be no more––it will be much less. How much less is not certain, but a figure in the area of 50,000 theoretically preventable deaths seems reasonable. This is not 200,000—but neither is it negligible.

TABLE 7 Economic Costs of Alcohol Misuse and Alcoholism in the United States, 1975

Item	Cost (billion $)
Lost production	19.64
Health and medical	12.74
Motor vehicle accidents	5.14
Violent crime	2.86
Social responses	1.94
Fire losses	0.43
TOTAL	$42.75

Source: NIAAA (1978, p. 31; prepared by Berry et al. 1977).

Economic Costs

Table 7 is a tabulation of how much alcohol abuse and alcoholism cost the United States in 1975. This $42.75 billion figure is not a "net cost" estimate; that is, it is not intended to represent the net reduction in gross national product resulting from alcohol use. It is a conglomeration of dollar values: discounted lifetime earnings attributable to about 69,000 people whose deaths in 1975 were estimated to be alcohol-related (25 percent of the $42.75 billion); lessened production as estimated by certain household wage differentials (35 percent); assignment of health care resources to alcohol-related problems (21 percent); and assorted other estimates. This collection of dollar values is held by its authors (Berry et al. 1977, Berry and Boland 1977) to be the "external cost" of alcohol abuse—paid not by the alcohol abuser but by the rest of society.

It is difficult to assign precise significance to this result. For one thing, a large fraction of the costs tabulated here do appear to be borne by the alcohol abuser: from 25 percent to 50 percent of the total amount cited. There are implicit assumptions about the labor supply in 1975—namely, that it was short—that are not consistent with the then-prevailing economic indicators (unemployment was at 8.5 percent). The transposition of "associated with alcohol" into "due to alcohol abuse," which is to say transformation of an estimated correlation into a direct cause, occurs in the discussion despite explicit cautions by the authors to the contrary.

There are some problems raised by this approach of estimating costs in absolute, autonomous terms. Economic cost estimates are usually related either to a specific program of action or to a specific annual budget. In the first instance, one might estimate the costs and benefits of a specific program to modify alcohol abuse. In the second instance, one would try to estimate the proportion of, for example, the federal budget or aggregate personal income, or capital stock, expended on, diverted by, or destroyed as a result of alcohol-related problems (Walsh 1979, Mäkelä and Österberg 1979). Absent from such contexts, a multibillion dollar cost figure is simply a large price tag on an empty box. It draws attention, but gives little guidance.

Behavior

In Table 8, approximately 1 in 10 drinking American adults (about 10 million people) are classified as "problem drinkers"; an additional 1 in 4 (25 million) as "potential problem drinkers"; the other 65 million or so drinkers reportedly had "no problems." What precisely do these rather ominous figures mean? In the original table (Johnson et al. 1977), the authors listed these categories as "frequent problem drinking symptoms," "potential symptoms," and "no symptoms." The most common of these symptoms ("taking 2 or 3 drinks at one sitting"—"sometimes but not often" or "frequently"; "going several days or weeks without taking a drink and then having several drinks at one time"—"sometimes but not often" or "frequently") are indistinguishable from criteria for moderate drinking patterns. The next most common symptoms ("talking

TABLE 8 Rates of Problem Drinking Among U.S. Drinkers, by Drinking Population, 1973–1975

	Percentages for Each Survey			
Drinking Population	March 1973	January 1974	January 1975	June 1975
All Drinkers				
No problems	64	70	65	63
Potential problems[a]	26	24	24	26
Problem drinkers[b]	11	6	10	10
Males				
No problems	57	66	62	57
Potential problems[a]	29	27	23	31
Problem drinkers[b]	14	8	15	13
Females				
No problems	74	77	70	73
Potential problems[a]	21	19	27	21
Problem drinkers[b]	5	4	3	6

[a] A potential problem drinker experienced two or three of sixteen problem drinking symptoms frequently or four to seven symptoms sometimes.
[b] A problem drinker experienced four or more of sixteen problem drinking symptoms frequently or eight or more symptoms sometimes.
Source: NIAAA (1978, p. 26; adapted from Johnson et al. 1977).

a lot about drinking"—"sometimes but not often" or "frequently"; "showing the effects of liquor more quickly than most people"—"sometimes but not often" or "frequently"; "taking a drink to feel better"—"sometimes but not often" or "frequently") are not, on their face, "problems." Sixty percent of the reported symptoms fall into one of the categories above. The authors warn that "this problem index is presented for comparison purposes over time and should not be used as an absolute definition of problem drinkers." As an index it simply demonstrates that there has been relative stability between 1973 and 1975 in reports of these drinking patterns.

These data really shed little light on the relationship between alcohol use and consequences. A few categories of symptoms (not the most common ones) are markedly more prevalent among heavier drinkers than other drinkers. But this report's conglomeration and renaming of categories obscures more than it enlightens. It is impossible to attach meaning to Table 8 beyond the indication of stability across time.

SOCIALLY IMPORTANT EFFECTS OF ALCOHOL USE

It is clear that there are problems with efforts to assay the effects of alcohol by toting up detached dollar values, "associated" deaths, and "symptoms." There are a number of alternative ways to present and organize information about these effects. No single way is right for all purposes, but my concern here is to highlight the socially important effects, to examine the degree to which alcohol is a principal cause, to roughly estimate the population affected, and to note what other generative factors besides alcohol use might be important and subject to modification.

Taking this view, I can identify five principal environments that, when combined with alcohol, produce effects. Each is best visualized as a system. First is the internal organic environment of the body, in W. B. Cannon's (1932) sense: its basic, enduring physiological forms and processes. Our principal interests here are the organs most sensitive in the long term to alcohol exposure: the heart and the liver.

The second environment is the personality. It is also internal to the person and one can even speak of it as based in an organ: the brain. But the physiological base does not suffice to complete our thinking about effects on the personality system. We need to use psychological concepts that we would not think of applying to hearts or livers.

The third system is external and the one that we so often think of as "the" environment: i.e., the physical one. It is of great importance that the physical environment we touch most directly is heavily transformed by human activity. The world of moving vehicles, building elevations, and concrete and floating surfaces is no less physical for being fabricated.

The fourth system, external like the last, is the system of intimate social relations, most particularly those we traditionally regard as primary and important to us: kin and coworkers or, to stress the long view, family and occupational careers.

Finally, there is the institutional environment, the relatively impersonal social matrix of secondary relations that we can ordinarily relate to only in abstract or remote terms, although it diffusely and sometimes very directly transfigures our lives. Most particularly, we are interested here in the broad public systems of responsibility for health care, civil safety, and economic well-being.

Physiological Effects

When the possibilities of sustained changes in physiological systems due to alcohol use are examined, very different effects come into view depending on whether the depth of intoxication, drinking patterns, or total consumption is considered. Due to the surprising capacity of the liver to rapidly convert alcohol to usable carbohydrate, the physiological effects of intoxication tend to be short-lived. The principal exception is intoxication that leads quite immediately to death. In examining drinking patterns, on the other hand, the primary effect involves physical dependence on alcohol, which induces vulnerability to the pathological syndrome of withdrawal or abstinence. The medical recognition and management of this syndrome, however, has relegated it to a fairly insignificant role in terms of physiological health—in contrast to the effects (which physiological alcohol addiction may accentuate or support) of drinking patterns on psychological and interpersonal matters. Finally, clear links between drinking patterns and physiological changes have been investigated most thoroughly in relation to total consumption. In this connection, rates of coronary heart disease and cirrhosis of the liver are by far the most prominent physiological aspects known at this time.

Deaths from alcohol overdoses, alone and in combination with other drugs, account for close to 10,000 deaths and perhaps 100,000 episodes of medical intervention each year. The finding of BAC in excess of 0.3 in autopsy toxicological findings is generally considered to strongly suggest the possibility that the death was the consequence of an acute alcohol reaction. Approximately 5,000 such deaths occur each year (Day 1977). In addition, approximately 5,000 fatal overdoses involving alcohol in combination with other drugs occur in the country each year, based on extrapolation from Drug Abuse Warning Network (DAWN) data (IMS America 1976). Since these reports indicate only measurable presence of blood alcohol and not specific BAC and do not speculate about relative mechanisms, it is difficult to know in how many of these deaths alcohol was a sufficient or necessary cause. The incidence of suicide within this group is not known, although in the alcohol-in-combination cases the death certificates estimate suicidal intent in about 40 percent of the cases.

Cirrhosis of the liver is a disease process known to have multiple causes, that are not all well understood (French 1971). Certain protein deficiencies are invariably followed by appearance of cirrhosis—the dissolution of liver cells and their replacement by scar tissue. Cytological studies on the effects of alcohol use have shown that there is a buildup of fatty yellow liver cells as a consequence of metabolizing alcohol. In laboratory studies of short-term high-level exposure to alcohol, mitochondria (essential organelles within liver cells) disintegrate. Both of these processes may contribute to liver cell death and fibrosis (cirrhosis). The supply of certain amino acids in the diet also appears to have a strong influence on how well liver cells resist deterioration; but this protective effect is by no means absolute (Lieber et al. 1971).

These microbiological findings are strongly supported by macroepidemiological studies (Schmidt 1977), which show that gross rates of cirrhosis deaths track shifting rates of total consumption in a population. Among clinical alcoholic populations in which consumption of 5 ounces of alcohol daily for long stretches of time is an approximate lower limit of alcohol use, a prevalence of cirrhosis damage of 8 percent has been reported, far in excess of the general population; another 25 percent suffer acute liver inflammation, generally regarded as a precursor to cirrhosis. In postmortem studies of individuals identified as alcoholics in New York City (Haberman and Baden 1978), the incidence of moderate-to-severe fatty changes in the liver exceeded the rate of cirrhosis by approximately the same proportions. In general, it is expected that a long period of exposure to high levels of consumption, on the order of 15 years or more, is necessary to bring the disease process to a life-threatening state; however, cessation of alcohol use can freeze the process.

The rate of liver cirrhosis death in the United States at several points, beginning in 1961, is reported in Table 9. It has been estimated that if no alcohol were consumed, the death rate from cirrhosis would approximate 3-4 per 100,000 (see Schmidt 1976, 1977; Skog 1979). The rise from 11.3 to 15.4 between 1961 and 1971 coincides with the 26-percent increase in per-capita U.S. alcohol consumption in this period.

TABLE 9 Statistics Relevant to Alcohol-Related Mortality in the United States, Death Rates per 100,000 by Year and Cause of Death

Cause of Death	1961	1971	1975	1979
Malignant neoplasms of buccal cavity and pharynx	3.5	3.7	3.8	3.6
Malignant neoplasms of respiratory system	23.1	35.3	40.7	46.6
Ischemic heart disease	274.6	326.1	301.7	283.6
Cerebrovascular disease	105.5	101.1	91.1	76.5
Atherosclerosis	19.3	15.2	13.6	13.0
Peptic ulcer	6.3	3.9	3.2	2.2
Cirrhosis	11.3	15.4	14.8	13.4
Motor vehicle accidents	20.8	26.3	21.5	23.3
Other accidents	29.6	28.6	26.8	23.7
Suicide	10.4	11.7	12.7	11.7
Homicide	4.7	9.1	10.0	10.0
Pneumonia	29.0	26.9	24.1	21.4

Sources: National Center for Health Statistics (1963, 1975a, 1978, 1980).

Since 1971, however, alcohol use has held steady while the cirrhosis death rate has declined to 13.6. The reason for this decline is simply not known. It is expected that cirrhosis death rate changes lag somewhat behind changes in total consumption, so we would have actually expected the increase to continue somewhat into the 1970s. However, general death rates in the United States have been declining steadily during the 1970s, and the decrease in cirrhosis death is in approximate line with this general decrease.

In the 35-54 age band, cirrhosis accounted for 6.5 percent of all deaths in 1979. Of the 1.93 million deaths in the United States in 1979, 30,000 were due to cirrhosis. At best guess, between 20,000 and 25,000 of these were a primary result of alcohol consumption. Studies in Ontario, Canada, which has a total consumption similar to the United States as a whole, lead to the estimate that roughly one-half of this last figure involved individuals whose total consumption is above the lower limit observed in clinical alcoholic populations (Schmidt 1977). We would therefore estimate that about 10,000-12,000 deaths from cirrhosis occur among clinical alcoholics in the United States, and another 10,000-12,000 among people who would most likely not meet clinical criteria for chronic alcoholism.

Alcohol use has also been investigated in relation to heart disease. The findings here are not clear-cut (see Klatsky et al. 1978, Hennekens et al. 1978). Very high consumption of alcohol has been linked to cardiomyopathy and hypertension; however, moderate consumption has been linked in a number of prospective and case control studies to reductions in ischemic heart disease, the major cause of death in the United States, claiming about 525,000 lives in 1979. Relative to non-drinkers, these studies indicate risk reduction of 20 percent to as much as 70 percent. However, the samples are specialized and the studies are inconsistent in regard to whether or when increases in consumption beyond two drinks per day cease being "protective."

From 1961 to 1971, while alcohol use on the whole increased 26 percent, the rate of death from ischemia increased 18 percent; but this change was confined to cohorts born before 1910—there was virtually no change in rates of coronary death for age brackets below 65. Since 1971, ischemic heart death has been declining dramatically in all age categories. This could be a lagged effect of earlier consumption changes, a result of improved cardiological care or general health, or the result of other causes entirely.

In summary, the major significant physiological effects of alcohol in the United States today are overdose, liver disease, and heart disease. Alcohol is involved in roughly 10,000 overdose deaths annually, half in combination with other drugs. It is difficult to place any boundaries around the precise population at risk for all of these overdoses, although it has been shown that in the combination deaths the sexes are evenly split and most were people under 30, whereas alcohol overdose deaths follow the general demographic profile of clinical alcoholic populations: mostly 30- to 55-year-old men. Alcohol is also involved in 20,000-25,000 cirrhosis deaths annually, of which about half occur in populations that probably meet appropriate criteria for clinical alcoholism. There is reason to believe that alcohol use may have a significant effect on coronary disease. If alcohol use were responsible for a 5-percent net decrease in coronary mortality, this would be 33,000 lives saved, and concomitant decreases in nonfatal heart attacks and other disabilities. The population of reduced risk in this instance would extend to 55 or 60 percent of the adult population of the United States.

Personality

It is likely that alcohol cannot create effects in the system of personality that cannot be created in other ways or that have not been observed in people who have never consumed alcohol. The primary clinical criterion by which alcoholism is recognized is "loss of control" over intake of alcohol (Jellinek 1960, DSM-III 1980). The concept of loss of control has considerable kinship with other important notions in abnormal psychology, particularly the notions of locus of control and of powerlessness or anomie. Both have been deemed pervasive aspects of modern life. In terms of clinical presentation, the psychiatric syndrome most often and most worrisomely associated with alcoholism is depression (Schuckit 1978) and particularly depressive states serious enough to entail risk of suicide.

The relation of alcohol use to suicide has been studied enough to develop some rough notions about its quantitative contributions (Aarens and Roizen 1977). It can be estimated that alcohol was present in about one-third of the annual 25,000 U.S. suicides, and about half of these drinking suicides could have been diagnostically ascertained as alcoholics. The prevalence among attempted suicides (estimated to number 200,000-400,000 per year) of such drinking problems as alcohol-related job loss, spouse loss or threat thereof, and self-identification as a problem drinker or alcoholic, all run approximately five times the rates to be expected in a demographically comparable group. There is considerable overlap between these alcohol-involved suicide figures and the alcohol-in-combination overdose deaths reported through the DAWN system.

If we attempt to examine the structure of personality by such simple measures as are available in wide-scale population surveys, we can make some estimate of the nonclinical impact of alcohol on personality in the United States. Insofar as evidence bears on this (Cahalan et al. 1969), it indicates that across the general span of drinkers, there is little difference (as there is little difference from abstainers) in either the structure of personal goals or degrees of satisfaction with the attainment of such. This is not to say that there are not unhappy or dissatisfied individuals among drinkers, as the numbers appealing to Alcoholics Anonymous and other treatment modalities clearly assure us. But if there are proportionately as many unhappy abstainers as unhappy drinkers, we should be especially hesitant to expect that alcohol itself has a significant independent impact on the "general happiness" or any index of it such as the rate of suicide.

When we examine changes in this suicide rate in Table 9, we note that the rate increased between 1961 and 1971, the period during which alcohol use increased; increased at an even more rapid rate between 1971 and 1975, the period during which total consumption was stable; and then decreased again in the latter years of the 1970s. These figures suggest that while alcohol may be involved in patterns of depression and suicide, the involvement does not have very much independent causal force.

Physical Environment

Because intoxication's most reliable correlate in behavior is a certain clumsiness in handling things and in moving about, accidental displacement of mass and energy is among the most serious aspects of drunkenness. Correlation of accident involvement with intoxication depth and, to some degree, patterns of intoxication, is best established. Studies of accident rate linkage and total consumption have not, as a rule, yielded significant relations (but see Cook, in this volume).

There are four principal types of accidents that concern us: motor vehicles, fires, falls, and drownings. The association between auto vehicle accidents and drinking is both the most notorious and largest—but it is also subject to a lot of misconception. The most common observation is that "50 percent of all driver fatalities are drunk drivers." Driver fatalities are not the only the fatalities associated with motor vehicles; about half of these fatalities are pedestrians and passengers. Moreover, fatalities are not the only consequence of motoring accidents. There are roughly 45,000 deaths, but there are also 1-2 million injuries, in 15-30 million crashes. Not all the "drunk" drivers in any of the well-known statistics are legally drunk—many were not perceptibly intoxicated except by sensitive measurement (BAC less than 0.05). At any time, about 10-20 percent of all drivers on the road have measurable BAC levels, so one would expect this fraction of accidents to involve alcohol, even though no causal significance might be attached to the fact.

Once all these factors are taken into account, if we ask how many bad effects of motor vehicle operations might theoretically be prevented if no one ever entered a driver's seat under *any* alcohol influence, we would estimate about 12,000 lives, 200,000 injuries, and about 1 million crashes—respectively 24 percent, 12 percent, and 6 percent of each of these effects (Reed, in this volume; see also Ouellet et al. 1977, 1978). Additional savings could be achieved if no pedestrian ever crossed a street while intoxicated—but this is a different matter from drunk driving and has more in common with the other types of fatalities.

Deaths in fires, falls, and drownings constitute a somewhat different type of problem. Accident deaths of these sorts approximately equal fatal motor crashes in number. In comparing changes in these across recent years, it is clear that the motor vehicle death rates have changed much more dramatically than other accident rates (most of which are for fires, falls, and drownings), particularly in the 15-44 age band. While there is evidence of alcohol involvement for each of these accident categories, centering in the 30- to 40-percent range for fatalities in the most careful studies, it is again true that a relatively smaller proportion can be attributed to alcohol, since measurable BAC is quite prevalent in the situations in which these accidents occur. The attributable fractions are probably below the range of those for auto crash effects (Aarens et al. 1977, Ouellet et al. 1977).

It is difficult to say how the risk of alcohol-caused accidents is spread through the population. In the case of accidents induced by drunk driving, there is a considerable mismatch between the older character of clinical alcoholic populations and the young male emphasis in the death statistics. In the other accident cases, there is significant difference in the distribution of rates across ages; the cirrhosis death rate climbs rapidly after 25, peaks in the 45-54 age band, and then declines, whereas other accidents occur at a steady rate between 15 and 44 and then rise progressively with each older age. The clinical alcoholic populations display higher accident mortality and morbidity, but are clearly not responsible for the bulk of alcohol-caused accidents.

Interpersonal Effects

The effects of alcohol on one's private relations with others, especially in the family, comprise a most important part of the effects of alcohol in society. We know from clinical histories the ways in which alcohol use can seriously aggravate problems that disturb families and mar work relations and career opportunities. But the very complexity of these long-term bonds makes it exceedingly difficult to know how and whether alcohol acts as an autonomous causal agent in such troubles. At the same time, we know that alcoholic beverages are an integral part of special family occasions ranging from shared evening cocktails to the wedding toast. How do we quantify the effects of alcohol as parts of such occasions? On the basis of current data, we simply cannot do so. This does not make them less real.

The interpersonal effects of alcohol that have drawn the most attention and provide the best opportunities for study are its putative links to belligerence and impoverishment. These, after all, are dramatic events and conditions; they are more likely to come to light beyond the family circle than the milder, but perhaps ultimately as tragic, breakdowns in trust, cooperation, and morale to which chronic alcohol dependence may contribute. Belligerent behavior in the family has become a prominent topic in the 1970s (Yahraes 1979) on which there are limited data available, and they are mainly on injurious assaultive behavior. At the current time, the most common view of researchers is that when alcohol is involved in familial assault, its role is to provide a rationale for justifying or excusing assaultive behavior (Straus et al. 1977). In studies of marital disputes, it appears that a past episode of drinking is as often a subject in a dispute as a current drinking episode is a precipitant (Epstein et al. 1977). In disputes serious enough to lead to calling of the police, drinking has been found to be about half as prevalent in cases of actual assault as in cases of no assault (Bard and Zacker 1974). This does not suggest that drinkers do not engage in belligerent family behavior, but it does argue that drinking may defuse as well as precipitate assaults. Data on causes of child abuse are sparse, but those available do not indicate elevated rates of "problem drinking" among families of physically abused children (Scientific Analysis Corporation 1976).

In none of these areas are longitudinal records yet available. It is doubtful that total consumption of alcohol, with which such records might be compared, would be of great significance in this connection. We know, for example, that the overall homicide rate, which largely reflects intimate violence, increased 94 percent, from 4.7 to 9.1 per 100,000 between 1961 and 1971, while total alcohol consumption increased 26 percent. Both rates have remained approximately level through 1980. But among black men between 15 and 54, whose total consumption of alcohol is less than that of white men, the homicide rate is 10 times as high, accounting for nearly half of all homicide victims. This suggests the need for great caution in ascribing a significant causal role to alcohol consumption in this connection.

Expenditure on alcoholic beverages has been a theme of long standing in the literature on temperance. Family complaints about money spent on alcohol do show up in surveys of drinking problems. It is nonetheless unlikely that the cost of alcohol as such can be as broad an issue as the possible effects of drinking on employability. The mean proportion of U.S. personal income spent on alcoholic beverages has declined steadily during the past 25 years, despite stable or increasing per-capita consumption, and now is lower than in any country in the industrial world.

The relations between drinking and occupational careers are doubtless complex. On one hand, consumption of alcohol is positively correlated to income; on the other hand, the heaviest drinking patterns are in households whose income distribution lies below the national average. There are occupations in which drinking is forbidden for long periods and others in which a pattern of frequent intoxication is virtually an occupational hazard (Trice and Roman 1979).

At the aggregate level, the period of increasing total consumption, 1961 to 1970, corresponds to the largest sustained boost in per-capita income in U.S. history. We do not know whether serious shifts in drinking patterns occurred during this expansion, and we do not know to what degree the economic gains were differentiated by drinking patterns.

In summary, while we have a long history of imagery that pits drinking against the family (Levine 1978), suggesting that it leads to abuse, neglect, and poverty, there is little evidence from the recent past of the United States to support this conception. Drinking may indeed be an important screen on which the inner drama and tension of family relations or career difficulties can be projected, but when we look beyond the clinical data there is little evidence to indict it for anything further.

Institutional Effects

In looking at the institutional effects of alcohol use, three distinct types of effects have attracted the greatest interest: public safety, health care delivery, and the economy. Each involves a formidable coalition of interests affected by alcohol around which a history of policy measures has risen.

The main concern in public safety has been public intoxication, most particularly intoxication of sufficient depth to create serious incapacity for walking or driving safely. The extent of the driving problem was discussed above; here we focus on the ambulatory problem. Public intoxication as an institutional issue has two components: one is the public nuisance aspect, that many citizens find the sight and sound of public drunkenness to be obnoxious; in the second place, drunks in public are vulnerable to crime and exposure. For both of these reasons, public drunkenness becomes a matter of public interest, and an institutional response is called for, mainly police and judicial involvement.

It has been noted across different countries that arrest rates for public drunkenness vary slightly inversely with total consumption of alcohol (NIAAA 1978). The "wetter" the country, the less concerned its citizens are to see their drunken fellows whisked into police vans (Mäkelä 1978; Room 1978). Since 1970 the trend in U.S. jurisdictions has been to decriminalize public intoxication, i.e., to delete criminal statutes governing inebriation and instead provide protective custody for drunks (Giffen and Lambert 1978). The net result has been to steadily decrease police contact with drunks and to shift responsibility for securing their safety from the jails and courts to special alcohol detoxification units. (It is still largely police who convey drunks to these units.)

Regarding the effect of alcohol use on health care delivery it has been reported that, special treatment programs aside, patients in hospitals tend either to be alcoholics or report alcohol-related problems more frequently than one would expect from prevalence in the population. Estimates of the prevalence of problem drinkers in general hospital populations run to 30 percent or more; however, it is difficult to attach credence to these estimates since the inclusion criteria are ill defined. As we have seen, sufficiently lax definitions of "problem drinking symptoms" can qualify 35 percent of drinkers (25 percent of adults) in "problem" categories. The types of specific problems that might lead heavier drinkers to excessive hospitalization include liver ailments, traumatic injuries, depression, and heart problems. But there are no studies that document excesses of heavy drinkers in any of these areas in hospital populations (Berry and Boland 1977).

In general, trends of hospitalization in recent years have been steadily downward. Rates of admission and average length of stay declined during the 1970s across virtually all categories of diagnosis. The impressive inflation of daily hospital costs has been used largely to capitalize high-technology medical equipment and to maintain overhead in the face of declining bed counts. There is little ground for thinking that any of these factors are especially influenced by alcohol use.

The final area of institutional effects is the economy. There are two respects in which drinking has economic impact. First, there is economic activity in the alcoholic beverage industry itself: manufacturing, distribution, and retail sales. Second, there is the leisure and recreational complex in which drinking has come to play a strong supportive role.

The alcoholic beverage industry is substantial and includes three tiers: distillers, brewers, and vintners; importation and distribution companies; and bars, taverns, liquor stores, groceries, restaurants, and other licensees selling drinks and packaged beverages directly to the public. Sales of alcoholic beverages at final retail outlets now amount to roughly $35 billion annually in retail trade. About one-third of this income is retained as government revenue. Beyond direct sales, alcoholic beverages have become an integral part of several institutions that serve U.S. consumers. Drink sales are a profit leader that provides important margins for an indeterminate number of general goods stores, groceries, and restaurants. Spectator and participation sports have become closely allied, via advertising and concession sales, to alcoholic beverages.

Over the past decades, drinking has largely shifted from public places to private homes, and drinking of wine and white spirits (vodka, gin, etc.) has become much more prominent (Gavin-Jobsen Associates 1978). While per-capita alcohol consumption has remained stable, stationary excise taxes on alcohol have retarded price growth, and therefore alcohol has become less expensive relative to other commodities (Cook, in this volume). The shrinkage in overall revenue share has not significantly affected the industry, since it has largely accrued as tax reduction, and government revenues have been compensated by income tax bracket inflation. Changes in the composition of demand for types of beverages have been accompanied by horizontal monopolization and acquisition among large manufacturing firms.

In summary, while the industry today does not by any means have the degree of economic importance that the vertical beer monopolies of the late 1800s had, its contribution to economic activity is considerable.

CONCLUSION

Scientific knowledge about alcohol use and its consequences, like our knowledge about nearly all human activity, has been strongly molded by our collective needs. Of course, this molding is only a first step in the process, for practical questions can only shape—not supply—their own answers. We have fair-to-good information about alcohol in its social context in relation to those effects about which modern societies have chosen to be intensely concerned. Where that concern is very recent, where it has simply not arisen, or where strong, high-quality research traditions that could be adapted for use have not been right at hand, our knowledge is poor or virtually nil.

Historically, interest in alcohol in the United States after repeal moved away from the terrain of family structures, market forces, and nonpathological personality, and instead came to focus on internal medicine, on abnormal psychology, and later on automobile casualties. The state of the art reflects this. The reemergence since 1960 of an increasingly sophisticated public health and social science research community committed to the study of alcohol problems should, by the end of the century, have brought things back into balance.

REFERENCES

Aarens, M., Cameron, T., Roizen, J., Roizen, R., Room, R., Schneberk, D., and Wingard, D. (1977) *Alcohol, Casualties, and Crime*. Final Report of Alcohol, Casualties, and Crime Project, NIAAA Contract No. ADM-281-76-0027 (reprinted August 1978). Berkeley, Calif.: Social Research Group.

Aarens, M., and Roizen, R. (1977) Alcohol and suicide. Pp. 466-524 in *Alcohol, Casualties, and Crime*. Final Report of Alcohol, Casualties, and Crime Project, NIAAA Contract No. ADM-281-76-0027 (reprinted August 1978). Berkeley, Calif.: Social Research Group.

Bales, R. F. (1944) The 'Fixation Factor' in Alcohol Addiction: An Hypothesis Derived from a Comparative Study of Irish and Jewish Social Norms. Unpublished Ph.D. dissertation, Harvard University.

Bales, R. F. (1946) Cultural differences in rates of alcoholism. *Quarterly Journal of Studies on Alcohol* 6:480-499.

Bard, M., and Zacker, J. (1974) Assaultiveness and alcohol use in family disputes: Police perceptions. *Criminology* 12(3):281-292.

Berry, R. E. Jr., Boland, J. P., Smart, S. and Kanak, J. (1977) The Economic Costs of Alcohol Abuse and Alcoholism—1975. Unpublished report prepared for National Institute on Alcohol Abuse and Alcoholism under Contract No. ADM-281-76-0016.

Berry, R. E. Jr., and Boland, J. P. (1977) *The Economic Cost of Alcohol Abuse*. New York: The Free Press.

Bruun, K., Edwards, G., Lumio, M., Mäkelä, K., Pan, L., Popham, R. E., Room, R., Schmidt, W., Skog, O.-J., Sulkunen, P., and Österberg, E. (1975) *Alcohol Control Policies in Public Health Perspective*. New Brunswick, N.J.: Rutgers Center of Alcohol Studies.

Bureau of the Census (1975) *Historical Statistics of the United States. Colonial Times to 1970. Bicentennial Edition*. Part 1. Washington, D.C.: U.S. Department of Commerce.

Cahalan, D., Cisin, I. H., and Crossley, H. M. (1969) *American Drinking Practices*. New Haven, Conn.: College and University Press.

Cahalan, D., and Room, R. (1974) *Problem Drinking Among American Men*. New Brunswick, N.J.: Rutgers Center of Alcohol Studies.

Cannon, W. B. (1932) *The Wisdom of the Body*. New York: Norton.

Day, N. (1977) *Alcohol and Mortality*. Paper prepared for National Institute on Alcohol Abuse and Alcoholism under Contract No. NIA-76-10(P). Berkeley, Calif.: Social Research Group.

deLint, J., and Schmidt, W. (1968) The distribution of alcohol consumption in Ontario. *Quarterly Journal of Alcohol Studies* 29:968-973.

DSM III; Diagnostic and Statistical Manual of Mental Disorders, 3rd Ed. (1980) Washington, D.C.: American Psychiatric Association.

Epstein, T., Cameron, T., and Room, R. (1977) Alcohol and family abuse. Pp. 526-573 in M. Aarens et al., *Alcohol, Casualties, and Crime*. Final Report of Alcohol, Casualties, and Crime Project, NIAAA Contract No. ADM-281-76-0027 (reprinted August 1978). Berkeley, Calif.: Social Research Group.

French, S. W. (1971) Acute and chronic toxicology of alcohol. Pp. 437-511 in B. Kissin and H. Begleiter, eds., *The Biology of Alcoholism, Volume 1: Biochemistry*. New York: Plenum Press.

Gavin-Jobsen Associates (1978) *The Liquor Handbook*. New York: Gavin-Jobsen Associates.

Gerstel, E. K., Harford, T. C., and Pautler, C. (1975) *Final Report: A Pilot Study of the Social Contexts of Drinking*. RTI Project No. 234-892, for Contract No. HSM-42-73-110 (NIA). Research Triangle Park, N.C.: Research Triangle Institute.

Giffen, P. J., and Lambert, S. (1978) Decriminalization of public drunkenness. Pp. 395-440 in Y. Israel, F. B. Glaser, H. Kalant, R. Popham, W. Schmidt, and R. G. Smart, eds., *Research Advances in Alcohol and Drug Problems*, Vol. 4. New York: Plenum Press.

Guttorp, P., and Song, H. H. (1977) A note on the distribution of alcohol consumption. *Drinking and Drug Practices Surveyor* 13:7-8.

Guttorp, P., and Song, H. H. (1979) A rejoinder to Skog. *Drinking and Drug Practices Surveyor* 14(6):29-30.

Haberman, P. W., and Baden, M. M. (1978) *Alcohol, Other Drugs and Violent Death*. New York: Oxford University Press.

Harford, T. C. (1979) Ecological factors in drinking. Pp. 147-182 in H. T. Blane and M. E. Chafetz, eds., *Youth, Alcohol, and Social Policy*. New York: Plenum Press.

Hennekens, C., Rosner, B., and Cole, D. S. (1978) Daily alcohol consumption and fatal coronary heart disease. *American Journal of Epidemiology* 107(3):196-200.

Hetzel, B. S. (1978) The implications of increasing alcohol consumption in Australia—A new definition of the alcohol problem. *Community Health Studies* 2(2):81-87.

Houthakker, H. S., and Taylor, L. D. (1970) *Consumer Demand in the United States: Analyses and Projections*. Cambridge, Mass.: Harvard University Press.

Hyman, M. H., Zimmerman, M. A., Gurioli, C., and Helrich, A. (1980) *Drinkers, Drinking, and Alcohol-Related Mortality and Hospitalizations: A Statistical Compendium*. New Brunswick, N.J.: Center on Alcohol Studies, Rutgers University.

IMS America (1976) *Project DAWN III: Drug Abuse Warning Network, Phase III Report*. Ambler, Pa.: IMS America, Ltd.

Jellinek, E. M. (1947-1948) Recent trends in alcoholism and alcohol consumption. *Quarterly Journal of Studies on Alcohol* 8:1-42.

Jellinek, E. M. (1960) *The Disease Concept of Alcoholism*. New Haven, Conn.: College and University Press.

Johnson, P., Armor, D. J., Polich, J. M., and Stambul, H. (1977) *U.S. Adult Drinking Practices: Time Trends, Social Correlates and Sex Roles*. A working note prepared for the NIAAA under Contract No. ADM-281-76-0020. Santa Monica, Calif.: The Rand Corporation.

Keller, M., and Gurioli, C. (1976) *Statistics on Consumption of Alcohol and on Alcoholism*. New Brunswick, N.J.: Journal of Studies on Alcohol, Inc.

Klatsky, A. J., Friedman, G. D., and Siegelaub, A. B. (1978) Alcohol use, myocardial infarction, sudden cardiac death, and hypertension. *Alcoholism: Clinical and Experimental Research* 3(1):33-39.

Knupfer, G. (1966) Some methodological problems in the epidemiology of alcoholic beverage usage: Definition of amount of intake. *American Journal of Physical Health* 56(2):237-242.

Ledermann, S. (1956) *Alcool—Alcoolisme—Alcoolisation*. Données scientifiques de caractre physiologique, économique et social, Institut National d'Études Démographiques, Travaux et Documents, Cahier No 29. Paris: Presses Universitaires de France.

Levine, H. G. (1978) The discovery of addiction: Changing conceptions of habitual drunkenness in America. *Journal of Studies on Alcohol* 39(1):143-177.

Lieber, C. S., Rubin, E., and DeCarli, L. M. (1971) Effects of ethanol on lipid, uric acid, intermediary, and drug metabolism, including the pathogenesis of the alcoholic fatty liver. Pp. 275-305 in B. Kissin and H. Begleiter, eds., *The Biology of Alcoholism, Volume 1: Biochemistry*. New York: Plenum Press.

Mäkelä, K. (1971) *Measuring the Consumption of Alcohol in the 1968-1969 Alcohol Consumption Study*. Helsinki, Finland: Social Research Institute on Alcohol Studies.

Mäkelä, K. (1978) Level of consumption and social consequences of drinking. Pp. 303-348 in Y. Israel, F. B. Glaser, H. Kalant, R. E. Popham, W. Schmidt, and R. G. Smart, eds., *Research Advances in Alcohol and Drug Problems*, Vol. 4. New York: Plenum Press.

Mäkelä, K., and Österberg, E. (1979) Notes on analyzing economic costs of alcohol use. *The Drinking and Drug Practices Surveyor* 15:7-10.

Mosher, J. (1979) Dram shop liability and the prevention of alcohol-related problems. *Journal of Studies on Alcohol* 40(9):773-798.

Mulford, H. A. (1964) Drinking and deviant drinking, U.S.A., 1963. *Quarterly Journal of Alcohol Studies* 25:634-650.

National Center for Health Statistics (1963) *Vital Statistics of the United States, 1961*. Washington, D.C.: U.S. Government Printing Office.

National Center for Health Statistics (1975a) *Vital Statistics of the United States, 1971*. Washington, D.C.: U.S. Government Printing Office.

National Center for Health Statistics (1975b) *Vital Statistics of the United States, 1972*, Vol. II. Washington, D.C.: U.S. Government Printing Office.

National Center for Health Statistics (1978) *Vital Statistics of the United States, 1975*. Washington, D.C.: U.S. Government Printing Office.

National Center for Health Statistics (1980) *Monthly Vital Statistics Report, Provisional Statistics, Annual Summary for the United States, 1979. Births, Deaths, Marriages, and Divorces*. DHHS Publication No. (PHS) 81-1120, Vol. 28, No. 13. Hyattsville, Md.: National Center for Health Statistics.

National Institute on Alcohol Abuse and Alcoholism (1978) *Third Special Report to the U.S. Congress on Alcohol and Health (Technical Support Document)*. Washington, D.C: U.S. Department of Health, Education and Welfare.

Ouellet, B. L., Romeder, J.-M., and Lance, J.-M. (1977) *Premature Mortality Attributable to Smoking and Hazardous Drinking in Canada. Volume I: Summary*. Ottawa, Canada: Long Range Health Planning Branch of the Department of National Health and Welfare.

Ouellet, B. L., Romeder, J.-M., and Lance, J.-M. (1978) *Premature Mortality Attributable to Smoking and Hazardous Drinking in Canada. Volume II: Detailed Calculations*. Ottawa, Canada: Long Range Health Planning Branch of the Department of National Health and Welfare.

Polich, J. M., Armor, D. J., and Braiker, H. B. (1980) *The Course of Alcoholism: Four Years After Treatment*. Santa Monica, Calif.: The Rand Corporation.

Popham, R. E., and Schmidt, W. (1978) The biomedical definition of safe alcohol consumption: A crucial issue for the researcher and the drinker. *British Journal of Addictions* 73:233-235.

Room, R. (1971) Survey vs. sales data for the U.S. *Drinking and Drug Practices Surveyor* 3:15-16.

Room, R. (1978) Evaluating the effect of drinking laws on drinking. Pp. 267-289 in J. A. Ewing and B. A. Rouse, eds., *Drinking*. Chicago: Nelson-Hall.

Rooney, J. F., Jr., and Butt, P. L. (1978) Beer, bourbon and Boone's Farm: A geographical examination of alcohol drink in the United States. *Journal of Popular Culture* 11(4):832-856.

Schmidt, W. (1976) *Effects of Alcohol Consumption on Health*. Toronto, Canada: Addiction Research Foundation.

Schmidt, W. (1977) The epidemiology of cirrhosis of the liver: A statistical analysis of mortality data with special reference to Canada. In M. M. Fisher and J. G. Rankin, eds., *Alcohol and the Liver*. New York: Plenum Press.

Schmidt, W., and Popham, R. E. (1975-1976) Heavy alcohol consumption and physical health problems: A review of the epidemiological evidence. *Drug and Alcohol Dependence* 1:27-50.

Schmidt, W., and Popham, R. E. (1980) Sex differences in mortality: A comparison of male and female alcoholics. Pp. 365-384 in O. J. Kalant, ed., *Research Advances in Alcohol and Drug Problems*, Vol. 5. New York: Plenum Press.

Schuckit, M. A. (1978) The identification and management of alcoholic and depressive problems. *Drug Abuse and Alcoholism Review* 1(4):1-8.

Scientific Analysis Corporation (1976) *Family Problems, Social Adaptation, and Sources of Help for Children of Alcoholic and Non-Alcoholic Parents*. San Francisco, Calif.: Scientific Analysis Corporation.

Skog, O.-J. (1971) *Alkoholkonsumets fordeling i befolkningen* (The Distribution of Alcohol Consumption in the Population). Oslo, Norway: National Institute for Alcohol Research.

Skog, O.-J. (1979) *Is Alcohol Consumption Lognormally Distributed*? SIFA Mimeograph No. 21. Oslo, Norway: National Institute for Alcohol Research.

Skog, O.-J. (1980) *Social Interaction and the Distribution of Alcohol Consumption*. SIFA Mimeograph No. 30. Oslo, Norway: National Institute for Alcohol Research.

Straus, M. A., Gelles, R. J., and Steinmetz, S. K. (1977) Violence in the family: An assessment of knowledge and research needs. In M. Van Stolk, ed., *Child Abuse, Its Treatment and Prevention: Interdisciplinary Approach*. Toronto: McClelland and Stewart.

Sulkunen, P. (1976) Drinking patterns and the level of alcohol consumption: An international overview. Pp. 223-281 in R. J. Gibbins et al., eds., *Research Advances in Alcohol and Drug Problems*, Vol. 3. New York: John Wiley.

Trice, H. M., and Roman, P. M. (1979) *Spirits and Demons at Work: Alcohol and Other Drugs on the Job*, 2nd ed. Ithaca, N.Y.: Cornell University.

Walsh, B. M. (1979) The economic cost of alcohol abuse in Ireland: Approaches and problems. *The Drinking and Drug Practices Surveyor* 15:3-6, 57-59.

Warburton, C. (1932) *The Economic Results of Prohibition*. New York: Columbia University Press.

Warner, K. E. (1978) Possible increases in the underreporting of cigarette consumption. *Journal of American Statistical Association* 73(362):314-318.

Wilkinson, R. (1970) *The Prevention of Drinking Problems: Alcohol Control & Cultural Influences*. New York: Oxford University Press.

Yahraes, H. (1979) Physical violence in families. Pp. 553-576 in *Families Today, A Research Sampler on Families and Children*, Vol. II. Rockville, Md.: National Institute of Mental Health.

The Paradox of Alcohol Policy: The Case of the 1969 Alcohol Act in Finland

DAN E. BEAUCHAMP

INTRODUCTION

The case of alcohol policy in the Nordic countries of Finland, Sweden, and Norway during the 1960s, which was intended to liberalize their alcohol control systems, is already an important and well-researched episode in the alcohol policy literature. This is especially true for Finland. Data are readily available on the relationship between altering the regulation of alcohol availability and changes in total consumption, in drinking for purposes of intoxication, in heavy consumption, and the like. An examination of the Nordic experience is important for the ways in which it can help frame our ways of thinking about alcohol policy and how alcohol policy is affected by larger forces for social and cultural change in advanced industrialized societies generally, not just for these small nations.

It is arguable that the results of these changes are of minor relevance to American policy. There are important structural differences between the Nordic countries and the United States: most commonly mentioned (Castles 1978, Hancock 1972, Eckstein 1966, Pesonen 1974) are their

Dan Beauchamp, a member of the panel, is at the Department of Health Administration, School of Public Health, University of North Carolina.

The author is grateful to Kettil Bruun of the Finnish Foundation for Alcohol Studies, and to members of the Social Research Institute of Alcohol Studies in Helsinki, whose invaluable assistance made this study possible.

parliamentary forms, elite autonomy, multiparty traditions, a long history of dominance by the Social Democrats in most countries, and a unique mode of government by consensus and consultation.[1]

The small size of the Nordic countries, especially Finland, makes them questionable candidates for comparative research. Finland's population is very small compared with that of the United States—roughly 4.5 million inhabitants as opposed to over 200 million in the United States. On the other hand, Sweden, Norway, and Finland bear important similarities to the United States. All have had a strong temperance tradition; Finland and Norway experienced prohibition; and the state monopoly system of alcohol control is similar to many state systems in the United States.

While there are always dangers in generalization from comparative research, Finnish alcohol policy seems important for the United States in two ways. First, the changes have been studied closely by internationally respected researchers. Their findings (which are summarized in this paper) constitute an important body of alcohol policy knowledge regarding alcohol availability and its control by governmental measures. The Finnish literature regarding the 1969 Alcohol Act is an invaluable stock of knowledge regarding the effectiveness of a small number of conventional policy instruments commonly found in advanced industrialized countries—such measures as tax policy, availability restrictions, and age limits. Thus, one goal in examining the situation in Finland is to develop a wider base of knowledge about substantive elements of alcohol policy as well as some general relationships between available policy instruments that any number of similar societies might consider, recognizing that this knowledge will not be free of specific contextual, historical, or cultural contingencies.

As a second aim, we seek to situate this Finnish policy change in the context of cultural changes occurring in mature industrial societies within the past several decades. We refer to the struggle between cultural modernization and cultural fundamentalism (Gusfield 1963) that typically arose as these societies rapidly completed processes of industrialization and moved toward postindustrial or advanced industrial patterns of dominant service sectors, advancing levels of affluence, consumption, urbanization, and so forth.

This maturation is apparently occurring throughout industrial societies, irrespective of differences in political systems and cultures. If this

[1] This dominance of the Social Democrats is less true for Finland than for Sweden and Norway. Also, the Social Democrats in all three Nordic countries suffered setbacks in the mid-1970s.

view is valid, then alcohol policy changes in Finland provide important insights not only into the impact of specific policy instruments such as taxation or availability, but also into the context of alcohol policy formation in societies, such as Finland and the United States, which contain both fundamentalist and modernist views toward alcohol, perhaps as well as an emergent postmodern view.

THE FINNISH ALCOHOL POLICY DEBATE

Norway, Finland, and Sweden have all had relatively restrictive systems for controlling the sale of alcoholic beverages during this century. Both Finland and Norway experienced an era of prohibition—in Finland it lasted from 1920 to 1933. Sweden only narrowly missed prohibition in a nationwide referendum.

All three societies have similar systems for control of alcohol. In each country there is a state monopoly for the manufacture and sale of most alcoholic beverages, except for beer of very limited alcohol content (in Finland, beer with less than 2.25 percent alcohol by weight). Generally speaking, until the liberalization legislation each country only permitted this very light beer to be distributed in shops and restaurants. Wine and spirits are sold through the state monopolies.

The state monopolies vary only in detail from country to country. In Finland, it is generally agreed, the monopoly (ALKO) is more independent from the Ministry of Social Affairs and Health. In both Finland and Sweden, the monopoly has in recent years been headed up by rather prominent political figures—the director in Finland served as a minister of social affairs and health in the early 1970s. The director of Vinmonopolet in Norway has been more a nonpolitical manager.

ALKO casts a formidable shadow across the Finnish economy. The revenues derived from ALKO constitute almost 8 percent of all state revenues. (The comparable figure for the United States is 2.2 percent of general federal revenues and roughly 1 percent of state and local revenues.) Furthermore, ALKO also has substantial restaurant and hotel holdings: as much as 10 percent of all hotel rooms in Finland are reportedly owned by the monopoly. ALKO is not just a retailing organization like the American state monopoly systems; it is also a substantial manufacturer of alcohol beverages.

ALKO's size and prominence in Finland seem to intensify real contradictions among the three statutory purposes of ALKO: to manufacture and sell alcoholic beverages, to raise revenues, and to hold down consumption. An important new factor for ALKO has also entered into alcohol policy within the last decade. This is the emergence of an "incomes policy" in Finland, a process of negotiation in which the government, the unions, and business attempt to adjust prices, wages, and salaries. The upshot is a rather direct pressure on ALKO from the trade unions to restrain price increases in order to moderate the impact of inflation.

The current director of ALKO, Pekka Kuusi, is a social scientist who joined ALKO shortly after World War II. One of his earliest research activities was the completion of one of the first national sample surveys in Finland, a survey of Finnish drinking habits (Kuusi 1948). His most ambitious alcohol policy research project was an experiment to measure the impact of making beer and wine more available in rural market towns. At that time all alcoholic beverages were prohibited in the rural communes as part of the original legislation establishing the postprohibition alcohol control system. In fact, this rural prohibition extends back to 1902.

The Alcohol Legislation Committee was formed by the Parliament in 1948, and its report in 1951 prompted wide public discussion. The anticipation of this policy debate was a key factor in the establishment by ALKO of the Finnish Foundation of Alcohol Studies in 1950 as well as in ALKO's support for Kuusi's study of rural prohibition. It was during this period that ALKO independently liberalized wine sales by no longer requiring identity cards for purchasing lighter wines (Kuusi 1957).

The changes recommended by the Alcohol Legislation Committee—the end of rural prohibition and the liberalization of beer—would overturn established traditions. Rural prohibition predated national prohibition and continued after its repeal. State monopoly stores were permitted only in urban areas or market towns; since 1902 no sale of alcohol had taken place in rural areas in Finland, which as late as 1960 encompassed 60 percent of the Finnish population. Thus, when the Alcohol Legislation Committee recommended in 1951 that beer (with no more than 3.5 percent alcohol by weight) be sold in retail shops and licensed restaurants, this modest recommendation stood to overturn an unbroken 50-year-old policy. (This change did not actually occur until 18 years later. The Alcohol Act was passed in 1968 and went into effect on January 1, 1969. In fact the act went beyond the 1951 recommendation, as we shall see below.)

THE FINNISH DRINKING STYLE

The way in which Finns define their alcohol problem—the Finnish drinking style—is intimately connected to the policy debate over making lighter beverages more available. An excerpt from a 1958 study by the eminent Finnish sociologist, Erik Allardt, might help clarify the Finnish view of their alcohol problems (Allardt 1958, p. 15):

> The consumption of alcohol in Finland is smaller that in most European countries. . . . More than half the adult population uses alcoholic beverages less often than once a month and the most common pattern is to use alcohol a few times in a year. In spite of this low rate of consumption drunkenness is a comparatively common phenomenon. In 1950 the rate of arrest for drunkenness was about ten times higher in Finland than in Sweden and three times higher than in Norway. There is evidence showing that Finnish immigrants in the United States in the early nineteen-twenties had rates of arrest for drunkenness higher than any other immigrant group. . . . Studies of Finnish drinking habits reveal that the Finns drink at irregular intervals, mainly on Saturdays and Sundays, and they are apt to drink great quantities at a time. The consumption of distilled spirits is heaviest, while the consumption of wines and brewed beverages is much smaller. The drinking of alcohol is to a great extent regarded as a means of getting drunk.

The Finnish drinking style is seen as the inclination to engage in drinking bouts for the explicit purpose of serious intoxication, interspersed with rather long periods of relatively little or no use of alcohol. Survey researchers in Finland have been mainly interested in how much an individual drinks on a particular occasion—with relatively little interest in constructing detailed typologies of drinkers. Consequently, in Finland there has until recently been very little interest in estimating the number of "alcoholics." By contrast, in the United States, the paradigmatic definition of alcohol problems has been the long-term and chronic ingestion of alcohol, particularly manifesting itself in deteriorating health, career, and family life among middle-aged men. Intoxication and public drunkenness are clearly alcohol problems, but the central problem of alcoholism is "loss of control" over alcohol, which for all practical purposes means heavy, chronic use of alcohol for very long periods of time. This is regarded as the problem of a distinct minority of individuals, although that minority is quite large in number due to the large size of the United States.

Until the late 1960s, Finnish aggregate or total consumption was among the lowest in the world. It still remains in the middle range despite the notable increases during the 1960s and the 1970s. Despite this modest overall consumption, researchers in Finland and elsewhere have commented on Finland's relatively high per-capita arrests for drunkenness. Southern European countries drank much more per capita than Finland yet had much lower arrest rates for drunkenness (Ahlström-Laakso and Österberg 1976). Thus, a point of view emerged that an increase in consumption, specifically of wine and beer, might lead to learning more moderate drinking styles and a decline in arrests for drunkenness.

THE FIRST TEST FOR THE DEBATE

The debate over the existing alcohol control system and the peculiarities of the Finnish drinking style provide the introduction to the first test of the effectiveness of liberalizing alcohol measures—the rural prohibition experiment of the early 1950s (Kuusi 1957).

The introduction to Kuusi's report on this field experiment addresses alcohol policy in the postwar period from the perspective of those individuals and groups who sought to modernize and develop Finnish society. Kuusi argued that the increase of state power to pursue collective security in many areas is a necessary attribute of modern social democracies. But Kuusi warns of the need to remove excessive restrictions in the social sphere, to reevaluate the necessity of burdensome traditional and moralistic restraints that may be counterproductive. Kuusi is clearly taking the cultural modernist view, referring to these restraints as categorical restrictions.[2]

Regarding the controversy over making alcohol available in rural areas and making certain forms of beer more available in retail shops, he lists six different hypotheses set forward regarding the results of such change:

> (1) the number of temperate persons will decline; (2) the frequency of drinking will increase; (3) a shift in the consumption patterns will occur to the advantage of the light beverages; (4) the use of illicit liquors will decline; (5) the quantities consumed at one sitting will decline; (6) drunkenness and breach of public peace will become more common.

Kuusi notes that those in favor of temperance tended to stress the first and second hypotheses—temperance will decline and drinking frequency will increase. The state alcohol monopoly favored the third and fourth hypotheses, predicting a shift to light beverages and a decline in the use of illicit beverages. Kuusi noted that the division of opinion on

[2] Kuusi (1957, pp. 2-3) argues that, while on balance the trend of modern social policy has been to stress social organization and collective approaches, it is often the case that the boundary line between the demonstrated needs of individuals for security from risk and their need for freedom must be carefully assessed. He goes on to suggest in rather veiled terms that sometimes this line can be drawn too heavily on the side of security and indeed may be done so to safeguard the interests of "certain groups."

the expectation of a decline in temperate persons and an increase in frequency of drinking was not so large as it might seem, since both sides agreed that this might happen, with those favoring change arguing that more drinking and more frequent drinking were actually occurring than the temperance groups seem to believe. The sharpest point of disagreement centered around the shift in consumption to light beverages.

The experiment took place in four dry market towns located in rural areas. Two were control towns: beer and wine only were made available through new state-owned stores. A survey was made in 1951 of the experimental and control towns before the opening of the stores and followed up with the measurements taken in the period 1952-1953 to determine the effects of the change (Kuusi 1957).

In general, Kuusi found that those who used alcohol very little were little affected by the opening of the experimental stores; the use of beer increased sharply in some areas (particularly in those communities in which drinking frequency was already higher); and the use of illicit liquor declined. Kuusi also found only slight indications of an increase in the amount of alcohol consumed at a given occasion. There was no appreciable increase in public drunkenness, and, indeed, Kuusi questioned the relation of the control system to the frequency of excessive drinking.

The results of the experiment were somewhat frustrated by changes in alcohol policy put in place by ALKO at the beginning of the experimental period. However, the results were interpreted by Kuusi as broadly supportive of the liberalization position, although he states this position very carefully and tentatively.

THE 1969 ALCOHOL ACT: THE SECOND TEST

It is not possible to reconstruct here the complex forces that led to the 1968 policy changes followed by the 1969 Alcohol Act, but for our purposes the broad outlines are of interest. The central point is that there were substantial forces for liberalization within Finnish society, and some of those in a strong position to influence policy were in favor of these changes. It is clear that sentiment strongly favoring a less restrictive alcohol policy grew during the 1960s among the leaders of ALKO, among the experts at the Finnish Foundation for Alcohol Studies, and among the general public. In the opinion of these experts, the key was the sharp social and demographic changes occurring in postwar Finland, accelerating during the early 1960s, and culminating in the election of 1966. The left generally did well in this election, and the new government was a coalition of the Center Party, the Social Democrats, and the party of the Finnish Communists.

This election was seen as a pivotal one, especially among young people in Finland, as it represented something of a break with the past, a strong commitment to growth, and a promise to liberalize Finnish society. A very popular book of this period was Pekka Kuusi's *A Social Policy for the Sixties: A Plan for Finland* (1964), which called for a comprehensive plan of social provision for the nonproducing sector (the aged, the poor, and the handicapped) in order to maintain expanding, aggregate economic demand. These ideas of development and deliberate stimulation of growth, coupled with a growing climate of permissiveness and a much younger population, seemed to collide with what was perceived as a very restrictive state alcohol policy.

Temperance organizations within Finland were opposed to these changes. The Nordic countries have had a rather strong temperance history. While membership has declined in the post-World War II period, this decline has been somewhat offset by the tendency of temperance members to be active in politics. Also, the temperance organizations have in a sense become part of the bureaucracy, in that they are subsidized by the State Ministry for Social Affairs and Health directly and through the operation of local temperance boards, which serve mainly as alcohol education agencies at the communal level.

The forces for liberalization easily carried the day, and the statutory changes that are of interest here are these: restrictions were abolished on the number of rural areas in which alcohol was permitted (in 1968 nearly 50 percent of Finland lived in areas in which no state store was permitted), the state alcohol monopoly's policy toward the licensing of restaurants was changed so as to permit alcohol (especially beer) to be sold in more restaurants, and the legal drinking age was reduced from 21 to 20 years for distilled beverages, 18 for light beer, and medium beer was allowed to be sold in retail shops.

THE AFTERMATH OF LIFTING RESTRICTIONS

What were the results of this much-debated, long anticipated change? As a parliamentary commission was later to report, the effects of these changes took virtually everyone by surprise. First of all, and perhaps most spectacularly, total per-capita consumption of alcohol in beverages increased by 46 percent (1.33 liters) during the first year. Most of this initial increase (1.17 liters) was accounted for by the 124-percent rise in beer drinking. The consumption of spirits also rose in 1969 by 0.15 liters of alcohol, then continued to rise at a steady 0.20 (approximately) liters per year through 1975; the rise in beer consumption continued at a much lower rate of 0.05 liters per year after 1969. By 1975 the use of

TABLE 1 Per-Capita Consumption of Alcoholic Beverages in Finland, 1951–1975, in Liters of Absolute Alcohol

Year	Distilled Spirits	Wines	Beer	Alcoholic Beverages, Total	Retail	Catering
1951	1.36	0.10	0.32	1.79	1.33	0.46
1957	1.15	0.24	0.33	1.79	1.29	0.43
1962	1.42	0.26	0.43	2.11	1.63	0.48
1968	1.43	0.51	0.94	2.88	2.19	0.69
1969	1.58	0.52	2.11	4.21	3.09	1.12
1972	2.19	0.63	2.28	5.10	3.69	1.41
1975	2.81	0.97	2.41	6.19	4.73	1.46

Source: Ahlström-Laakso and Österberg (1976, p. 9). Reprinted by permission.

beer had increased a cumulative 156 percent (0.147 liters) since 1968; spirits 96 percent (1.38 liters), wine 87 percent (0.46 liters) (see Table 1). Total consumption reached 6.19 liters in 1975. In recent years alcohol consumption has apparently stabilized. The sustained increase in strong alcoholic beverage use in the intervening years is given an interesting interpretation by the Finnish researchers, as we shall see below.

The changes in availability of medium beer with the new law were radical and sudden. The number of applicants for licenses to retail medium beer was considerable, and ALKO granted a license to almost every applicant. The total number of monopoly stores, medium beer shops, and full-licensed restaurants also increased sharply during the first 3 years of the change in the system. These changes are summarized in Table 2.

According to a survey by Mäkelä (1971), the estimated number of drinking occasions during which blood alcohol of drinkers reached at least the 0.10 level increased by about 25 percent. Another researcher

TABLE 2 Finnish Distribution Network for Medium Beer, 1969–1975

	Number of Outlets			
Year	Retail Outlets	Rural Retail Outlets	Service Outlets	Rural Service Outlets
1969	17,431	9,878	2,716	1,195
1971	15,560	9,034	3,406	1,647
1973	13,550	7,374	3,319	1,524
1975	11,965	6,965	3,086	1,393

Source: Ahlström-Laakso and Österberg (1976, p. 6). Reprinted by permission.

TABLE 3 Per-Capita Alcohol Consumption, Finland and Sweden

Per-Capita Alcohol Consumption	Finland		Sweden	
	Liters	Percentage	Liters	Percentage
Total				
1951	2.58		3.49	
1962	3.01		4.03	
1973	7.41		6.73	
Increase				
1951–1962	0.43	16.7	0.54	15.5
1962–1973	4.40	146.2	2.70	67.0

Source: Mäkelä and Österberg (1976, p. 6).

estimated an increase in heavy consumers of alcohol from 61,000 in 1961 to 91,000 in 1969, whereas the increase would be likely to have stopped at 71,000 had the law remained unchanged (Purontaus 1970, cited in Bruun et al. 1975, p. 81).

The sharp increase in consumption during this period in Finland was put in a larger historical perspective by Ahlström-Laakso (1975, p. 1):

> In Finland and also obviously in other countries the concern over the increased consumption among young people is connected with the general growth of alcohol consumption. After abolishing the Prohibition Act in 1933, less than a litre of pure alcohol was consumed a year in Finland. The two-litre limit was broken after three decades in the first half of the 1960's. From the year 1968 to the year 1973, the consumption of alcohol again doubled. This took five years; in other words, one-sixth the time that it took for the previous consumption to double. Finns consumed 5.6 litres of pure alcohol per capita in 1973.

This dramatic rise in consumption during the 1960s and early 1970s was not restricted to Finland; it happened in most industrial countries, especially those with a relatively low aggregate consumption (Sulkunen 1976). But the percentage rise in Finland exceeds that of any comparable country (Keller and Gurioli 1976). Mäkelä and Österberg (1976) have compared Finland with Sweden during the period 1951-1973. Their research reveals that Finland's growth in per-capita consumption paralleled Sweden until the Finnish liberalization in 1969; after this change the Finnish level rose to a point higher than that of Sweden during the early 1970s (see Table 3).

MORE RECENT SURVEY REPORTS

Jussi Simpura of the Social Research Institute of Alcohol Studies provides us with a very complete picture of changes that occurred in 1969

TABLE 4 Data on Previous Week's Use of Alcohol by 15- to 69-Year-Old Finns in 1968, 1969, and 1976

	Year	Women	Men
Alcohol consumer (% of total)	1968	57	87
	1969	65	91
	1976	80	91
Weekly drinking instances per alcohol consumer[a]	1968	0.5	0.9
	1969	0.8	1.5
	1976	0.7	1.4
Average consumption of alcohol in a single instance of drinking (centiliters 100-percent alcohol)[a]	1968	3.0	7.3
	1969	3.2	6.2
	1976	4.4	8.4
Average weekly consumption of alcohol (centiliters 100-percent alcohol, all respondents)[a]	1968	0.88	5.96
	1969	1.66	8.37
	1976	2.52	10.32

[a] The results are based on drinking instances within the week preceding the interview and represent a typical autumn week of the year.
Source: Simpura (1978). Reprinted by permission.

and subsequently. Simpura in 1976 essentially replicated the two earlier national surveys conducted by Mäkelä in 1968 and 1969. The few key findings for the alcohol policy debate in Finland were (Simpura 1978): from 1968 to 1975, abstinence rates among women fell sharply and, while there was some slight decline for men abstainers in the first year of the new legislation, the number of male abstainers remained constant thereafter (see Table 4).

The mean or average drinking instance changed in a rather surprising way. Not unexpectedly, the average number of drinking instances increased rather sharply in the first year; this number then declined (very slightly) over the next 6-year period.

Recall that from the standpoint of the ongoing alcohol policy debate in Finland, an increase in drinking frequency was regarded by many as a sign that drinking was becoming more moderate and that lighter drinks (such as beer) were being consumed.

Simpura (1978) speculates that there actually was an increase in drinking instances in 1976, but this was not revealed in the survey because medium beer had in the intervening period become defined as a nonalcoholic drink (much like the light beer that was available in Finland in retail shops prior to the change). Thus, he believes respondents were less likely to report the drinking of medium beer in their responses. We will discuss this change further below, but the researchers at the Finnish Foundation are persuaded that medium beer instances are not accurately reflected in the 1976 survey.

There is a decline from 1968 to 1969 in the average amount of alcohol consumed in a single instance of drinking, as the advocates of the change had hoped. However, there was a subsequent rise in the average amount of alcohol consumed on a given occasion beyond the 1969 level. In this longer term (through 1975), the Finnish style of drinking heavily on occasion seems not to have been reduced by the 1969 Alcohol Act.

Finally, as Table 4 indicates, there has been a strong and sustained increase in average weekly alcohol consumption during this period, from 5.96 to 10.32 centiliters of alcohol per week for men, and from 0.88 to 2.52 centiliters for women.

One of the more interesting findings of these surveys is the estimate of the number of heavy consumers, defined by Finnish researchers as persons exceeding 1,000 centiliters of alcohol in annual consumption (this is the Finns' estimate of consumption that produces increased risk of cirrhosis). Although there was a rather sharp increase from 1968 to 1969, this figure has remained rather constant since 1969 for men, despite the increase in per-capita consumption during this period. This is especially significant, since a central tenet of the distribution-of-consumption thesis is the strong, positive temporal relationship posited between per-capita consumption and the proportion of heavy consumers.

Simpura (1979) has attempted to determine whether the increase in consumption is attributable mainly to the decline in abstinence, the increase in frequency of drinking, or the amount consumed per occasion. His equation regresses the share of alcohol consumers within the population, the annual frequency of drinking instances per consumer, and the average volume consumed per occasion, on the annual consumption of alcohol. His estimates suggest that "(t)he growth in average consumption in a single instance is the overriding factor increasing consumption among men, and it is also of greater importance for women than the other chief factor, the increasing number of alcohol consumers" (1979, p. 11).

Another perspective on how the new law affected Finnish drinking patterns was explored by asking what proportion of drinking occasions were for the explicit purpose of intoxication. Simpura (1978) reports that in 1969, 58 percent of men's consumption was for the purpose of intoxication, rising to 65 percent in 1976. The corresponding trends for women were from 26 percent in 1969 to 39 percent in 1976.

MEDIUM BEER

The new legislation produced a marked increase in the consumption of medium beer during the first year. Consumption of medium beer then began to stabilize and even decline over the next several years. Perversely (recalling the thesis of the reformers), the consumption of strong liquors began to increase steadily.

To refer again to Table 1 for the changes in per-capita consumption of the three principal forms of beverage alcohol, the overall increase in beer consumption was sharp in the first year and was somewhat less from 1969 to 1975 (2.11 to 2.41). But an analysis of the purchase of medium beer (as opposed to strong beer) shows that its consumption actually declined from 1970 onward. The slight overall increase in the consumption of beer is accounted for mainly by the increased purchase of strong beers in ALKO shops and in restaurants.

The speculation of researchers at the Finnish Foundation for Alcohol Studies is that medium beer sold in retail shops and available in many new restaurants was no longer seen as a true alcoholic beverage. The paradoxical effect was to first increase the consumption of medium beer among consumers, and then to have them move on to the use of stronger beverages purchased in ALKO shops and elsewhere.

PUBLIC DRUNKENNESS

Österberg (1979) and others have sought a more precise understanding of the impact of these legislative changes on alcohol-related problems or damages such as public drunkenness, alcohol-related crimes, highway crashes, etc. Space will permit consideration only of public drunkenness, which was of special significance in the alcohol policy debate, being seen as a key indicator of the Finnish drinking style. Per-capita arrests for drunkenness fell at a fairly constant rate from 1950 to 1958. There was a slight increase in per-capita drunkenness arrests between 1958 and 1967, but the 1967 rate was still smaller than that of 1950. In the next 2 years, arrests for public drunkenness fell by 10 percent, with this change being largely accounted for by a change in the legal codes that removed criminal sanctions for drunkenness. This was a period of generally rather lax arrest activity in Helsinki and elsewhere. It was also the period during which the new legislation took place. We do not, therefore, have an unambiguous picture of the legislation's immediate impact on public intoxication or arrests. From 1969 to 1975, arrests for per-capita drunkenness doubled.

Österberg is careful to point out the many perils in using arrest statistics, particularly national aggregates. He notes that after the changes in legislation arrest policies became more stringent again in Helsinki in 1970, then relaxed somewhat again in 1972. To attain more precision, Österberg examined arrest patterns for Helsinki alone and found that they conformed generally with the national picture. Of course, arrests in Finland's largest city would dominate the national statistics. Nevertheless, arrest practices for one jurisdiction should be more uniform and changes should be more easy to isolate.

Österberg then examined arrest rates as a function of per-capita consumption of alcohol and found that there is a steady overall decline from 1950 to 1975. The arrest rate for 1975 is approximately half that of 1950. If one examines the post-1968 period, a somewhat less critical evaluation of the effects of liberalization can be offered. The liberalization period interrupted a decline from the high point of 1950, with the period of 1969 to 1975 remaining fairly constant (arrest rates as a function of per-capita consumption).

COMPARISONS OF SWEDEN AND FINLAND

Mäkelä and Österberg (1976) have compared Sweden with Finland during the period 1951 to 1963. Both acknowledge the central importance of price and income during the period affected but argue that factors beyond real price changes are necessary to explain the greater increase of alcohol consumption in Finland.

If one assigns the 1951 price level a value of 100, the real price for alcohol in Finland fluctuated in a narrow range between 105 (1970) and 95 (1968). A similar stability in real price for all alcoholic beverages occurred in Sweden, but there were important changes in the relative prices of the different types of beverages in both countries—in general, the relative price of spirits increased.

There are important differences between the two countries in disposable income and consumer expenditure trends. In Sweden, as Table 5 indicates (Mäkelä and Österberg 1976), relative increases in consumer expenditures per capita were greater in Finland, especially during the period 1968-1973. The authors note that between 1965 and 1973, "alcohol's share of Finnish consumer expenditures increased from 4.2% to 7.6%. In Sweden the equivalent values are 4.3% and 5.3%" (p. 37). "The real expenditure on alcohol per capita was 3.5 times greater in

TABLE 5 Annual Percentage Increase in Overall Consumer Expenditures Per Capita

	1951–1962	1962–1973	1962–1968	1968–1973
Finland	3.4	3.7	2.1	5.7
Sweden	2.4	2.0	3.0	0.8

Source: Based on Mäkelä and Österberg (1976, p. 37).

TABLE 6 Total and Per-Capita Increases in Monopoly Stores in Finland and Sweden, 1951–1973

Year	Finland	Sweden
Number of Stores		
1951	80	241
1957	90	260
1962	101	275
1968	132	292
1973	184	309
Inhabitants Per Store (in thousands)		
1951	29	51
1957	28	48
1962	28	45
1968	27	35
1973	26	26

Source: Based on Mäkelä and Österberg (1976, p. 38).

1973 than in 1959, real consumer expenditures per capita had during the corresponding time increased by approximately 75%" (p. 37).

Mäkelä and Österberg argue that both the rise in overall consumer spending and the 1969 Alcohol Act combined to accentuate the increase in alcohol's share of consumer expenditure in Finland. They note in particular the steady rise in the number of state stores (ALKO outlets) per capita since 1950. This process accelerated in the 1960s before the passage of the new act. In 1969 there was a tremendous expansion in the network of restaurants in which alcohol was available (see Table 6). In 1951 there were 51,000 inhabitants per state store, or 80 stores in all of Finland. By 1973 there were 184 stores, or 1 per 26,000. In Sweden the change was less marked, from 241 stores in 1951 (1 store per 29,000 inhabitants) to 309 stores in 1973 (1 per 26,000) (see Table 6).

During the same period the percentage of expenditures for alcohol consumed in restaurants declined sharply in Sweden, from 14 percent in 1951 to 6 percent in 1973. In Finland, on the other hand, this percentage increased from 19 percent in 1951 to 26 percent in 1973. This change occurred mostly after 1968, parallel with the issuing of new restaurant licenses for the sale of alcohol after the 1969 Alcohol Act. At the present time there is rough parity between the two countries, with Finland having 3.4 licenses per 1,000 inhabitants and Sweden having 3 licenses per 1,000 inhabitants.

Mäkelä and Österberg sum up by arguing that the creation within a short period of a dense distribution network in Finland surely has been an independent contributor to the greater increase in per-capita consumption there. There were never large rural areas in Sweden with the kind of prohibition that existed in Finland until the late 1960s. Also, the expansion of restaurants and ALKO shops—a process that began before the 1969 act—along with the strong increase in consumer expenditures per capita helped to accentuate the changes in Finland.

EVALUATING THE IMPACT OF THE 1969 ALCOHOL ACT

Finnish researchers at the Finnish Foundation for Alcohol Studies (and other key figures such as ALKO officials) appear unanimous on three central points regarding the impact of the 1969 act. First, the changes following the 1969 act were more dramatic and much more negative than most people had predicted. Second, there is some consensus that the worsened situation in the mid-1970s regarding alcohol problems cannot fairly be attributed to the 1969 Alcohol Act alone. On this second point, there is substantial agreement at the Finnish Foundation for Alcohol Studies that, while the act clearly had an independent and significant impact, the 1970s would likely have experienced a worsening of many problems without it. Finally—there is less consensus on this point from the temperance movement—the 1969 act was a more or less inevitable product of well-nigh irresistible forces for cultural change in Finnish society.

The researchers point in particular to at least three crucial factors that must be understood about Finnish society before we can assess the impact of these legislative changes: first, the widespread demographic, social, and economic changes in Finnish society since World War II; second, important changes in drinking styles and behavior (particularly rates of abstinence) already under way in 1968 (these began slowly in the 1950s and accelerated during the 1960s); and third, the changes in the political and cultural climate during the 1960s; in particular the election of 1966 brought in a major change in government. A coalition between the Center Party and the Social Democrats also marked the first participation of the Finnish Communist Party in government since the late 1940s.

While we review the first two of these changes, it is worth noting that these forces for change in Finnish society have a very familiar ring. These changes—although at a much later time and with a more rapid pace—mirror the cultural struggles that seem common to advanced industrial societies in which local, rural, and traditional values conflict with an emerging order that stresses consumption, leisure, and emancipation of the individual from community restraints.

This process of modernization has certainly left its traces in alcohol policy debates throughout Western societies. Indeed, as Gusfield (1963, 1968) has suggested, the modern temperance movement in the United States was shaped in many ways by the upheavals caused first by industrialization and finally by the maturing of the American economy and the associated emergence of the modern consumer ethic to challenge the traditional work ethic. This is the difference between what Gusfield calls "cultural fundamentalists" and "cultural modernists."

Gusfield's categories are helpful in framing the alcohol policy debate in Finland, sharply illuminating the research and findings of the Finnish Foundation and other groups active in alcohol policy. Gusfield defines the two central types of modernists and fundamentalists as follows (1963, p. 136):

> The cultural fundamentalist is the defender of tradition. Although he is identified with rural doctrines, he is found in both city and country. The fundamentalist is attuned to the traditional patterns as they are transmitted within family, neighborhood, and local organizations. His stance is inward, toward his immediate environment. The cultural modernist looks outward, to the media of mass communications, the national organizations, the colleges and universities, and the influences which originate outside the local community. Each see the other as a contrast. The modernist reveres the future change as good. The fundamentalist reveres the past and sees change as damaging and upsetting.

As Gusfield (p. 142) points out, alcohol has figured prominently in this central cultural struggle within the Western industrialized societies, particularly those with a strong Protestant tradition. As Gusfield argues, the culture of mass society has been a culture of "compulsive consumption, of how to spend and enjoy, rather than a culture of compulsive production, of how to work and achieve" (p. 142).

As Gusfield and many others take great care to point out, the distinction between fundamentalism and modernity can be worked too hard (Gusfield 1967, Vidich and Bensman 1958, Bendix 1964). There is a tendency among some writers to speak of modernity as meaning roughly industrialization, mass society, posttraditional society, etc. But Gusfield is talking specifically about the rejection of Protestant values of hard work, frugality, and abstemiousness, which have buttressed development of Western industrial societies. Modernist values are based on the rejection of the work-and-save values of the industrial revolution and extol instead the values of leisure, personal consumption, and the emancipation from limits in matters of personal conduct (e.g., smoking, the use of alcohol and, lately, other drugs, and sexual conduct).

Finnish society provides a rather striking glimpse of the rapidity with

TABLE 7 Distribution of the Finnish Labor Force in 1950, 1960, and 1970

	1950	1960	1970
Agriculture and forestry	46	36	20
Industry and construction	27	30	34
Commerce, transport, and services	27	34	46

Source: Ahlström-Laakso and Österberg (1976, p. 9). Reprinted by permission.

which these changes may occur. As Ahlström-Laakso and Österberg indicate (1976, p. 9), and as Table 7 shows, the shift toward urban areas is striking:

In 1951 about 67 percent of Finns lived in the countryside, while the corresponding share in 1960 was about 60 percent and in 1973 about 43 percent. Migration has been very heavy. Moreover, there has been a rapid movement in the countryside from remote villages to population centers, and in the cities from the center to the suburb. Migration has been connected with changes in the vocational structure of population. . . .

The comparison between Finland, Sweden, Great Britain, and Hungary for the period 1950 to 1970 for the changing mix of the agricultural, industrial, and service sectors is even more revealing (see Table 8). These data show that the shift from primary to secondary and tertiary employment occurred in Finland at much the same time as opposed to other Western countries.

As was noted above, this period was also marked by a dramatic increase in real incomes. Ahlström-Laakso and Österberg report (1976, p. 10) that private consumption expenditure per capita increased by about 44 percent in 1951-1962 in real terms and by about 52 percent in 1962-1973. Disposable income also increased very rapidly during the

TABLE 8 Distribution of the Work Force for Selected European Countries, 1950, 1960, and 1970

	Agriculture			Manufacturing			Services		
	50	60	70	50	60	70	50	60	70
Finland	46	35	20	27	31	34	27	34	46
Sweden	21	14	8	41	46	40	38	40	52
Great Britain	5	4	3	48	49	46	47	47	51
Hungary	54	38	25	22	34	45	24	28	30

Source: These data were furnished by the Research Group of Comparative Sociology, University of Helsinki, Helsinki, Finland.

1960s and early 1970s, the 5-day work week was introduced during this period, and social legislation substantially improved the social security of the average Finnish citizen.

THE RISE OF THE "WET GENERATION"

Pekka Sulkunen of the Social Research Institute has reanalyzed (1979a, b) the four basic Finnish national drinking surveys undertaken by Kuusi in 1948, Mäkelä in 1968 and 1969, and Simpura in 1976 (see Kuusi 1948), relying primarily on abstinence rates and age at first drink as indices of major cultural change. What emerges is a revealing picture of a society undergoing a major cultural and social transformation in a very short period of time and its impact on drinking styles.

At the outset Sulkunen (1979a, p. 2) warns against the temptation to resort to vague notions of modernization or growth to explain the rapid increase in per-capita consumption of alcohol and associated problems in Finland. Sulkunen challenges the idea that increased leisure time, increased income, or urbanization necessarily or automatically brings in its train increased consumption of alcohol (1979a, pp. 1-2). Sulkunen's statements are reminiscent of the cautions that students of modernization and postindustrialism have advanced about the use of such terms (Gusfield 1968, Bendix 1964). Modernization or the occurrence of a postindustrial shift toward a service economy is not a polar extreme from industrialism. The matter is more complicated.

Sulkunen's very detailed and exhaustive analysis of these four surveys constitutes a retreat from his earlier work on the growth of consumption among western societies. His key finding about the postwar period in Finland has been the emergence of a "wet generation." This became especially noticeable in the 1960s. However, there is evidence that these changes, especially drinking at younger ages, began much earlier.

Table 9 reflects the declining rates of abstinence between 1946 and 1976. Sulkunen argues that this decline began earlier than the 1969 Alcohol Act, yet one is struck by the sharp decline in female abstinence after 1969. Table 10 demonstrates that the change in drinking habits has been most noticeable among white-collar groups and increased greatly during this period. Table 11 illustrates that the average age at first drink has declined appreciably and that it is occurring equally early for rural and urban drinkers.

Sulkunen's conclusions are complex and interesting. He argues that strong economic growth during the 1960s may have given impetus to the increase in consumption during that period, but that the more important changes were generational and cultural in nature. The generations born

TABLE 9 Last Drinking Occasion of the Finnish Population over 19 Years, in 1946, 1968, 1969, and 1976

	Percent of the Population			
	Never	Month Ago or More	More Recently	N
Total Population				
1946	19	30	51	2,891
1968	19			1,663
1969	15			1,570
1976	9	20	70	2,601
Women				
1946	31	33	36	—
1968	32			420
1969	25			399
1976	15	25	60	1,326
Men				
1946	5	26	69	—
1968	4			1,243
1969	3			1,171
1976	4	16	80	1,275

Source: Sulkunen (1979a, p. 11). Reprinted by permission.

after World War II present a sharp rate of change in contrast with the generations that preceded in terms both of abstinence rates and age of initial alcohol use.

An extended quotation by Sulkunen (1979a, pp. 36-37) helps clarify this point:

Drinking by adolescents seems, in the light of the preceding observations, to be a phenomenon that began to emerge among those who were born in 1916-1925 and who reached their adolescence after the war (approximately 1930-1946). Kuusi (1948, 111) estimates that probably more than 25% of the males in this age group had their first drink in the army or in the war.

Commenting on these data, Kuusi wondered in worry: "Is it too early now to assess whether this practice of boozing at a premature age will be found as a transitional aftermath of the war, or will it develop into a permanent custom? In any case, this is a phenomenon that deserves continuous attention" (Kuusi, 1948).

After thirty years, the answer is somewhat distressing. Boozing at a premature age did not only develop into a permanent custom. It developed into a practice of boozing in childhood. In fact, in 1976 drinking at the age of 15-19 was no less common that at the age of 20-29, if measured by the proportion of those who had ever taken a drink.

TABLE 10 Last Drinking Occasion of Finnish Population over 19 Years by Occupational Position, in 1946 and 1976

Occupational Position	Never	Month Ago or More	More Recently
Unclassified or unknown			
1946	—	—	—
1976 (N = 117)	9	24	67
Farmers			
1946	22	33	45
1976 (N = 415)	21	26	53
Farm workers, etc.			
1946	23	32	45
1976 (N = 57)	14	16	70
Industrial workers			
1946	17	28	55
1976 (N = 1,033)	9	20	71
Lower white-collar			
1946	16	30	54
1976 (N = 670)	5	19	76
Upper white-collar			
1946	10	24	66
1976 (N = 309)	3	16	81

Source: Sulkunen (1979a, p. 14). Reprinted by permission.

Sulkunen does not so much argue that this "wet" generation drank more on average than did the earlier generation but rather that it tended to be more likely to drink and to begin drinking at an earlier period.

Sulkunen pays particular attention to the decline of abstinence among older women in Finland (1979a). He argues that there is tentative evidence that the younger group in turn encouraged the older persons to drink, especially older women. Sulkunen regards these differences in drinking styles as truly generational rather than age differences.

Sulkunen's assessment of the meaning of declining abstinence rates and a sharp increase in the numbers who use alcohol at an early age can be summarized as follows: Any evaluation of the 1969 Alcohol Act and its consequences must take into account a post-World War II sea change in the general cultural and political climate regarding alcohol. The 1969 Alcohol Act is the culmination of postwar values. Abstinence began to decline sharply from the start of this period especially during the 1960s, among young women and teenagers of both sexes. Sulkunen does not deny that the 1969 act had an independent impact on this trend toward earlier drinking and lower rates of abstinence. In fact, if one carefully

TABLE 11 Changes in Age of First Drink for Finns in 1946 and 1976 by Rural-Urban Residence

	1946		1976	
Age at First Drink	Rural	Urban	Rural	Urban
–15	8	8	24	26
16–17	19	22	42	46
18–20	42	47	68	74
21–	68	79	89	92
Never	20	14	11	8
Don't know	12	7	—	—
TOTAL	100	100	100	100
N	—	—	1,056	1,545

Source: Sulkunen (1979a, p. 32). Reprinted by permission.

scrutinizes the data that Sulkunen relies on, one can note that the declines in these abstinence rates became more marked and noticeable in the period following the 1969 act.

One might quarrel with some details of Sulkunen's interpretation. Clearly the impulse toward liberalization found widespread support among the postwar generation. Yet it is unlikely that this group had a decisive political voice in the mid-1960s, even if one acknowledges the great increase in participation of the young in that election. We must remember that this period was one during which many factors contributed to a more liberal climate, and that many key figures in the alcohol policy debate favored these changes.

There is room for disagreement as to what weight to assign to cultural changes in drinking habits, the rise in consumer expenditure for alcohol, and the increased availability brought by the 1969 act, in evaluating the subsequent rise in consumption and associated problems. But the principals in the alcohol policy debate see all three as factors, differing only in which they would emphasize.

SUMMARY AND CONCLUSIONS

The most striking fact to an outsider speaking with various participants in alcohol policy in Finland is the consensus they present on the debate's central issues. Officials of ALKO and the Department of Social Affairs and Health, Finnish Foundation researchers, and leaders of the temperance organizations appear to agree about several key issues.

First, there is widespread consensus that the impact of the 1969 act was on balance not favorable and that, while many factors beyond the act must be taken into account in explaining the rise in consumption in Finland after 1969, the act itself was directly responsible for a significant fraction of the increase in problems.

Paradoxically, there is at least tacit acceptance by almost everyone that, while the legislative changes were probably introduced too abruptly and with too much optimism, there was little likelihood that these changes were avoidable by any government during this period. There is the possibility, however, that despite support for liberalization during the post-World War II period, these changes might have been introduced more gradually. There were similar changes in the other Nordic countries, and the move toward liberalization and the left in Finland was very popular with a young and very vocal generation whose values were taken up by the major parties in Finland. Reforms of the existing alcohol system were one of those changes that enjoyed wide popularity among the general public.

It is important to note that there is also something of a consensus on just why these adverse consequences occurred. A collective process of social learning (Heclo 1974) seems to have occurred among those responsible for or attentive to alcohol policy.

Perhaps the most widely shared area of interpretation is the addition hypothesis. Nearly everyone on the Finnish alcohol policy scene now agrees that it is not so easy to simply replace existing drinking patterns or structures with new ones, particularly by the process of enlarging the availability of lighter beverages. Thus the earlier optimism of the Finnish Alcohol Legislation Committee, which in 1951 recommended making lighter beverages more available to stimulate the substitutions of beer and wine for spirits, seems to have been the central casualty of the aftermath of the 1969 act. Further, there is endorsement by all of the principals that per-capita consumption must itself be a central feature of Finnish alcohol policy. The parliamentary committee formed to analyze the impact of the liberalization episode took the central goal to be not only controlling consumption but also lowering it (Report of the Alcohol Committee 1978). The main instruments for this goal are to be pricing policy and restrictions on availability (mainly Saturday closings for ALKO and stricter licensing of beer establishments).

In a larger sense, the Finnish case represents something of a shift in the conflict between Gusfield's cultural modernism and cultural fundamentalism, and perhaps the emergence of a new perspective in this longstanding conflict.

The postwar period in Finland had speakers for both viewpoints. The research community at the Finnish Foundation and the policy makers at ALKO generally endorsed a cultural modernist viewpoint. They not only argued for liberalization of alcohol control policy, but they also voiced this support in terms of the changing cultural values and climate of the times (see Kuusi 1957, pp. 1-3). The cultural fundamentalists resisted these moves, sometimes in terms of the traditional values of abstinence, at other times in terms of the researchers who wrote darkly of the Finnish hereditary type—types who were given over to compulsive, explosive drinking that all too often resulted in violent outbursts.

The disputants have moved somewhat closer to one another in this debate. It is misleading to depict this episode as one involving contending interest groups of roughly equal size, especially today. By and large the groups who are concerned with alcohol policy are quite small in Finland, and alcohol policy issues—while enjoying wide coverage in the press—have become defined as a secondary problem in Finnish society. The terms of the debate narrowed considerably under the new consensus; the issues became much more matters of technique and detail rather than broad policy.

It seems that all sides have recognized (with different degrees of regret) that times have fundamentally changed in Finland. Most parties seem to sense the permanence of postindustrial society and its rejection of temperance and abstinence.

At the same time, all sides now realize that modernity is not a process of replacing old values and structures with new and improved ones. Rather modern and traditional values regarding alcohol tend to coexist cheek-by-jowl with one another; traditional patterns of drinking are not replaced, but endure and are even magnified. One is reminded of Joseph Schumpeter on the process of economic modernization (cited in Bendix 1964, p. 10):

> Social structures, types and attitudes are coins that do not readily melt. Once they are formed they persist, possibly for centuries, and since different structures and types display different degrees of ability to survive, we almost always find that actual group and national behavior more or less departs from what we should expect it to be if we tried to infer it from the dominant forms of the productive process.

The dominant form of the productive process to which Schumpeter refers can be for our case that the shift is a predominance of rural to industrial occupational structures undergirded and overlaid by the values of localism, religious fundamentalism, and an ethic of work productivity, thrift, and abstemiousness. This shift, which has been occurring for quite a number of decades in many Western societies (and for a very much shorter period in Finland), is defined as a marked decline in agricultural labor, then stabilization of the industrial sector, and finally rapid growth in the service sector (happening in Finland almost simultaneously). The traditional values are ultimately displaced by the values of consumption, leisure, and emancipation from restrictions on such matters as drinking, sexual conduct, and other areas of private life.

The current period in Finland represents something of a retreat from the heady optimism of the 1960s and not just for alcohol policy. In fact, the rate of growth has slowed considerably, and the general debate concerning the wisdom of rapid, continuous growth espoused by nearly all parties during the 1960s is heard in Finland as elsewhere. Further, there is assimilation by the younger generation (and likely not just because they are becoming older and the normal age-related decline in rates of alcohol consumption has begun to set in).

These changes have brought something of a cultural reassessment and social retrenchment. A central tenet of cultural modernism is that modern men and women should be free from the burden of tradition. Though there are clear differences between societies, the triumph of the individual from moral restraints has been a broad cultural theme of the West. This triumph occurred at the same time as general social security provisions for the entire society were being strengthened, and society's share of consumption in the form of taxes and control of economic activity increased dramatically. Nevertheless, for the modernist the individual was to be free of restrictive social norms aimed at closely shaping his or her personal life-style.

It would be a serious mistake to see the shift in Finland regarding alcohol policy as a return to earlier traditional positions. Those who supported the liberalization measures have on balance prevailed. The principle of individual freedom from onerous and restrictive alcohol policy measures anchored in conflicts over societal status has been firmly established.

The old traditional-modern split seems to be in a process of replacement by one that represents different issues in postindustrial culture. In that culture, the fundamental value of individual autonomy in personal life-style choices remains solidly established, and the goal of closely determining by legal mechanisms styles of drinking has been firmly rejected. The central goals for alcohol policy are to shift from proscribing drinking that is seen as harmful to relying on macro policy and regulatory mechanisms to prevent harmful aspects from worsening.

One of the surprising implicit areas of agreement that can be faintly discerned in Finland is the gradual recognition of the benefits of the Finnish style of drinking. One of the virtues of episodic drinking is that it minimizes adverse impact on productivity and also minimizes the effects of more sustained or chronic ingestion on the liver and other physical structures. This view was expressed both by the temperance movement representatives and alcohol policy researchers.

This is an interesting development. It represents a modified acceptance of the goal of freedom for the individual regarding matters of alcohol use and abandonment of the attempt to directly shape the modal pattern of drinking in Finland. The traditional and predominantly rural values that stress abstemious if not abstinent conduct have begun to disappear. (The parliamentary committee did, however, suggest that as an educational goal the government should attempt to reduce "admiration for alcohol, and especially drunkenness.") At the same time, the belief that simply liberalizing measures (and the disappearance of the earlier stern attitude) would result in an improved situation for alcohol problems has gone by the board.

What remains is the commitment to liberating individuals from an excessively restrictive alcohol policy (e.g., rural prohibition) as a basic principle and, paradoxically, the reappearance of the necessity for broad aggregate control mechanisms. There is now a growing emphasis on health and safety issues, which apparently is enlarging the previous preoccupation with the Finnish drinking style. The attempt to find a better marriage between alcohol and larger economic policy, especially income policy, is also a new central part of the alcohol policy debate.

The question of medium beer represents something of a problem for alcohol policy makers, with the temperance groups anxious to find some way to retreat from the current widespread availability. But one senses that medium beer has become a symbolic issue; other than a general tightening of controls over licensing, the removal of medium beer from the retail shops will probably not occur. This debate in the Parliament continues at this writing, with the trend seeming to be one in which the government is seeking to gain more control over the policy-making apparatus of ALKO, primarily by strengthening the functions of the administrative board. The future of alcohol policy in Finland seems to be an attempt to sort out the new boundaries between personal freedom and interest in Finnish society so that problems with alcohol can be held to as low a level as the recognition of this freedom (and other values) permits.

IMPLICATIONS FOR THE UNITED STATES

At the first glance, the debate in Finland over alcohol policy seems remote and over issues that do not concern the United States. The federal government's power over alcohol policy is a great deal more limited here than in Finland, and the debate here is conducted in different terms. The evidence that liberalization did have a discernible impact on alcohol problems is important for the growing discussion of alcohol policy in the United States, but it does not in itself constitute decisive evidence. Even in monopoly states there is little opportunity to avail ourselves of these same policy measures.

The principal value of the Finnish experience, beyond the light it sheds on general issues regarding alcohol availability and other control measures, may well lie in providing a new conceptual context for the ongoing debate regarding alcohol and alcohol policy in the United States.

In the United States, policy measures to restrict the availability of alcohol or to influence per-capita consumption are seen by some as a retreat from the values of cultural modernism and a return to rural, traditional viewpoints—viewpoints currently characterized as "neoprohibitionism." This characterization might well have applied if these issues had been raised in the period following repeal. Then a call by the "drys" for a return to, if not prohibition, at least a much more restrictive control system for alcohol could fairly be interpreted as a victory for cultural fundamentalism and the "drys."

But almost a generation has passed since repeal, and the post-World War II generation in the United States grew to maturity protesting United States involvement in Vietnam, advocating civil rights, and urging less restrictive measures for drugs such as marijuana. Alcohol policy per se was off the public agenda in this period of our postindustrial society.

What we have now, however, is the entry of several new factors in the debate that may well fuel reconsideration of the postrepeal alcohol policy model. What has arisen in postindustrial society is a fundamental concern with the quality of the environment, the social costs of unrestricted growth, inflation, and unchecked consumption. A very broad and diverse body of support has grown up around the ecology movement, and there seems to be little diminution of support for this issue even as its economic impact becomes more manifest. Likewise, the impact of inflation and expensive and depleting energy sources have called into question assumptions about the modal personality of postindustrialism, about consumption as a central and relatively unrestricted right. In the area of health policy specifically, there is increased awareness of the limits of clinical health care and increasing attention to lifestyle and environmental hazards.

This means that in postindustrial culture the privacy and anonymity of the individual has undergone some pressures for alteration. It is true that this debate still contains many lines of the older traditional-modernist conflict. The life-style debate in health policy (Kass 1975, Knowles 1976), for example, is often challenged by some as one more example of attempts by public health reformers to conduct campaigns of moral reform.

The dominant view appears to be that there is an appropriate place for reasonable limits over life-style threats to health. These limits are likely to be established along the lines of the struggle for consumer or environmental protection: strong advocacy against certain features of industrial organization, especially pollution, manufacturing of hazardous products, and advertising practices.

This emerging concern with the quality of life in postindustrial society involves a struggle to find a new language for appropriate limits to consumption behavior, while preserving the gains of postindustrial society in abandoning harmful or counterproductive moralistic restraints. In the preliminary skirmishes between those who advocate a reexamination of alcohol policy and those opposed, the first group attempts to persuade the second that they are not party to a general programmatic return to traditional systems restraint, but instead are part of a general trend within the larger society for heightened concern for environmental protection, safety in transportation, and consumer protection.

There is no evidence currently available that this particular development has occurred in Finland, nor is it necessary for this argument. The shifts in alcohol policy in Finland can be seen as preliminary skirmishes in a larger cultural struggle to define the postindustrial values, particularly for alcohol. In the earlier period in Finland, the struggle between traditional and modern values predominated. In the most recent period, this tension has begun to fade (but not to disappear). There seems to be a readiness at least for a discussion of limits to alcohol consumption. This indicates the reliance on tax measures and broad regulatory policy (licensing, limiting ALKO's marketing pressures, education, etc.), but rejection of measures like personal identity cards or a system of sharply restricted availability. At the same time there is heightened appreciation of the health benefits of the traditional Finnish style of drinking and a deepened respect for the inability of available and acceptable social policy measures to influence this radically.

To Americans this suggests that there is significant value in retaining our own present structure of alcohol consumption. Ours is an enormously pluralistic structure that shows great variation from region to region and ethnic group to ethnic group. Nevertheless, a fundamental attribute of the structure is a modal drinker who either drinks not at all or very little indeed.

To retain this overall pattern it cannot, however, be treated as an attempt to closely shape the life-styles for drinkers. That approach, which is the heritage of the cultural fundamentalism, must be rejected, and the modernist emphasis on freedom from close restraint and individual choice must be firmly upheld. Nevertheless, the government and others involved in policy should respect the value of this pattern of alcohol consumption and try to discourage modes of drinking that are markedly divergent.

The policy implication of this for the United States is that a cornerstone of alcohol policy might well take the form of attempting not only to minimize adverse consequences from alcohol consumption, but also to discourage drinking models that attempt to undermine the dominant or modal pattern of limited drinking. This should always be undertaken as an explicit attempt to work out in a new way an appropriate model for alcohol use that is congruent with the values and dominant images of postindustrial culture.

REFERENCES

Ahlström-Laakso, S. (1975) Changing Drinking Habits Among Finnish Youth. Report No. 81 from the Social Research Institute of Alcohol Studies. The State Alcohol Monopoly, Helsinki, Finland.

Ahlström-Laakso, S., and Österberg, E. (1976) Alcohol policy and the consumption of alcohol beverages in Finland in 1951-1975. *Bank of Finland Monthly Bulletin* No. 7. Helsinki: Bank of Finland.

Allardt, E. (1958) Drinking norms and drinking habits. Pp. 7-109. in E. Allardt, T. Markkanen, and M. Takala, eds., *Drinking and Drinkers: Three Papers in Behavioral Sciences*. Stockholm: Almquist and Wiksell.

Bendix, R. (1964) *Nation-Building and Citizenship*. New York: Anchor.

Bruun, K., Edwards, G., Lumio, M., Mäkelä, K., Pan, L., Popham, R. E., Room, R., Schmidt, W., Skog, O.-J., Sulkunen, P., and Österberg, E. (1975) *Alcohol Control Policies in Public Health Perspective*. Helsinki: Finnish Foundation for Alcohol Studies.

Castles, F. G. (1978) *The Social Democratic Image of Society*. London: Routledge and Kegan Paul.

Eckstein, H. (1966) *Division and Cohesion in Democracy*. Princeton, N.J.: Princeton University Press.

Gusfield, J. (1963) *Symbolic Crusade: Status Politics and the American Temperance Movement*. Urbana, Ill.: University of Illinois Press.

Gusfield, J. (1967) Moral passage: The symbolic process in public designations of deviance. *Social Problems* 15:175-188.

Gusfield, J. (1968) Prohibition: The impact of political utopianism. Pp. 257-308 in J. Braeman, R. Bremner, and D. Brody, eds., *Change and Continuity in Twentieth Century America: The 1920's*. Athens, Ohio: Ohio State University Press.

Hancock, M. D. (1972) *Sweden: The Politics of Postindustrial Change*. Hinsdale, Ill.: Druden Press.

Helco, H. (1974) *Modern Social Politics in Britain and Sweden*. New Haven, Conn.: Yale University Press.

Kass, L. (1975) Regarding the end of medicine and the pursuit of health. *Public Interest* 40:11-42.

Keller, M., and Gurioli, C. (1976) *Statistics on Consumption of Alcohol and Alcoholism*. New Brunswick, N.J.: Journal of Studies on Alcohol.

Knowles, J. (1976) Individual reponsibility. In John Knowles, ed., *Doing Better and Feeling Worse*. New York: Norton.

Kuusi, P. (1948) *Suomen Viinapulma Gallup-Tutkimuksen Valossa*. Helsinki: Otava.

Kuusi, P. (1957) *Alcohol Sales Experiment in Rural Finland*. Publication No. 3 of the Finnish Foundation for Alcohol Studies. Helsinki.

Kussi, P. (1964) *Social Policy for the Sixties: A Plan for Finland*. Helsinki: Finnish Social Policy Association.

Mäkelä, K. (1971) Measuring the Consumption of Alcohol in the 1968-1969 Alcohol Consumption Study. Social Research Institute of Alcohol Studies, Helsinki.

Mäkelä, K., and Österberg, E. (1976) Alcohol consumption and policy in Finland and Sweden, 1951-1973. *Drinking and Drug Practices Surveyor* 12:4-45.

Österberg, E. (1979) Indicators of Damage and the Development of Alcohol Conditions . . . 1950-1975. January, 1979 mimeo for the International Study of Alcohol Control Experience, Finnish Foundation for Alcohol Studies, Helsinki.

Pesonen, P. (1974) Finland: Party support in a fragmented system. Pp. 271-314 in R. Rose, ed., *Electoral Behavior: A Comparative Handbook*. New York: The Free Press.

Poikolainen, K. (1979) Increase in Alcohol-Related Hospitalizations in Finland 1969-1975. Mimeo, Department of Public Health Science, University of Helsinki.

Purontaus, J. (1970) Cost-Benefit Analysis and the Finnish Alcohol Policy. Unpublished Master's Thesis, University of Helsinki.

Report of the Alcohol Committee. (1978) English Summary of Committee Report. Prepared by Oy ALKO Ab Information Service.

Simpura, J. (1978) The Rise in Aggregate Alcohol Consumption and Changes in Drinking Habits. The Finnish Case in 1969 and 1976. Manuscript, The Social Research Institute of Alcohol Studies, Helsinki.

Simpura, J. (1979) Who Are the Heavy Consumers of Alcohol? International Study of Alcohol Control Experiences. Mimeo, March 1979, the Finnish Foundation for Alcohol Studies, Helsinki, Finland.

Sulkunen, P. (1976) Drinking patterns and the level of alcohol consumption: An international overview. Pp. 223-281 in K. J. Gibbons, et al., eds., *Research Advances in Alcohol and Drug Problems*, Vol. 3. New York: John Wiley.

Sulkunen, P. (1979a) Abstainers in Finland 1946-1976. A Study in Social and Cultural Transition. Report No. 126, August 1979, the Social Research Institute of Alcohol Studies, Helsinki, State Alcohol Monopoly, Helsinki, Finland.

Sulkunen, P. (1979b) Drinking Populations, Institutions and Patterns. Part II: Individual Drinking Patterns. Draft mimeo, July 31, 1979, the Social Research Institute of Alcohol Studies, State Alcohol Monopoly, Helsinki, Finland.

Vidich, A. J., and Bensman, J. (1968) *Small Town and Mass Society*. Princeton, N.J.: Princeton University Press.

The Effect of Liquor Taxes on Drinking, Cirrhosis, and Auto Accidents

PHILIP J. COOK

INTRODUCTION

Alcoholic beverages have been taxed at a relatively high rate throughout the history of the United States. During the last 20 years, however, taxes on beer, wine, and liquor have increased more slowly than the overall price level. The result has been a substantial reduction in the price of alcoholic beverages relative to other commodities. Federal and state alcohol tax policies during this period have thus had the effect of providing an economic incentive for increased drinking. Since alcohol consumption is a contributing factor in the etiology of highway accidents, violent crime, suicide, cirrhosis, and a number of other causes of injury and death, it is possible that the downward trend in the relative price of alcoholic beverages has had the effect of reducing Americans' life expectancies and increasing morbidity.

Does the rate of alcohol taxation in fact have an important influence on rates of morbidity and mortality? There is almost no direct evidence on this question in the social science literature, although the potential importance of alcohol taxation as a public health policy instrument has

Philip J. Cook, a member of the panel, is at the Insitute for Public Policy Studies, Duke University.

Research assistance for the project was provided by Kent Auberry, Andrew Pescoe, R. J. Plummer, and Robert Schmitt. This draft reflects several helpful comments on an earlier version from Charles Clotfelter and Michael Murray.

been discussed in several recent scholarly presentations.[1] The most controversial aspect of this question is whether changes in the price of alcohol influence the drinking habits of heavy drinkers; this group accounts for the bulk of alcohol-related problems, and it is widely believed that the drinking habits of this group are insensitive to price.

In this paper, I review available evidence on the price elasticity of demand for alcohol and present a new statistical analysis that tends to support the view that liquor consumption is moderately responsive to price in the United States. More important, I am able to demonstrate with a high degree of certainty that increases in the tax rate on spirits reduce both the auto fatality rate and the cirrhosis mortality rate. The virtually inescapable conclusion is that the demand for alcohol by heavy drinkers is responsive to price.

The next section is a brief history of alcohol taxation, prices, and consumption in the United States since 1950. A review of the econometric literature follows on the price elasticity of demand with new results for the 1960-1975 period using a "quasi-experimental" technique. The next section reviews available results relating alcohol consumption levels to mortality from certain causes and then presents new findings on the impact of alcohol prices on cirrhosis and auto fatalities. The following section discusses the use of excise taxes on alcoholic beverages as part of a public health strategy for reducing alcohol abuse, and the final section presents concluding observations.

PRICES AND TAXES

While the average prices of beer, wine, and distilled spirits have been increasing during the last two decades, the rates of increase are less than the overall inflation rate. The statistics in Table 1 demonstrate that the price of spirits, relative to the average price of all other consumer goods (as measured by the consumer price index [CPI]), has declined 48 percent since 1960. During the same period, the relative price of beer has declined by 27 percent and wine by about 20 percent.

We would expect that price reductions of this magnitude, especially when coupled with the substantial increases in average real disposable income during this period, would result in increased consumption. In fact, average consumption of ethanol per person (aged 15 and over) increased 29 percent between 1960 and 1971; since then average consumption has remained roughly constant at about 2.7 gallons of ethanol

[1] See Bruun et al. (1975); Popham et al. (1976 and 1978); Medicine in the Public Interest (1979).

TABLE 1 Average Prices/Pint of Ethanol, 1960–1980

	Current Prices (dollars)			Adjusted Prices (1980 dollars)		
Year	Beer and Ale	Wine	Distilled Spirits	Beer and Ale	Wine	Distilled Spirits
1960	6.48	—	7.79	17.05	—	20.50
1961	6.50	—	7.84	16.89	—	20.39
1962	6.53	—	7.87	16.78	—	20.22
1963	6.59	4.02	7.96	16.73	10.21	20.21
1964	6.64	4.01	7.99	16.67	10.07	20.05
1965	6.70	4.03	8.01	16.56	9.96	19.80
1966	6.81	4.05	8.06	16.34	9.73	19.35
1967	6.93	4.10	8.16	16.14	9.56	19.01
1968	7.12	4.26	8.27	15.95	9.54	18.54
1969	7.30	4.44	8.36	15.47	9.41	17.73
1970	7.54	4.79	8.57	15.16	9.63	17.22
1971	7.82	5.02	8.68	15.01	9.64	16.67
1972	7.89	5.21	8.86	14.67	9.69	16.48
1973	8.01	5.55	8.91	14.01	9.71	15.60
1974	8.78	6.04	9.05	13.87	9.55	14.30
1975	9.72	6.32	9.31	14.09	9.16	13.50
1976	9.95	6.46	9.46	13.63	8.85	12.97
1977	10.10	6.64	9.59	12.93	8.51	12.27
1978	10.66	7.29	9.98	12.69	8.67	11.88
1979	11.77	7.95	10.39	12.60	8.50	11.12
1980 (Jan.)	12.40	8.27	10.74	12.40	8.27	10.74

Note: Current prices were calculated as follows. *The Liquor Handbook*, p. 24 (see Gavin-Jobson Associates, Inc., 1978), gives data on consumer expenditures and volume purchased for beer, wine, and spirits in 1976. These data yield estimates of the average price per pint of each type of beverage ($0.45, 0.94, and 4.06, respectively). In 1976, beer averaged 4.49 percent alcohol, while wine was 14.6 percent and spirits 42.9 percent (calculated from data in National Institute of Alcohol Abuse and Alcoholism (1978, Table 3, p. 6)). Therefore, 1 pint of alcohol was contained in 22.3 pints of beer, or 6.85 pints of wine, or 2.33 pints of spirits. These figures for the number of pints of beverage per pint of alcohol were multiplied by the average prices per pint of beverage to obtain the 1976 figures in the first three columns. Prices in other years were derived from these 1976 prices by use of the Bureau of Labor Statistics price indexes for beer, wine, and spirits (unpublished).

per year. Figure 1 depicts the recent history of consumption rates for beer, wine, and spirits.

Nominal consumer expenditure for alcoholic beverages increased from $13 billion to $32 billion between 1960 and 1975 (DISCUS Facts Book 1977, p. 26), but this represents an increase in real terms (controlling for inflation) of only 36 percent. The percentage of total consumer expenditures accounted for by alcoholic beverages declined from 3.7 percent to 3.0 percent during this period. Thus the decline in the

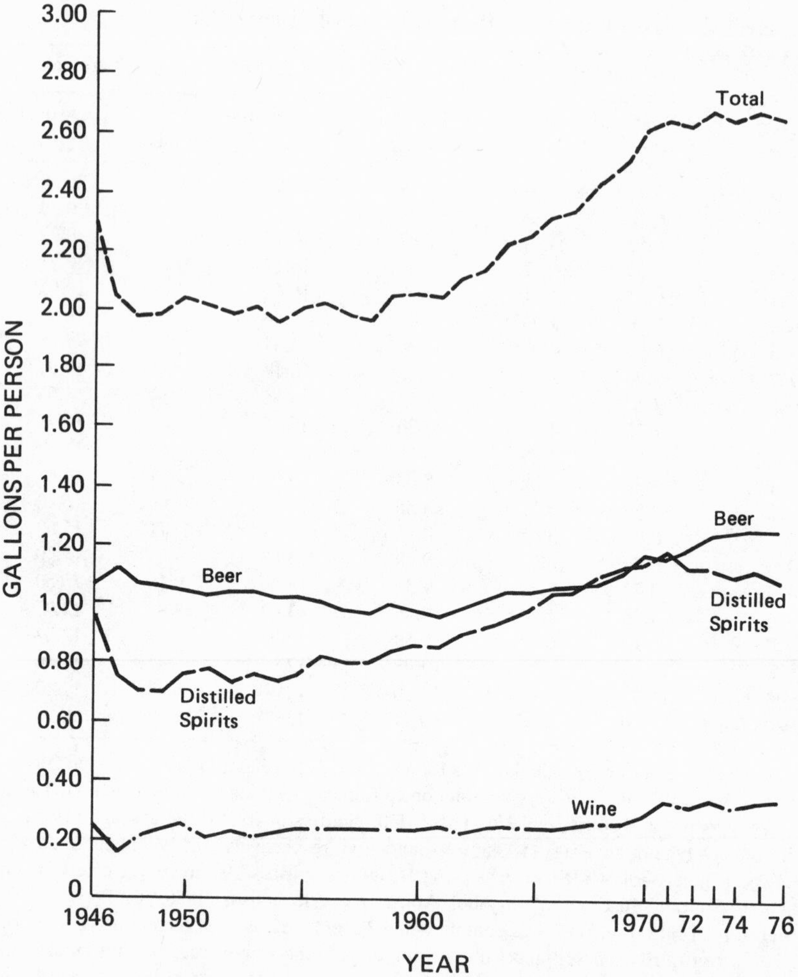

FIGURE 1 Trends in per-capita ethanol consumption in U.S. gallons, based on beverage sales in each major beverage class in the United States, 1946-1976.

real cost of all types of alcoholic beverages in recent years has been accompanied by increased consumption but a reduced importance in the typical consumer's budget.

What accounts for the secular decline in the relative prices of alcoholic beverages? Part of the answer, particularly in the case of distilled spirits, is that excise tax rates have not kept up with inflation.

The last three columns of Table 1 were derived from the first three columns, converted to "1980 dollars" by the consumer price index. For example, according to the CPI, 1960 dollars had 2.63 times as much purchasing power as 1980 dollars. Therefore, 1960 prices were converted to 1980 dollars by multiplying in each case by 2.63.

ALCOHOL BEVERAGE TAXATION

Alcoholic beverages are subjected to a complex array of taxes and other controls that affect retail prices. Distilled spirits are taxed more heavily than beer and wine; taxes on distilled spirits include a federal excise tax of $10.50 per proof gallon and state and local taxes and fees that averaged $5.55 per gallon in 1975 (DISCUS 1977a) and import duties on foreign products. Beer and wine are also taxed by all levels of government. As a result, various direct taxes and fees account for about one-half of retail expenditures for spirits and about one-fifth of retail expenditures for both beer and wine (DISCUS 1977a, p. 2).

The states have legislated a considerable degree of government control on alcoholic beverage prices and sales. Nineteen states have a legal monopoly over the wholesale trade in spirits, and all but two of these also monopolize the retail trade in spirits. Most of these monopoly states also require that wine be sold only in state stores, and several have included beer as well.[2] In monopoly states, then, prices are set by administrative fiat. In the remaining states, retail distributors must be licensed by the state, and in most of these states the distributors are subject to fair trade controls on pricing. The decline in real prices of alcoholic beverages, then, directly reflects choices made by state legislatures and regulatory agencies. More fundamentally, however, it is clear that these choices have been influenced by costs. A major component of the cost of distilled spirits is the federal excise tax; the fact that it has remained constant at $10.50 per proof gallon for the last 30 years has greatly contributed to the decline in real cost of this type of beverage. If this excise tax had been "indexed" to keep up with inflation since 1960, it would now stand at about $28 per proof gallon; assuming this tax increase had been passed along to the consumer with a 20-percent markup, the real price of spirits would only have declined by about 22 percent since 1960, in contrast with the actual decline of 48 percent.

Thus, prices of alcoholic beverages are controlled by legislation and government agencies to a considerable extent. Tax and regulatory decisions in this area influence patterns of consumption, which in turn influence the public health. The remaining sections of this paper are devoted to developing evidence on the magnitudes of such effects and discussing their implications for pricing policy.

[2] Details on state regulations are given in DISCUS (1977b).

THE DEMAND FOR ALCOHOLIC BEVERAGES

The prices of alcoholic beverages have had a downward trend since 1960 (compared with the overall price level), while during much of this period average consumption increased. Was the decline in price responsible, at least in part, for the increase in drinking during the 1960s? More generally, how much do economic variables—prices and income levels—influence drinking habits? This question has motivated a number of empirical studies, both in North America and Europe. My review of this literature is limited to a few of the best of these studies based on U.S. or Canadian data. The new results presented below are based on recent data on prices, income, and consumption in the United States. I begin the review with a brief summary of the economic theory and terminology useful to understanding the empirical studies.

NOTES ON THE ECONOMIC THEORY OF CONSUMPTION

Economic theory demonstrates that an individual's rate of purchase (quantity demanded) for any commodity that he or she consumes will be influenced by the price of the commodity, the prices of related commodities, and by his or her purchasing power (wealth or income). The relationship between the consumer's quantity demanded and these economic variables can be characterized as a mathematical function of the follow form:

$$q_i = D\left(\frac{P_1}{P}, \ldots, \frac{P_i}{P}, \ldots, \frac{P_n}{P}, \frac{Y}{P}\right), \qquad (1)$$

where

q_i = the quantity demanded of commodity i,

$P_1, \ldots, P_n$ = prices of the various commodities available, to the consumers, including P_i (the "own price"),

P = a price index, such as the consumer price index,

Y = the consumer's income.

Adjusting prices and income for the overall price level P in this fashion is justified because only relative prices and income matter in determining demand. For example, a uniform increase of 10 percent in all prices and income would have no effect on demand.

Individuals have different tastes, and two consumers facing identical prices and income may differ considerably in the mix of commodities they buy; the mathematical form of the function (1) differs among individuals. However, economic theory predicts that consumers will be alike in their qualitative response to a price change; an increase in price will reduce the quantity demanded.[3] It would be surprising if alcoholic beverages proved to be exceptions to this basic principle of economics.

Economics offers some useful terminology for characterizing the shape of a demand function:

(1) If the quantity demanded of a commodity increases with income, the commodity is termed *normal*; otherwise it is *inferior*. Cheap wine of poor quality may be an example of an inferior commodity.

(2) If two commodities are typically consumed together, one enhancing the utility derived from the other, they are termed *complements*. More precisely, two commodities are complements if a reduction in the price of one increases the demand for the other. Beer and whiskey are complements for people who consume their alcohol in the form of boilermakers. For most people, however, we might expect beer, wine, and liquor to be *substitutes*, meaning that a reduction in the price of any one of these three would increase the demand for the other two.

(3) Economists usually express the responsiveness of demand to prices and income in terms of *elasticities*. For example, the "own price elasticity of demand" is defined as the percentage change in quantity demanded resulting from a 1-percent change in own price. The "income elasticity of demand" is defined analogously.

If the own price elasticity is between zero and minus one, demand is *inelastic*; if less than minus one, it is *elastic*. It can be demonstrated mathematically that if the demand for a commodity is inelastic, then an increase in its price will result in an increase in expenditure on that commodity; if demand is elastic, expenditure will fall in response to price increase.

THE ECONOMETRIC APPROACH TO ESTIMATING PRICE AND INCOME EFFECTS

In practice, estimates of price and income effects for alcoholic beverages have been based on aggregate data rather than data on individual consumption decisions with respect to individual brands of alcoholic beverages:

(1) Adequate data on consumption and income are usually not avail-

[3] A sufficient, but by no means necessary, condition for this prediction is that the commodity is "normal" (i.e., quantity demanded tends to increase with income).

able for individuals or households. Most demand estimates have been based on data for average consumption, income, and price levels for the entire populations of geographic units—states, provinces, or even entire countries.

(2) Most studies have aggregated the scores of varieties of alcoholic beverage into three categories: liquor, wine, and beer. Quantity is measured by volume of liquid within each category and price as some sort of average or index of the prices of all the brands included in the category. This type of aggregation conceals interbrand substitution that may result from price changes. For example, an increase in the average price of liquor may induce consumers to substitute cheaper for more expensive brands, thereby perhaps maintaining their volume of liquor consumption despite the general increase in prices.

Equation (1) is too general (in a mathematical sense) to be estimated. In practice, the demand function that has been estimated has a form similar to the following:

$$Q = a + b_S\left(\frac{P_S}{P}\right) + b_B\left(\frac{P_B}{P}\right) + b_W\left(\frac{P_W}{P}\right) + c\left(\frac{Y}{P}\right) + d_1X_1 + \ldots + d_mX_m, \tag{2}$$

where

Q = per-capita consumption of spirits (or beer or wine),

P_S, P_B, P_W = prices of spirts, beer, and wine, respectively,

Y = per-capita disposable income,

P = consumer price index,

$X_1, \ldots, X_m$ = other variables thought to influence consumption.

Some studies have used the logarithm of these variables instead of estimating the linear form of the equation. Other mathematical transformations are also possible. The "correct" mathematical form is not known; any particular form chosen by the econometrician is at best an approximation of the correct form.

Given the appropriate data, it is possible to estimate the parameters of (2) using standard econometric techniques (e.g., regression analysis). The resulting estimates for b_S, b_B, and b_W measure the effect of these prices on quantity demanded. In a demand equation for spirits, we would expect b_S to be negative, b_W and b_B to be positive (assuming they are substitutes for spirits), and c to be positive (assuming spirits is a normal commodity). The econometric estimates of these parameters serve as a test of these qualitative hypotheses. These estimates are never precise; the regression analysis provides a basis for calculating the range of statistically reasonable values for each parameter (confidence intervals).

There are a variety of statistical problems encountered in attempting to develop reliable estimates of price and income effects. These problems are introduced and discussed in the course of the literature review below.

RESULTS FROM ECONOMETRIC STUDIES

Of all published econometric studies of alcoholic beverage demand using data for the United States or Canada, the most noteworthy is by Johnson and Oksanen (1977). They estimated separate demand equations for beer, wine, and spirits, using panel data for 10 Canadian provinces for 15 years (1956-1970). Each of their demand equations includes the following independent variables: prices for beer, wine, and spirits relative to the consumer price index; real income per adult; the dependent variable (consumption) lagged 1 year; and variables representing ethnicity, schooling, religion, and strikes affecting alcoholic beverage sales. Several estimation techniques were used. Since they yielded virtually identical results, I will limit my discussion to one of these techniques: ordinary least squares with separate dummy variables for each province. The following points should be taken into account in considering their results:

(1) The authors assume that prices are exogenous, because "In the Canadian institutional setting, prices are established by government agencies" (p. 114). The possibility that the price policies of government agencies may be influenced by demand conditions is not considered.[4]

[4] Presumably retail price decisions are influenced by prices charged by manufacturers and importers, which will in turn be influenced by demand conditions. Demand should play a particularly important role in determining the price of aged whiskey and wine, which are in more or less fixed supply in the short run. If observed prices are influenced by demand conditions, rather than being an exogenous determinant of quantity demanded (as Johnson and Oksanen, and most other econometric studies, have assumed), then the parameter estimates in their demand equations will be biased and inconsistent. This is an example of a widely recurring problem in statistical analyses of data generated by a "natural" process characterized by complex causal interactions among the variables.

It should also be noted that Johnson and Oksanen use data on *average* prices of alcoholic beverages. It would be preferable to use price index data in this type of study, but alcoholic beverage price indexes are not available for Canadian provinces.

(2) The authors avoid the problem of developing a statistical explanation for the large cross-sectional (interprovince) differences in consumption because of the nature of their data. All equations are estimated in first difference form, which of course is impossible when only cross-sectional data for 1 year are available.[5] This is important because the large cross-sectional differences in consumption levels are probably more a reflection of cultural differences than of differences in prices and income (Simon 1966). The use of the first difference form reduces the importance of finding adequate empirical proxies for these cultural differences. The same point applies to the problem of cross-sectional differences in nonprice regulations of beverage sales. Of course, to the extent that "culture" and nonprice regulations change over time, it is still necessary to control for them. But these intertemporal changes are likely to be small relative to cross-sectional differences, at least for the 15-year period under consideration here.

(3) The authors' specification includes dummy variables for each province. These variables are meant to capture province-specific trends in the underlying determinants of consumption behavior not otherwise specifically accounted for in the demand equation.

(4) The inclusion of all three prices in each demand equation permits estimation of the cross-price effects on demand, as well as the own price effect.

(5) The inclusion of lagged consumption in the specification is justified by the possibility that alcohol consumption behavior exhibits some degree of habit formation, and hence the full, long-run response to a change in price or income does not emerge immediately. Including this variable creates certain econometric problems, which are discussed by the authors.

The key results of this analysis are summarized in Table 2. The short-run demand for spirits is elastic with respect to own price, but beer and wine are inelastic. Income changes have little or no effect on demand. The cross-price effects (not shown in this table) tend to be small. These estimated effects suggest that spirits and beer are complements, whereas wine and beer are substitutes—but the evidence from the Johnson-Oksanen study is not very strong on this issue.

[5] By way of illustration, the first difference form of the equation

$$y_t = a + bX_t + ct,$$

is

$$(y_t - y_{t-1}) = b(X_t - X_{t-1}) + c.$$

TABLE 2 Estimated Elasticities for Spirits Demand in Canada

Beverage Type	Short-Run Own Price Elasticity	Long-Run Own Price Elasticity	Short-Run Income Elasticity
Beer	−0.26	−0.29	−0.02
Spirits	−1.13	−1.70	0.10
Wine	−0.68	−1.36	0.01

There have been no published studies using U.S. data that are comparable in scope and quality to that of Johnson and Oksanen. A major limitation on studying U.S. alcoholic beverage demands is the lack of state-level price data for beer and wine. One exception is Hogarty and Elzinga (1972), who were able to obtain data on beer prices for two brands; these data were entered into the public record as a result of an antitrust suit by the U.S. Department of Justice. They used these annual state-level data for the period 1956-1959 to estimate the following equation:

$$\text{Log } B = a + b \text{ Log } P_B + c\left(\frac{1}{Y}\right) + d \text{ Log } F, \tag{3}$$

where

B = beer consumption per adult,

P_B = price of beer,

F = percent foreign born,

Y = per-capita income.

Their results imply a price elasticity of −0.9 and an income elasticity of 0.4. They also experimented with including a spirits price variable in their equation but rejected it because the result implied that beer and spirits are complements—a result that they thought highly unlikely. This study makes no special use of the panel structure of the data, and the specification is highly inadequate to the task of controlling for non-economic influences on beer consumption.

Price elasticities of spirits consumption estimated from U.S. state-level data have been questioned because of problems with official data on liquor sales. These data are based on reports by wholesalers to the tax authorities and differ from actual liquor consumption by state residents in three potentially important respects: (1) "moonshine" liquor is of course not included; (2) wholesalers may underreport their sales in order to evade state taxes; and (3) some consumers may "import" their liquor from an adjacent state if price differences between the two states make this activity worthwhile.

If a state raises its tax on liquor, the resulting reduction in reported sales may thus exaggerate the true reduction in consumption by state residents due to either an increase in moonshining, an increased propensity of wholesalers to conceal sales from tax authorities, or increased purchases in adjacent states (the "border effect"). These problems may also be present for beer and wine sales, of course.

Two empirical studies (Waler 1968 and Smith 1976) have attempted to estimate the magnitude of the "border effect" in spirits commerce. Waler (1968) reports on the basis of his analysis of 1960 cross-sectional data for 42 states that ". . . the price coefficient becomes negligible and insignificant when interstate effects are included, suggesting that the large and significant price coefficient obtained in the misspecified model is completely spurious" (p. 858). Waler's method for taking account of border effects is complex and problematical, and his specification of the demand equation is highly inadequate (his only explanatory variables are price and income). His conclusion therefore is in doubt.

Smith (1976) estimates his liquor demand equation from 1970 cross-sectional data on 46 states. One of his variables is the price of liquor in the adjacent state with the lowest price—a variable that he believes should "control for" the border effect. This interpretation is dubious, since relative prices in all adjacent states are relevant to determining the net magnitude of the border effect for any one state.[6]

Smith also attempts to allow for underreporting of liquor sales by wholesalers. He finds that the tax elasticity of demand is larger than the elasticity with respect to net price (price net of tax) and interprets this difference in elasticities as an indication that an increase in the state tax increases the propensity of wholesalers to underreport. These results are critiqued by Hause (1976), and it is clear that Smith's interpretation is tenuous. In any case, Smith's specification is highly simplistic—he, like Waler, makes no attempt to control for interstate differences in culture, ethnicity, religion, or nonprice regulations on alcohol sales and consumption. We are left without reliable estimates of the importance

[6] Note that a state may be a net exporter across one border and an importer across another. The magnitudes of cross-border flows depend on the price differences, the number of people who live on the higher-priced side of each border, and the effort devoted to enforcing the laws against this sort of cross-border purchase.

of the border effect, underreporting by wholesalers, and other difficulties with state data.

There have been several econometric studies of the demand functions for alcoholic beverages based on aggregate time series data for the United States. Niskanen (1962), for example, estimated both demand and supply functions for spirits, beer, and wine, using observations of 22 years (1934-1941 and 1947-1960). The use of U.S. aggregate data eliminates the border effect as a concern. A major problem is that the small number of observations necessitates use of very simple specifications; the list of explanatory variables in Niskanen's demand equations is limited to prices and two measures of consumer purchasing power. Presumably, there were other factors having an important influence on drinking patterns that changed between 1934 and 1960: Niskanen's omission of these other factors (whatever they might be) almost certainly causes his estimates of price elasticities to be biased. Nonetheless, these estimates are widely cited—in part because they are dramatically large. He found the demand for spirits to be highly price elastic (-2.0), while beer was -0.6 and wine around -0.7. Another econometric analysis of spirits consumption based on time series data for the United States, by Houthakker and Taylor (1966), is cited for the opposite reason: their estimate of price elasticity was slightly *positive*, though not significantly different from zero in a statistical sense.

My conclusion from reviewing econometric studies of alcoholic beverages is that there are no reliable estimates for the price elasticities of demand based on U.S. data.[7] Recent studies, summarized above, should serve to sensitize researchers to some of the econometric estimation problems that are peculiar to this class of commodities.

QUASI-EXPERIMENTAL ANALYSIS OF THE DEMAND FOR LIQUOR

The econometric approach to estimating the price elasticity of demand for alcoholic beverages has dominated this empirical literature, but it is not the only available estimation technique. Julian Simon (1966) introduced a "quasi-experimental" approach and argued persuasively that it offers a basis for more reliable estimates than regression analysis of cross-sectional or time series data.

Simon analyzed 23 cases in which the price of liquor increased by

[7] A number of other reviews of the econometric literature are available. Bruun et al. (1975, pp. 74-78) and Medicine in the Public Interest (1979, pp. 64-68) present nontechnical summaries. More detailed reviews are in Lau (1975) and Ornstein and Levy (ca. 1978).

more than 2 percent from one year to the next in a state. Apparently he limited his study to cases in which such a price change was induced by an increase in the state liquor tax. For each such case, he calculated the apparent price elasticity of demand, using the following procedure:

(1) Calculate the following formula as a measure of the proportional change in the state per-capita liquor consumption resulting from the price increase:

$$\frac{(C^{a}_{t+m} - C^{a}_{t-m})}{C^{O}_{t-m}} - \frac{(C^{O}_{t+m} - C^{O}_{t-m})}{C^{O}_{t-m}}, \quad (4)$$

where

C^{a}_{t+m} = per-capita liquor consumption in state *a* (the "trial" state) for a 12-month period beginning 3 months after the tax change;

C^{O}_{t+m} = comparison states' per-capita consumption during a 12-month period beginning 3 months after the tax change;

and so forth. The "before" and "after" periods are each 12 months long, chosen so as to leave a 7-month gap around the month of the price change. Simon used two groups of "comparison states": if the price change was in a monopoly state, then the group of comparison states consisted of all other monopoly states except those that had a contemporaneous price change; if the price change was in a license state, then the group of control states consisted of all other license states except those with a contemporaneous price change. Simon experimented with several methods for averaging consumption changes across control states.

(2) The number calculated from the above formula was divided by the percentage change in price; the resulting ratio is then an estimate of the price elasticity of liquor demand.

(3) The median of the 23 estimates of price elasticity thus derived was Simon's choice for a point estimate of the price elasticity. His median was −0.79.

Simon's method shares some of the problems discussed above in reference to the econometric studies. In particular, he makes no attempt to control for the border effect or for possible changes in the accuracy of reporting of sales by wholesalers. (He does take the moonshining problem into account by excluding states with a high incidence of moonshining from his analysis.) It is possible, then, that his elasticity estimates are biased.

There are two great virtues of Simon's technique: first, since the price changes he studies are induced by changes in the tax rate, it is reasonable to suppose that these changes are truly exogenous to the market and not induced by changes in demand. The "causal ordering" problem is thus circumvented—unless it can be argued that state legislatures and regulatory agencies take market conditions into account when setting the tax rate. Second, Simon's method for controlling for nonprice influences on liquor demand—his use of control states—is probably more reliable than the econometric technique of specifying one or more variables in the regression equation to reflect and control for these influences. It is not clear that these influences are adequately measured by any short list of socioeconomic variables. Simon's technique is justified by the presumption that nonprice factors that influence consumption in one state in any given year will also be influential in other states in that year. As long as the sample of trial states is representative of the United States as a whole, then Simon's approach yields an unbiased estimate.

Simon's price elasticity estimate of −0.79 seems reasonable. Since he has only 23 observations, and these observations have a rather high variance, his point estimate is not precise. He reports that "We may say with 0.965 probability that the mean of the population from which this sample of elasticity estimates was drawn is between −0.03 and −0.097" (p. 198).

A REPLICATION

I have replicated Simon's study using state tax changes occurring between 1960 and 1975. I have limited this study to license states and excluded states in which the tax change was less than $0.25 per gallon. Alaska, Hawaii, and the District of Columbia were also excluded. The resulting sample contains 39 observations.

The formula I used to calculate the proportional change in liquor consumption induced by these tax changes differs in a number of minor respects from that used by Simon: (1) The "before" and "after" comparison is based on the calendar years preceding and following the year of the tax change without regard to the month in which the change occurred. This approach simplifies computations considerably. (2) Instead of using some sort of average of proportional changes in consumption for some group of states as a control, I have in all cases used the median proportional change in consumption for the year in question as a basis of comparison. The median is calculated from the 30 license states (of the 48 contiguous states) in each year. Experiments with other approaches to choosing a basis of comparison (e.g., using the median of the entire list of 48 states in each year, or the proportional change in total U.S. consumption as the basis for comparison), demonstrated that this choice makes little difference to the results.

The formula I used can be written as follows:

$$D_t^s = \frac{C_{t+1}^a - C_{t-1}^a}{C_t^a} - \underset{s}{Median}\left\{\frac{C_{t+1}^s - (C_{t-1}^s)}{C_t^s}\right\}, \tag{5}$$

where

C_t^a = total liquor consumption divided by the population aged 21 and over in year t and state a (the state incurring the tax change).

My results are reported in Table 3. Of the 39 observations, 30 of the net consumption changes, D_t^a, calculated from this formula were negative. This is very strong evidence that an increase in tax reduces reported liquor sales in a state, other things being equal. One test of the statistical significance of this result is the following: The null hypothesis is that price has no effect on consumption. Under this assumption, the probability of obtaining 30 or more negative changes out of 39 trials is less than one-tenth of a percent. This nonparametric "sign" test is a strong indication that the null hypothesis should be rejected in favor of the alternative that an increase in price reduces consumption.

An estimate of the apparent price elasticity of demand can be developed by converting each of the statistics in Table 3 ("net change in liquor consumption") by the proportionate change in price caused by the tax change for each observation. To calculate the proportionate change in price, I multiplied the tax increase by 1.2 (i.e., I assumed a 20-percent markup on tax) and divided by the average retail price of liquor for the appropriate state and year.[8] The median of the resulting

[8] The prices used for this calculation were taken from data in various issues of *The Liquor Handbook* on "retail prices of leading brands." Average prices for each state and year were calculated by ANOVA (analysis of variance), with main effects for each state and each brand. Separate ANOVAs were run for each year. The coefficients on the "state effects" were used as a measure of average price. (The null category for brand was Bacardi rum.)

To calculate the markup on tax changes, I proceeded as follows: For each year, price changes for all 30 license states were calculated and standardized by subtracting that year's median price change. The resulting net price changes for states with tax changes in that year were divided by the tax changes. The median of the 39 ratios calculated in this fashion was 1.1875, suggesting a "typical" markup of 18.75 percent. This estimate was rounded to 20 percent in the elasticity calculations reported here.

TABLE 3 Effects of Changes in State Liquor Tax Rates

Year	State	Tax Change[a] ($)	Net Change in Liquor Consumption[b] (%)	Rank[c]	Net Change in Auto Fatality Rate[b] (%)	Rank[c]	Net Change in Cirrhosis Death Rate[b] (%)	Rank[c]
1961	Conn.[d]	1.00	− 8.7	2	13.2	28	− 9.0	6
	Mo.[d]	0.40	− 2.5	10	− 7.2	5	− 5.7	10
	Nev.[d]	0.60	− 6.0	5	−39.2	1	−16.3	3
1963	Fla.[d]	0.33	− 0.1	14	− 2.7	12	9.7	28
	Nebr.	0.40	− 7.8	1	2.8	21	− 7.1	7
	N.J.	0.30	− 1.9	7	4.7	23	− 3.3	12
	N.Y.[d]	0.75	− 7.4	2	− 0.0	15	2.3	21
	S.D.[d]	0.50	− 6.8	3	−17.2	4	−15.1	1
	Tenn.	0.50	0.3	17	2.0	19	8.7	26
	Wis.	0.25	1.6	22	− 4.3	8	−12.6	3
1964	Ga.	0.50	9.8	29	− 2.0	12	2.2	22
	Kans.	0.30	− 7.0	3	− 6.1	7	− 7.7	8
1966	Mass.[d]	0.70	− 6.9	3	− 4.2	10	− 3.0	9
1967	Calif.	0.50	− 4.6	6	− 6.4	4	−14.7	2
	Tenn.	1.50	−11.3	2	− 3.0	11	8.9	25
1968	Ariz.	0.56	0.8	19	10.3	26	5.7	22
	Fla.[d]	1.23	− 9.9	1	8.0	24	4.9	20
1969	Conn.[d]	0.50	− 9.2	1	− 6.3	6	− 0.9	12
	Del.	0.50	0.3	16	4.5	25	− 6.2	7
	Ill.[d]	0.48	− 4.8	6	− 5.7	7	− 3.2	9
	Mass.[d]	0.41	14.8	30	5.3	27	− 2.2	10
	Minn.	0.755	− 4.4	8	− 9.0	4	− 0.2	14
	Nev.	0.50	− 2.6	11	− 1.7	12	11.4	23
	N.J.	0.50	− 8.8	2	− 3.2	10	0.8	17
	R.I.[d]	0.50	− 4.3	9	−26.1	1	−17.9	2
1970	Ky.	0.64	− 0.5	13	0.3	16	6.7	22
	La.	0.82	− 0.6	12	− 1.5	14	− 1.0	13
1971	Del.	0.60	−10.7	3	−26.8	1	3.2	21
	Okla.	1.60	− 7.3	4	− 2.1	11	−11.2	3
	Minn.[d]	0.90	− 3.2	10	1.1	21	7.1	25
	Mo.[d]	0.80	− 7.1	5	0.3	17	− 9.2	4
	S.D.	1.75	1.6	21	19.6	30	3.3	22
	Tex.[d]	0.32	− 3.2	11	− 0.5	14	− 3.4	11
	Wis.	0.35	2.9	23	0.3	18	− 1.3	14
1972	Nebr.	0.40	− 2.3	11	−20.2	1	1.5	17
	N.J.	0.50	− 7.2	6	− 1.7	11	− 8.8	5
	N.Y.	1.00	−10.0	3	− 3.3	7	− 9.9	2
1974	Ariz.	0.50	− 4.3	3	− 2.2	9	− 3.9	9
1975	Mass.[d]	0.69	0.2	16	—	—	—	—

[a] A legislated change in the state tax on distilled liquor, expressed in dollars per gallon.
[b] The changes in consumption, auto fatality rate, and cirrhosis death rate are proportional changes net of the corresponding change for the median state in that year.
[c] Based on a ranking of the 30 license states (excluding Alaska and Hawaii) on the basis of the proportional change in the given rate.
[d] An increase in the tax on beer was enacted in the same year.

price elasticity estimates for the 39 observations was −1.6. This result implies that a 10-percent increase in the price of spirits in a state results in a 16-percent reduction in quantity purchased in that state. This is of course a point estimate, which is subject to statistical error. Given the distribution of the 39 elasticity estimates, it can be shown that there is a 95-percent chance that the true elasticity is less than −1.0 (i.e., that the demand is elastic with respect to price).

My point estimate of −1.6 is considerably larger than Simon's estimate of −0.8. The difference may be partly the result of Simon's using a larger assumed markup. He is not precise about what rate of markup he used, merely noting that it was the "customary retail" rate (which is surely higher than 20 percent). If I had assumed an 80-percent markup, for example, my elasticity estimate would have been reduced to −1.07. However, as explained in the preceding footnote, a 20-percent markup appears to be the appropriate assumption for the retail liquor industry.

It should be noted that the interpretation of the resulting elasticity estimate is complicated by the fact that in 16 of the 39 instances in which states increased the liquor tax, there was a contemporaneous increase in the state tax on beer. If beer is a substitute for liquor, then the effect of these contemporaneous changes will be to bias the liquor price elasticity upward. For this reason, the estimated price elasticity of demand for liquor may understate the true price elasticity.

For the procedure outlined above to give a valid, unbiased measure of the effect of tax changes on liquor sales, it is necessary that state tax changes not be systematically related to historical trends in consumption in the state. It is possible that the decisions of state legislatures to change liquor taxes are systematically related to trends in state sales (or tax revenue collections). Suppose *arguendo* that legislatures tend to raise taxes in response to an unsatisfactorily slow growth in liquor tax revenues. Then states that raise their taxes will be a biased "sample" of all states, the bias being in the direction of relatively slow growth in consumption. The statistical procedure reported above would then yield misleading results; the bias in the measure of price elasticity would be negative, yielding an exaggerated notion of the degree to which reported consumption was responsive to price. To test for this and related possibilities and more generally to test whether the state with tax changes can reasonably be viewed as a representative sample of all states, I calculated the net proportional growth rate in liquor consumption (the same measure as reported in Table 3) for all states that had tax changes during the sample period, but for a period that preceded the year of the tax change by 2 years. Consistent with the notation defined in formula (5) above, this statistic is denoted D^{a}_{t-2}. If tax changes occur "at random" among states and are not influenced by recent history in liquor consumption, then we would expect that it would equally be likely for any one of these values to be positive or negative. The results are consistent with these predictions: 21 of 36 states (58 percent) for which it was possible to do this calculation[9] had negative values, while 15 were positive. The probability that this difference would occur by chance alone is 40 percent. These results are compatible with the claim that tax changes are exogenous events with respect to consumption trends and that the states with tax changes are representative of all states during the sample period.

THE EFFECT OF TAX INCREASES ON AUTO AND CIRRHOSIS FATALITIES

INTRODUCTION

The association between drinking and excess mortality from a variety of causes has been thoroughly documented. Excess drinking has been particularly strongly implicated as a major causal factor in auto accidents and cirrhosis of the liver. If an increase in the price of liquor reduces average consumption, it is reasonable to suppose that an increase in the liquor tax rate may reduce the mortality rate due to these causes. Nevertheless, there are a number of reasonable doubts about this conjecture, which are summarized here in three questions.

While the evidence presented in the previous section convincingly demonstrates that an increase in the liquor tax rate reduces reported liquor sales, the effect on actual consumption of liquor may be smaller or nonexistent: the differences between reported sales and actual consumption, as discussed above, are the result of moonshining, underreporting by wholesalers, and the "border effect." Is average liquor consumption responsive to changes in the price of liquor?

Even if actual consumption of liquor is reduced as a result of tax (price) increases, consumers may maintain their average level of alcohol consumption by substituting beer or wine for liquor. (Liquor only accounts for about 40 percent of beverage alcohol consumed in the United States.) Is average alcohol consumption responsive to changes in the price of liquor?

Even if an increase in the price of liquor does reduce total alcohol consumption, this reduction will have no effect on cirrhosis and auto fatalities if it is limited to moderate drinkers. Studies of price effects on average alcohol consumption give no indication of the distribution of

[9] Since no data were available for 1958, it was not possible to include the three states that had tax changes in 1961.

these effects among different types of drinkers. Is average alcohol consumption of heavy drinkers responsive to changes in the price of liquor?

There is no need to respond to these questions directly. The quasi-experimental method of measuring the price elasticity of liquor consumption, developed in the previous section, can also be used to measure the effect of tax increases on cirrhosis and auto accident fatality rates. The resulting estimates can virtually speak for themselves. If changes in liquor prices have the expected effect, it will be revealed by this method, and we can then draw the appropriate conclusions about the three questions. I report my results after a discussion of the literature on cirrhosis. No discussion of the relationship between drinking and auto fatalities is presented, since David Reed's paper in this volume has a thorough discussion of this issue.

CIRRHOSIS AND DRINKING

Cirrhosis is a disease of the liver in which the capacity of the liver to cleanse the blood and perform its other functions is reduced due to scarring of the liver tissue. Most people who die of cirrhosis have a long history of heavy drinking.[10] The liver is capable of processing a moderate level of alcohol intake without sustaining any damage. The cirrhosis disease process reflects a repeated "overload" of alcohol in the system.[11] If the cirrhosis victim reduces consumption, then the scarring process will be slowed or stopped and his or her life will ordinarily be prolonged.

Cirrhosis Mortality Rates as an Indicator of Heavy Drinking

A number of researchers have suggested that the cirrhosis mortality rate for a population is a good indicator of the fraction of the population that is drinking "heavily" (Skog, forthcoming). As such, the cirrhosis mortality rate can be used to compare relative incidence of heavy drinking in different populations or to measure trends in the incidence of heavy drinking for a single population. The main difficulty with using cirrhosis mortality in this fashion is that the current cirrhosis mortality rate reflects not only the current incidence of heavy drinking, but also

[10] Not all cases of fatal cirrhosis are related to alcohol. Schmidt (1977) estimated that the death rate in Canada from causes other than excess drinking is about 4/100,000. If this "base rate" of alcohol mortality is applicable to the United States, then almost three-quarters of cirrhosis deaths are alcohol-related.

[11] Schmidt (1977) reports evidence that drinkers who consume as little as three ounces of ethanol per day for long periods have a heightened risk of cirrhosis. A large percentage of those who die of alcohol-related cirrhosis have a drinking history that is more moderate than that of a clinical alcoholic population.

the trend in heavy drinking during the preceding 15-20 years (Jellinek 1947); a change in drinking habits in a population will not be fully reflected in cirrhosis mortality for this period of time. However, the short-run response of cirrhosis to heavy drinking is not necessarily negligible. We can imagine a population to have a "reservoir" of cirrhosis victims whose disease has progressed to a greater or lesser extent (Schmidt and Popham, forthcoming); if this group changes its drinking patterns, there will be some effect on the cirrhosis mortality rate within a short time.

To the extent that cirrhosis mortality rates do serve as an indicator of the incidence of heavy drinking, they are of considerable value in alcohol research. A number of social and medical problems besides cirrhosis are related to heavy drinking.[12] If it can be demonstrated that a particular alcohol-related policy is effective in reducing cirrhosis, then it would be expected that this intervention is also having the effect of reducing other problems associated with heavy drinking.

Previous Studies Relating Price to Cirrhosis Deaths

Seeley (1960) calculated intertemporal correlations between an alcohol price variable and the cirrhosis death rate for Ontario and for Canada, using annual data for 1935-1956. His "price" variable was an index representing the price of a gallon of beverage alcohol, divided by average disposable income. His work was extended to other countries and time periods by Popham et al. (1976). The reported correlations are typically close to 1.0. The authors' intepretation of these findings is that consumption levels by heavy, cirrhosis-prone drinkers are responsive to price.

There are two main problems with these studies. First, the "price" variable confounds the price of alcohol with income. The studies reviewed in the section above on demand for alcoholic beverages have consistently found that average consumption responds differently to changes in income and changes in price. Johnson and Oksanen (1977) in particular found that drinking in Canada was highly responsive to price but unresponsive to income. The second problem is that these correlation results may well be the result of "third-cause" variables, not included in the analysis, that are responsible for trends in both price and in the incidence of heavy drinking.

Historical "experiments" with large changes in price and conditions

[12] See Polich and Orvis (1979) for an analysis of the relationship between consumption level and the incidence of a variety of alcohol-related problems in a sample of U.S. Air Force personnel.

of availability provide another source of evidence on the degree to which heavy drinkers are responsive to such environmental factors. Prohibition is an obvious case in point. Warburton (1932, p. 240) found that alcoholic beverage prices during Prohibition were three to four times higher than before World War I. Cirrhosis death rates reached their lowest level of the 20th century shortly after World War I and remained constant at this low level (7-8 per 100,000) throughout the 1920s (p. 213). Furthermore, the drop in cirrhosis death rates was apparently greater for the relatively poor than for others, a result that reinforces the notion that high prices were at least in part responsible for the reduction in the prevalence of heavy drinking during this era:[13] Warburton (p. 239) reports that the cirrhosis death rate for industrial wage earners fell further than for city residents as a whole. Terris (1967, p. 2077) reports that the age-adjusted cirrhosis death rates dropped further for blacks than for whites between 1915 and 1920 and that these rates preserved their new relative position through the 1920s.[14]

Conclusions

Most cirrhosis deaths are the result of many years of heavy drinking. Cirrhosis is to some extent an "interruptable" disease process, so that a reduction in consumption on the part of a cirrhosis victim, even one whose condition is quite advanced, will extend life expectancy. The cirrhosis death rate may be a reasonably good indicator of the incidence of heavy drinking in a population. Previous research has provided some evidence to the effect that an increase in the price of alcohol will reduce the incidence of heavy drinking and the cirrhosis death rate, although this evidence is by no means decisive or compelling.

RESULTS OF A QUASI-EXPERIMENTAL STUDY

The nature of and justification for the quasi-experimental approach to studying the price elasticity of demand for liquor was explained above in the section on demand for alcoholic beverages. The same approach is used here to measure the short-term effect of changes in the liquor

[13] Economic theory and common sense both suggest that the price elasticity of demand for a normal commodity tends to be relatively high for households in which expenditures on the commodity constitute a relatively large fraction of their budgets.

[14] A related bit of evidence is given in Terris (1967). He notes that, in England and Wales in 1950, the cirrhosis death rate increased strongly with socioeconomic class, unlike in the United States. His explanation is that "spirits have been taxed out of reach of the lower social classes in the United Kingdom, where only the well-to-do can really afford the luxury of dying of cirrhosis of the liver" (p. 2086).

TABLE 4 Analysis of Tax-Related Changes

Rank	Net Change in Liquor Consumption	Net Change in Auto Fatality Rate	Net Change in Cirrhosis Death Rate
1–5	16	8	9
6–10	8	8	9
11–15	6	9	6
16–20	4	4	3
21–25	3	5	9
26–30	2	4	2
Percent below median	76.9%	65.8%	63.2%
Sign test: prob-value[a]	<0.001	0.037	0.072

[a] Suppose that a price change had no effect on liquor consumption. Then each observation on the net change in consumption associated with a tax change would have a probability of 0.5 of being negative. This is the null hypothesis. The "prob-value" reported in the first column is the probability that 30 or more observations out of the 39 "trials" would be negative if the null hypothesis were true. The prob-values in the second and third columns are defined analogously. These probabilities were calculated using the normal approximation to the binomial distribution, applying the continuity correction. These procedures and terminology are found in Wonnacott and Wonnacott (1977).

tax on the death rate due to cirrhosis and auto accidents.[15] This study uses the same 39 observations as the consumption study with one exception: data did not permit inclusion of the 1975 tax change in Massachusetts.

The results are reported in Tables 3 and 4. The formula used to calculate the "net change in auto fatalities" is strictly analogous to the formula used in the consumption study, with the auto fatality rate replacing liquor consumption per capita. The formula used to calculate the "net change in the cirrhosis death rate" is a bit more complicated. It can be written as follows:

$$\frac{\sum_{i=1}^{3} D^a_{t+i} - \sum_{i=1}^{3} D^a_{t-i}}{D^a_{t-1} + D^a_t + D^a_{t+1}} - \underset{s}{Median}\left\{\frac{\sum_{i=1}^{3} D^s_{t+i} - \sum_{i=1}^{3} D^s_{t-i}}{D^s_{t+1} + D^s_t + D^s_{t-1}}\right\}\lambda, \tag{6}$$

where

D^a_{t+i} = the cirrhosis death rate in the "trial" state i years after the tax change,

[15] Mortality rates for cirrhosis and auto accidents were calculated from frequency counts published in National Center for Health Statistics (1975, Table 1-13) and related tables in previous editions. Annual state population estimates were taken from Bureau of the Census (1971, 1978).

and so forth. This formula permits delayed effects of the tax change on cirrhosis to be taken into account.

The results of this analysis are summarized in Table 4. About 66 percent of the "net change" observations in the case of auto fatalities were negative. If tax changes in fact had no effect on auto fatalities, we would expect that only about 50 percent of these observations would be negative. The probability that 66 percent or more would be negative given the null hypothesis of "no effect" is less than 4 percent. Therefore, we can conclude with considerable confidence that a liquor tax increase tends to reduce the auto fatality rate.

About 63 percent of the "net change" observations in the case of cirrhosis deaths were negative. The probability of this high a fraction of negative values by chance alone is about 7 percent. It appears likely, then, that increases in liquor tax reduce the cirrhosis death rate.

As in any statistical study, these findings do not offer definitive proof of anything. However, the preponderance of the evidence certainly supports the conjecture that the price of liquor is one determinant of the auto accident and cirrhosis death rates. The quasi-experimental technique employed here minimizes problems of interpretation and in particular minimizes doubts concerning the causal process that underlies the results.

I conclude, then, that despite the questions posed at the beginning of this section, there is good reason to believe that the incidence of heavy drinking responds to liquor price changes of relatively small magnitude. The magnitudes of these responses are highly uncertain but can be estimated using the same techniques as were employed in estimating the price elasticity of demand. I converted the "net change in auto fatality rate" and the "net change in cirrhosis death rate" statistics (from Table 3) into price elasticities. The median of these price elasticities is -0.7 for auto fatalities and -0.9 for cirrhosis deaths. It seems entirely reasonable that these elasticity estimates turn out to be less than the price elasticity of demand for distilled spirits.

EVALUATING ALCOHOLIC BEVERAGE TAXATION

Alcoholic beverage prices have a direct effect on the prevalence of chronic excess consumption and the prevalence of the various problems caused by chronic excess consumption. There are three sources of evidence for this conclusion: (1) numerous studies, including my own (see the section on demand for alcoholic beverages), have found that per-capita consumption of alcoholic beverages is responsive to price changes. It is possible but unlikely that this observed responsiveness of aggregate drinking to price is due entirely to light and moderate drinkers;[16] (2) large changes in price associated with the adoption of Prohibition and other "natural experiments" of this sort have been associated with large reductions in the cirrhosis mortality rate and other indicators of the prevalence of excess consumption; and (3) small increases in the tax rate for spirits appears to reduce cirrhosis and auto fatality rates.

Each of these pieces of evidence is subject to legitimate scientific doubt. Nonetheless, I believe that, taken together, they provide a strong case for the proposition that an increase in the price of alcoholic beverages will reduce the prevalence of excess consumption and the incidence of the various problems caused by chronic excess consumption. The magnitude of the effect that could be generated by, say, a 20-percent increase in the alcoholic beverage price level is highly uncertain, although it appears likely that the effect of such an increase would be measured in terms of thousands of lives saved per year and billions of dollars of savings in medical and related expenses. Since the prices of alcoholic beverages are currently and historically controlled to a considerable extent by government policy, it is appropriate to view alcoholic beverage prices as public health policy instruments. This conclusion is empirical, rather than normative—it is by no means equivalent to concluding, for example, that it would be a good thing to raise the federal excise tax on alcohol or that higher prices are better than lower prices. A complete evaluation of a change in alcohol price policy requires consideration of other effects in addition to those related to public health. In particular, the distributive effects of a tax-induced increase in price should be considered, as should the loss in consumer benefits associated with low alcoholic beverage prices. These two types of concerns are discussed below.

INCIDENCE

The distribution of alcohol consumption levels among individuals is very diffuse and skewed to the right. This characterization is valid for every population group that has been studied (Bruun et al. 1975). Roughly speaking, one-third of U.S. adults abstain, one-third drink very lightly (up to three drinks per week), and the remaining third account for most

[16] There is considerable evidence that the consumption levels of the median drinker and the, say, 90th percentile drinker are closely related, as demonstrated by comparing population groups that differ widely in per-capita consumption (Bruun et al. 1975). Hence it would appear that the consumption levels of the "typical" drinker and the "heavy" drinker are subject to the same environmental and cultural influences, and/or that drinking patterns are interdependent or "contagious."

of the total consumption.[17] More precise characterizations of drinking distributions can be calculated from two recent studies. The Rand survey of drinking practices in the U.S. Air Force (Polich and Orvis 1979) found that 10 percent of the surveyed population (including abstainers) consumes 51 percent of the alcohol; 10 percent of the drinking population consumes 47 percent of the total alcohol.[18] DeLint and Schmidt's (1968) study of bottle purchases from government stores in Ontario found that 10 percent of consumers purchased 42 percent of the total alcohol.[19] Because the distribution of consumption has this property of concentrating a high percentage of consumption among a relatively few people, the incidence of alcohol taxation is necessarily very unequal. Whether this degree of inequality is good or bad depends on one's perspective. Three questions, reflecting three rather different normative perspectives on the incidence issue, are posed and discussed below.

How is the incidence of alcohol taxation related to the consumer's ability to pay?

Almost $10 billion in direct taxes and fees on alcoholic beverages was collected by all levels of government in 1976 (DISCUS 1977a). In most jurisdictions this revenue was not earmarked for specific programs but rather is used to help finance a wide range of governmental activities. One traditional standard in the public finance literature is that such general public revenues should be collected on an "ability to pay basis"; households with equal incomes should make equal contributions, and tax contribution should increase with income. By this principle, taxes on alcoholic beverages clearly receive low marks. Households with equal incomes pay vastly disparate alcohol taxes, depending on their alcohol consumption levels.

How much of a burden does alcohol taxation impose on members of poor households?

The answer to this question is not known, but it is useful to outline the relevant issues. First, an increase in alcohol taxes is disadvantageous for adult individuals and household heads who drink; they would not voluntarily choose to pay higher prices. Their dependents may be made better off, however, depending on the response of the household's drinking members to the price change. An increase in the taxes on alcoholic beverages can either increase or reduce the total expenditures of poor

[17] See report of the panel, pp. 27-28.

[18] Calculated from statistics reported in Polich and Orvis (1979, Appendix E). In making the calculations, I used the midpoints of each interval in Table E10 and assumed a mean of 20.0 for the top, open-ended interval.

[19] Beer and on-premise consumption were not included in this study.

households on alcohol, thereby leaving more or less money for food, clothing, and shelter. For households whose demand for alcoholic beverages is elastic (price elasticity less than -1.0), an increase in price will cause a reduction in total expenditure on drinking, while expenditures will increase for other households. Surely poor households differ considerably among themselves with respect to price elasticity of demand. However, the evidence above suggests that the average household's demand for spirits, at least, is quite elastic; furthermore, poor households would tend to be more elastic than higher income households. Therefore, for a high but unknown percentage of poor households, an increase in alcohol taxation should reduce expenditures for alcoholic beverages. Furthermore, it is quite possible that a tax-induced reduction in drinking in households that are at the high end of the drinking distribution may lead to reduced medical expenditures and increased earnings from employment.

Is the incidence of alcoholic beverage taxation related to the benefits received from government?

An alternative to the "ability to pay" standard is the "benefits" standard, which states that the distribution of tax liability should be closely related to the distribution of benefits received from government programs. To a large extent medical care, alcoholism treatment, minimum income maintenance, and other social services are provided and financed by government. The various health and social problems associated with alcohol consumption place expensive demands on these services. These problems are highly concentrated among the same group that pays the bulk of alcohol taxes—the chronic excess consumers. It is clear, then, that there is fairly close positive association between the amount of an individual's alcohol tax contribution and the expected value of government services consumed by the individual for alcohol-related problems. We view alcohol taxes as analogous to insurance premiums that are calibrated to one determinant of risk—the average rate of alcohol consumption—just as health and life insurance premiums are adjusted for age.

We could label this view of the alcohol tax the "drinker should pay" standard. This standard suggests that it is appropriate to set alcohol taxes at a level such that tax revenues are equal to the total government-financed costs of alcohol-related problems. Or we could choose to go further, by the same justification, and structure taxes so that drinkers collectively pay the total bill for the alcohol-relted externalities, including private costs borne by other individuals (e.g., we all pay higher premiums for private health and life insurance policies because some insured people drink unhealthy or unsafe amounts). Aside from the

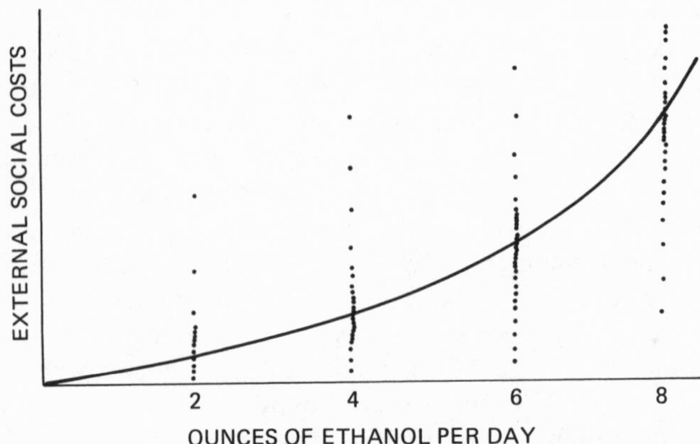

FIGURE 2 Hypothetical relationship between individual's consumption and resulting social harm.

problem of actually calculating the social costs of drinking, these quantitative standards are vulnerable to a major objection: the social harm of drinking is not proportional to the rate at which an individual consumes alcohol, so that a tax that is proportional to consumption will not be strictly proportional to alcohol-related harms. Figure 2 illustrates a hypothetical relationship that depicts average social cost increasing with consumption. The relationship is not strictly proportional because it incorporates two reasonable assumptions: (1) average social cost increases disproportionately with consumption; and (2) individuals at the same consumption level differ widely with respect to the external harm caused by their drinking, due perhaps to differences in personality, metabolism, drinking patterns, and so on. A tax that is proportional to the ethanol content of alcoholic beverages will then result in light drinkers, and the "safer" heavy drinkers, paying more than "their share" of the total bill for alcoholic-related social costs. Whether this arrangement is deemed better or worse than paying these costs from general tax revenues is a matter of preference. In my judgment, the "drinker should pay" principle is not sufficiently compelling in itself to justify high taxes on alcoholic beverages. But it should be kept in mind that high taxes reduce the social costs of drinking in addition to providing a mechanism for financing these costs.

COST-BENEFIT ANALYSIS

Besides providing a source of government revenue, taxes on alcohol influence the volume of total sales and the distribution of that volume among individual drinkers. If alcohol were not taxed, the price of alcoholic beverages would be too low because it would not reflect the negative externalities of drinking. The Pigovian principle requires that the tax on an externality-generating activity be set equal to the difference between the marginal social cost of the activity and its marginal private cost—an approach long advocated by economists for controlling environmental pollution. The objective of this type of tax is to "internalize" the external costs of the activity, thereby giving agents the incentive to curtail the activity in question to the appropriate level (i.e., the level at which every unit of the activity is valued at least as highly as the true social cost of that unit of activity). The normative force of this principle is undermined in the case of drinking by the fact (illustrated above in Figure 2) that the social cost of a drink differs depending on who consumes it and under what circumstances. Therefore, an increase in the tax on alcoholic beverages will deter some drinking that is socially worthwhile (the value to the consumer exceeds the social cost) as well as some that is not worthwhile. Given this situation and ignoring the distributional issues discussed above, the appropriate tax rate should be chosen by comparing costs and benefits at each tax level.

The marginal social benefit of a tax increase is equal to the value of the reduction in negative externalities that will result from reduced consumption, plus the additional tax revenue obtained. The marginal social cost of a tax increase is equal to the value of "consumers' surplus"[20] lost as a result of the tax. In principle, the tax rate is "too low" if the additional benefit of an increase exceeds the loss in consumers' surplus. It should be clear that it is very difficult to implement this principle due to the considerable uncertainty about the actual magnitudes of these theoretical constructs. But this discussion will perhaps serve as a useful framework for further empirical research in this area.

CONCLUSION

Public enthusiasm for government restrictions on drinking peaked in the early years of this century when many states and eventually the nation adopted Prohibition. Since the repeal of Prohibition in 1933 there has been a more or less steady decline in government restrictions on availability. Perhaps the most important aspect of this trend in recent

[20] Consumers' surplus is defined as the maximum amount consumers would be willing to pay for their current consumption level, minus the amount they actually are required to pay. This difference is positive because consumers value inframarginal units of the commodity at more than their price, as reflected in the fact that demand curves have negative slope.

years has been the rather sharp decline in the prices of alcoholic beverages (relative to average prices of other commodities) caused in large part by the failure to increase taxes commensurate with the inflation rate. While the public remains concerned about the "alcohol problem," there is a widespread belief that restricting availability is not an effective strategy for combating this problem. For example, a recent study by Medicine in the Public Interest (1979) concluded that state legislators are "generally skeptical about the effect of regulations, including taxation, on the incidence, patterns, or circumstances of use" (p. 31). On the basis of the evidence reported above, it appears quite likely that the legislators' view is incorrect—taxes do reduce total consumption and in particular reduce those portions of total consumption associated with auto fatalities and liver cirrhosis. If correct, these findings suggest that legislators should view alcohol taxation as a policy instrument for combating alcohol-related problems and not just a source of revenue. I am not advocating that taxes be raised—there are costs as well as benefits to raising taxes and the evidence presented above is far short of a complete cost-benefit analysis of a tax change. Rather the message of my results is that the benefits do exist and should be taken into account.

REFERENCES

Bruun, K., Edwards, G., Lumio, M., Mäkelä, K., Pan, L., Popham, R. E., Room, R., Schmidt, W., Skog, O-J., Sulkunen, P., and Österberg, E. (1975) *Alcohol Control Policies in Public Health Pespective*. The Finnish Foundation for Alcohol Studies, Vol. 25. Finland: Aurasen Kirjapaino, Forssa.

Bureau of the Census (1971) Population estimates and projections. *Current Population Reports*, Series P-25, No. 460. Washington, D.C.: U.S. Department of Commerce.

Bureau of the Census (1978) Population estimates and projections. *Current Population Reports*, Series P-25, No. 727. Washington, D.C.: U.S. Department of Commerce.

deLint, J., and Schmidt, W. (1968) The distribution of alcohol consumption in Ontario. *Quarterly Journal of Studies on Alcohol* 29(4):968-973.

DISCUS (1977a) *DISCUS 1977 Tax Briefs*. Washington, D.C.: Distilled Spirits Council of the United States, Inc.

DISCUS (1977b) *Summary of State Laws and Regulations Relating to Distilled Spirits*, 22nd ed. Washington, D.C.: Distilled Spirits Council of the United States, Inc.

DISCUS Facts Book (1977) *Beverage Alcohol Industry: Social Attitudes and Economic Development*. Washington, D.C.: Distilled Spirits Council of the United States, Inc.

Gavin-Jobson Associates, Inc. (1978) *The Liquor Handbook*. New York: Gavin-Jobson Associates, Inc.

Hause, J. C. (1976) Comment. *Journal of Law and Economics* 19(2):431-435.

Hogarty, T., and Elzinga, K. (1972) The demand for beer. *Review of Economics and Statistics* 54(May):195-198.

Houthakker, H. S., and Taylor, L. D. (1966) *Consumer Demand in the United States, 1929-1970*. Cambridge, Mass.: Harvard University Press.

Jellinek, E. M. (1947) Recent trends in alcoholism and in alcohol consumption. *Quarterly Journal of Studies on Alcohol* 8:1-42.

Johnson, J. A., and Oksanen, E. H. (1977) Estimation of demand for alcoholic beverages in Canada from pooled time series and cross sections. *Review of Economics and Statistics* 59(1):113-118.

Lau, H-H. (1975) Cost of alcoholic beverages as a determinant of alcohol consumption. In R. J. Gibbins et al., eds., *Research Advances in Alcohol and Drug Problems*, Vol. II. New York: John Wiley & Sons.

Medicine in the Public Interest, Inc. (1979) *The Effects of Alcohol-Beverage Control Laws*. Washington, D.C.: Medicine in the Public Interest, Inc.

National Institute of Alcohol Abuse and Alcoholism (1978) *Third Special Report to the U.S. Congress on Alcohol and Health*. Washington, D.C.: U.S. Department of Health, Education, and Welfare.

National Center for Health Statistics (1975) Deaths from 69 selected causes: United States, each division and state—1975. *Vital and Health Statistics of the United States*. Washington, D.C. U.S. Department of Health, Education, and Welfare.

Niskanen, W. A. (1962) *The Demand for Alcoholic Beverages: An Experiment in Econometric Method*. P-2583. Santa Monica, Calif.: Rand Corporation.

Ornstein, S. I., and Levy, D. (ca. 1978) Price and income elasticities of demand for alcoholic beverages. Graduate School of Management, University of California, Los Angeles.

Polich, J. M., and Orvis, B. (1979) *Alcohol Problems: Patterns and Prevalence in the U.S. Air Force*. Santa Monica, Calif.: Rand Corporation.

Popham, R. E., Schmidt, W., and deLint, J. (1976) The effects of legal restraint on drinking. In B. Kissin and H. Begleiter, eds., *Social Aspects of Alcoholism*, Vol. 4. New York: Plenum Press.

Popham, R. E., Schmidt, W., and deLint, J. (1978) Government control measures to prevent hazardous drinking. In J. A. Ewing and B. A. Rouse, eds., *Drinking*. Chicago: Nelson-Hall.

Schmidt, W. (1977) The epidemiology of cirrhosis of the liver: A statistical analysis of mortality data with special reference to Canada. In M. M. Fisher and J. G. Rankin, eds., *Alcohol and the Liver*. New York: Plenum Press.

Schmidt, W., and Popham, R. E. (forthcoming) Skog's "lagged" consumption variable: A comment on liver cirrhosis mortality as an indicator of heavy alcohol use. *British Journal of Addiction*.

Seeley, J. R. (1960) Death by liver cirrhosis and the price of beverage alcohol. *Canadian Medical Association Journal* 83:1361-1366.

Simon, J. L. (1966) The price elasticity of liquor in the U.S. and a simple method of determination. *Econometrica* 43(1):193-205.

Skog, O.-J. (forthcoming) Liver cirrhosis mortality as an indicator of heavy alcohol use: Some methodological problems. *British Journal of Addiction*.

Smith, R. T. (1976) The legal and illegal markets for taxed goods: Pure theory and an application to state government taxation of distilled spirits. *Journal of Law and Economics* 19(2):393-429.

Terris, M. (1967) Epidemiology of cirrhosis of the liver: National mortality data. *American Journal of Public Health* 57:2076-2088.

Waler, T. (1968) Distilled spirits and interstate consumption effects. *American Economic Review* 58:853-863.

Warburton, C. (1932) *The Economic Results of Prohibition*. New York: Columbia University Press.

Wonnacott, T. H., and Wonnacott, R. J. (1977) *Introductory Statistics for Business and Economics*, 2nd ed. New York: John Wiley & Sons.

Reducing Alcohol Abuse: A Critical Review of Educational Strategies

JOHN L. HOCHHEIMER

INTRODUCTION

Public concern about the control of alcohol abuse by individuals has had varying levels of intensity and focus throughout American history. The zenith probably occurred during the years immediately preceding the passage of the 18th Amendment to the U.S. Constitution, when advocates of temperance placed sufficient political pressure on the federal and state governments to embark the nation on the "noble experiment." Although the overall effects of Prohibition were mixed, the public came to believe that government-mandated abstinence was not a viable policy. To induce people to moderate their drinking behavior, it was thought

John L. Hochheimer is a graduate student at the Institute for Communication Research, Stanford University.

This paper was written with the guidance of Nathan Maccoby, Director, Institute for Communication Research.

I would like to thank Judith A. Courtney of McCann-Erickson, Inc., San Francisco; Dean Gerstein of the National Research Council; John Kaplan of the Stanford University School of Law; Robin Room of the Social Research Group at the University of California; John Pierce of the Institute for Communication Research at Stanford University; and Larry Wallack of the Social Research Group at the University of California for their help, comments, suggestions, and patience throughout the construction and various revisions of this paper.

For Janet K. Alexander, who is dearly missed.

that a program of education was needed to supplement government regulation of alcohol.

Many policy makers now perceive alcohol abuse to be a public health problem that requires responsible input from the medical and behavioral sciences. They believe that if the public can come to understand some of the personal and social problems associated with alcohol, this understanding will modify people's concerns about the consequences of their own alcohol consumption and ultimately change their drinking behavior. This model (knowledge leads to attitude change, which leads to behavior change—K-A-B) has been the foundation of most of the public information campaigns that have been launched against alcohol abuse. This paper examines attempts to educate people so that they can voluntarily choose healthy alcohol consumption behaviors.

Why should we bother with attitudes at all, since what we are really after is behavior change? Have there not been 30 years of attempts to modify drinking behaviors through the mass media with no demonstrable effect (Blane 1976, Blane and Hewitt 1977, Goodstadt 1978)? Is that not sufficient evidence against an educational approach?

The basic premise of this paper is that it is possible for people to consciously choose to change their behaviors if they are effectively educated. This requires information not only about some of the problems associated with alcohol abuse, but also about how people can change behaviors associated with consuming alcohol. Although it may be true that previous education campaigns focused on various aspects of alcohol abuse have had minimal, if any, effect (Blane 1976), as this paper shows, their apparent failures were due, not to the lack of viability of the approach, but to insufficient attention to some of the principles of mass persuasion and social learning theory that have been applied with some degree of success in related areas.

In brief, the argument is as follows: Many people used to (and still do) believe that mass education campaigns do not work because audiences are not interested. But studies show that people will listen and respond if (1) the campaign aims to provide practical education—something people believe they can readily use and (2) the campaign operates by means of tested communications principles.

In order to set practical goals for mass education campaigns, we need to have a good idea of how people behave and why or how behavior changes with education. "Hierarchies of effects" identified by communications theorists provide a series of models of behavior change that may be put into operation and tested for applicability to specific instances. These hierarchies involve knowledge, attitudes, and behavior.

Research shows that knowledge and attitudes can be influenced by information programs through the mass media,[1] while a combination of mediated and interpersonal persuasion works best in carrying change through all three levels, thus making the change most likely to stay in force. The Stanford Heart Disease Prevention Program (SHDPP), which this paper discusses in detail, was developed with these principles in mind. The program was based, in part, on McGuire's (1968) matrix of persuasive communication. This matrix, by including receiver and target variables, forces us to formulate a clear picture of *who* it is that we envision doing (or thinking or feeling) *what* as a result of the campaign. Principles of social learning and group influence tell us how interpersonal and institutional factors may be organized to reach campaign targets and receivers. The source, message, and channel parts of the matrix force us to envision how mass media can be used to advance these goals. Throughout a campaign, a smooth integration of the design (planning), program (implementation), and feedback (evaluation) functions is crucial to success.

In examining mass education campaigns aimed at alcohol consumption behaviors, we see that the absence of design research, good feedback, and clear targets vitiated their effectiveness. When viable objectives for an effective alcohol campaign are compared with other, successful health education experiences (for example, SHDPP and certain behavior modification programs for smoking and weight reduction) the resemblances encourage investment in alcohol education. A properly staged series of projects in which emphasis is placed on affecting both knowledge and attitudes is a sensible step at this time.

PUBLIC INFORMATION CAMPAIGNS: THE CLASSIC DEBATE

For many years, the prevailing judgment of the social science community was that any public information campaign was doomed to almost certain failure. The literature was replete with campaigns that foundered on the rocks of audience noncompliance, believed to be due to three factors: (1) "chronic know-nothings," (2) de facto selectivity, and (3) selective exposure to new information.

Hyman and Sheatsley (1947), in describing "Some Reasons Why Information Campaigns Fail," noted that a large proportion of the population was not familiar with any particular event, despite the strength

[1] Some modest behavioral effects can be initiated through the media as well (see Maccoby and Farquhar 1975 and Maccoby et al. 1977).

of the messages and the breadth of their distribution. The lack of reaction by such "chronic know-nothings" to the specific messages of the information campaigns they surveyed was ascribed to "apathy." Hyman and Sheatsley argued that, although the various media disseminating the information (at that time radio, newspapers, and magazines) were adequate to ensure the broadest possible coverage, and the messages themselves were sufficient to have some noticeable effect on the audience, "there is something about the uninformed which makes them harder to reach no matter what the level or nature of the information" (p. 413). It was felt that the apathy of these people was insurmountable and, most likely, the creators of the messages could do nothing to overcome it.

The literature often reports biases in the composition of voluntary audiences of public information campaigns (Sears and Freedman 1967). This "de facto selectivity" has been noted as far back as Lazarsfeld and his colleagues' classic study (1948) of voting behavior: "Exposure is always selective; in other words, a positive relationship exists between people's opinions and what they choose to listen to or read" (p. 164). Similar conclusions are reached by Lipset (1953), Schramm and Carter (1959), and Star and Hughes (1950).

Members of the audience who avail themselves of the information disseminated by a campaign were thought to be predisposed to seek information congenial with their existing attitudes. People apparently allow themselves "selective exposure" to new information, filtering it through existing cognitive structures and retaining or interpreting just that information which is congenial and supportive of previous attitudes. If new information conflicts with established beliefs, no message may be able to motivate the desired behavior change (Cartwright 1949).

Klapper (1960) concluded that the mass media changed attitudes minimally and functioned primarily to reinforce previously held views. As recently as 1976, in evaluations of the "Feeling Good" television series (Swinehart 1976) and of campaigns to persuade people to use seat belts (Robertson 1976), social scientists still argued that people will not learn and that the reasons for the communicators' failure to persuade resided within the individuals themselves. The focus has not generally been turned around on the producers to find out whether poor message construction, poor use of the media, or inadequate evaluation procedures might have been the true culprits.

Sears and Freedman's (1967) reanalysis of data, however, from some of the previously mentioned campaigns found the case for de facto selectivity and selective perception of information to have been greatly overstated. For example they show that in the Lazarsfeld et al. (1948) study, looking at the data from one perspective, only the Republicans had been selectively exposed, while viewing it from a different perspective, only the Democrats had been selectively exposed. A reanalysis of data from a massive campaign in Cincinnati to "sell" the United Nations (Star and Hughes 1950) shows that education, not previous orientation, was the best predictor of who got the most from this campaign. Selective exposure may exist, but it is only one explanation of the campaign effects.

Sears and Freedman found similar flaws in the assumption that people prefer only new information that agrees with their existing predispositions. They cite five studies showing some preference for supportive information, eight studies showing no preference, and five studies showing preference for nonsupportive information. They concluded that "[t]he available evidence fails to indicate the presence of a general preference for supportive information" (p. 208).

Sears and Freedman did find evidence that people prefer information that they expect to serve a practical purpose as opposed to less useful information. Past history of exposure to a particular issue being presented was also found to be a significant factor. Their work supports the view that education, information utility, and previous exposure are only three of many factors that interact to influence people involved in any campaign.

Mendelsohn (1973) goes further, arguing that discussions of the failure of public information campaigns have focused on the deficiencies of the audience to explain the presumed lack of effect, rather than on the creator of the message, the content, or the media. "With rare exceptions," says Mendelsohn, "mass communications researchers have been documenting and redocumenting the by now obvious fact that when communicators fail to take into account fundamental principles derived from mass communications research, their efforts will be generally unsuccessful. . . . In short, very little of our mass communications research has really tested the effectiveness of the application of empirically grounded mass communications principles simply because most communications practitioners do not consciously utilize these principles" (p. 51). He suggests that those involved in all phases of campaigns work together in the planning, implementation, and evaluation of campaigns in order to incorporate principles of communication effectiveness derived from research. In this way, those who create the campaign can come to understand and use social science principles, while social scientists can realize some of the constraints, norms, legal requirements, etc. of those who work regularly with the media.

The "National Driver's Test" is an example of this kind of effective integration. It had been established that most drivers tended to ignore the more than 300,000 messages disseminated each year about bad drivers, since 8 in 10 considered themselves to be good drivers (Mendelsohn 1973). A new approach was necessary to overcome this indifference to traffic safety messages. Viewers would be given the opportunity to become aware of possible deficiencies in their driving behavior and would be given information on how to correct the deficiencies they discovered.

A TV program was created with three specific objectives: (1) to overcome public indifference to traffic hazards that may be caused by bad driving; (2) to make bad drivers cognizant of their deficiencies; and (3) to direct viewers who became aware of their driving deficiencies into driver improvement programs in their respective communities. Mendelsohn notes that not only were the objectives realistic and specific, but they also lent themselves to objective post hoc evaluation.

A massive national publicity campaign for the program was mounted. The program aired on the CBS television network prior to the Memorial Day weekend, when concern for traffic safety was greater than usual.

The results were impressive. Over 30 million people watched the program (making it one of the highest rated public affairs programs to that date). Subsequently, CBS received mail responses from almost 1.5 million people. Of greatest interest is that, according to the National Safety Council, about 35,000 people enrolled in driver improvement programs shortly following the telecast. This represented an estimated threefold increase over previous annual enrollment.

Those involved with the National Driver's Test concluded that innovative programming, assisted by adequate promotion, can whet the public's appetite for useful information, overcoming the alleged apathy. Also, it is apparent that reasonable goals are amenable to rigorous evaluation and effective accomplishment.

We know that it is possible to wage an effective campaign for public education through the media. What is known about the effectiveness of campaigns aimed specifically at alcohol abuse, however, is "for the most part . . . highly tentative" (Blane and Hewitt 1977, p. 15).

ALCOHOL ABUSE CAMPAIGNS

The same factors generally found by Mendelsohn (1973) to characterize ineffective public education campaigns permeate campaigns about alcohol as well. Haskins (1969), for example, reviewed studies on the development and evaluation of mass mediated campaigns aimed at drinking and driving (among other safety campaigns). He found the research on which campaigns and their evaluations were based to be scanty. The formative research[2] that was available relied solely on laboratory experiments and verbal measures. Evaluative studies were found to lack control groups and statistical analyses. He found that the message strategies relied heavily on negativistic, threatening fear appeals—an approach that may not necessarily be the most successful.[3] Haskins reaches the same conclusion as Mendelsohn: campaigns should make "systematic use . . . of known communications research, accompanied by appropriate pre-testing research at various stages of the development process" (p. 65).

Concerning the formation of alcohol abuse campaigns, only two areas have been studied (Blane 1976). One seeks to ascertain how receptive different segments of the potential audience would be to various messages pertaining to alcohol use. The National Highway Traffic Safety Administration (NHTSA) tried to assess the extent and conditions of adolescent and adult involvement with alcohol. Those involved in social or business occasions in which alcohol was served at least once in the last 3 months were considered at highest risk for being in a drinking-driving situation: they comprised 54 percent of adults, according to the study. Of these, 43 percent (about 23 percent of the total sample) felt that they may have been driving while intoxicated once in the previous year. Of these people, 75 percent had taken some action to prevent the situation from occurring.

From survey data, Grey Advertising (1975a) was able to identify and classify four basic groups according to the degree and nature of countermeasures they were willing to perform to evade or prevent drinking and driving. "Social conformers" are primarily upper-class persons who would take the least obtrusive actions (e.g., serving food at parties and/or driving intoxicated guests home). "Cautious pre-planners" tend to be older women of lower socioeconomic status (SES) and are more likely to prevent drinking and driving. "Legal enforcers" were found to be older, middle-SES people who would be likely to use legal countermeasures, such as calling the police, as a way of attempting to prevent intoxicated people from driving. "Aggressive restrainers" tend to be younger men who would be willing to physically restrain their friends from driving under the influence of alcohol.

In their first evaluation of the formative data, Grey Advertising

[2] Early informal measures of effects that can be used to plan future messages, channel strategies, etc.

[3] Only under certain conditions is strong fear arousal the most effective use of fear appeals (Chu 1966).

(1975a) recommended that instead of using a shotgun approach to their campaigns as they had done in the past, NHTSA would do better to aim at drinking adults, rather than the entire population. They further recommended that NHTSA develop different strategies to sanction giving and receiving help in alcohol-related situations for "social conformers" (motivating conformity) and "aggressive restrainers" (motivating friendship).

In a second evaluation, Grey Advertising (1975b) suggested a basically similar approach for adolescents. NHTSA messages should be aimed at drinking youth as well as those adults in a position to influence them. Grey also advised that alcohol-related messages should attempt to establish a peer group norm that to give or receive aid when one is impaired by alcohol is acceptable behavior.

In another study, the Addiction Research Foundation utilized sample surveys in Ontario, Canada, in 1975 to find how much concern there was about alcohol-related problems and what remedies the public perceived to be adequate (Gillies 1975, Gillies et al. 1976a,b,c). A "Social Policy and Alcohol Abuse Survey" was conducted in order to "ascertain the attitudes of a cross-section of Ontario adults toward existing and hypothetical regulatory measures related to the use of alcohol" (Gillies 1975, p. i). In general, the survey discovered a high degree of concern about the incidence and consequences of alcohol abuse and a variety of opinions about what should be done about it.

There has also been some formative research on the development of messages for alcohol-related campaigns. Flynne and Haskins (1968), for example, wanted to find out if the statement of behavioral intention (confidence in driving after drinking) was indicative of actual willingness to perform the behavior. With such knowledge, campaign evaluations could be designed to determine whether messages aimed at reducing the unwarranted confidence of an intoxicated driver in his or her ability to drive were successful. Flynne and Haskins found that real-world observation (rather than laboratory testing) of drinking behavior was necessary for effective pretesting of messages. Real-world driving behavior differed significantly from behavior manifested under experimental conditions; hence, any means of measuring the effects of the campaign other than direct observation of real-world behavior would be invalid.

This finding applies to a program that was designed to eliminate indifference about drinking-driving messages; it resulted in the film, "A Snort History." The didactic approach that had characterized most such messages was avoided, because tests had shown that high-fear appeals had a boomerang effect. The 6-minute film told how alcohol affects judgment in that drinkers tend to become more optimistic about their driving ability precisely when pessimism would be more prudent (Mendelsohn 1973).

Of those viewers tested, 43 percent reported the film left them feeling concerned about the effects of alcohol on driving; 30 percent said that they would consider changing their ideas regarding safe driving (Mendelsohn 1973). Real-world observation would determine how much of this cognitive shift converts directly (without further training) into changed behavior.

Admittedly, these are less positive results than those from the "National Driver's Test." "A Snort History" does show, however, that information can be transmitted through different media with some positive effect, holding the attention of the audience while getting the message across. (In fact, the quality of "A Snort History" was good enough to have it placed as a short subject in a Denver first-run motion picture theater, where it was not subject to diffusion of impact by conflicting messages.) There may be a study or two that further refines what has been determined by these studies; aside from what has been presented here, however, there has been no further analysis of—to paraphrase Lasswell—what do you want to say to whom, how effectively do you want to say it, and through what channel.

Several assessments of program evaluation techniques (Douglas 1976, Driessen and Bryk 1972, Kinder 1975) have criticized the lack of methodological rigor. Some of the problems mentioned are: overly simplistic before-and-after designs; inability to control for confounding factors; the use of a variety of techniques in one campaign so that the effects of any one could not be singled out; and failure to consider the possible impact of a Hawthorne effect[4] among people being studied.

Blane (1976) found that few evaluations of public education campaigns related to alcohol abuse have been methodologically sound. Typically, a simple before-and-after design is used to assess changes in attitudes, knowledge levels, and behavior of a target audience. Changes are attributed to the intervention of the campaign, but due to the typical lack of control groups in the experiments, conclusions are weakened because of unmeasured factors that may have confounded the results.

Blane and Hewitt (1977) have listed the following deficiencies in campaign evaluation strategies:

(1) Lack of integration of evaluation into the overall campaign design, which prevents a symbiotic relationship between evaluators and designers of the campaign. A two-way flow of information is necessary so the

[4] Enhanced desirable behavior because of the knowledge that one is being observed.

program can be developed based on what is known from research and so feedback from the evaluative process can guide future refinement of the campaign.

(2) Lack of sufficient precampaign testing of message content and appeal, which should be conducted with suitable samples from within the target audience.

They make the following recommendations for improvement:

(1) Statistical analyses of results rather than merely reporting before-and-after percentages of variables as being something meaningful.

(2) Close coordination of campaign objectives and evaluation design.

(3) Long-term evaluation to assess other than immediate campaign effects.

(4) Use of unobtrusive and nonreactive measures.

(5) Determination of actual exposure of the target group to the campaign as well as correlation of these data with data concerning changes in attitudes, knowledge, and behavior.

(6) Attention to determining the possible negative, counterproductive side effects of the campaign.

Given the lack of both formative research and sufficient evaluation, it is no wonder that previous public education campaigns aimed at reducing the incidence of alcohol abuse have had such inconclusive results. Proper use of the mass media for effective dissemination of messages is a multifaceted process that requires a great deal of planning, evaluation, and willingness to replan during the campaign if necessary.

I have shown that an integrated approach can be helpful in certain campaigns. Rather than focus on other alcohol-specific campaigns to see how they might have been changed to be more effecive, I next review the elements of a well-designed campaign and what can be done to best implement them. I pay particular attention to health-related campaigns in which these factors have been manipulated to advantage.

THEORETICAL PRINCIPLES OF BEHAVIOR CHANGE

MODELS OF EFFECTS

If we have learned anything from the experiences of those who have attempted to implement campaigns in the past, it is that good intentions and the utmost earnestness of convictions are insufficient catalysts for engineering meaningful change in behavior. We have had no adequate way of determining whether previous educational strategies were appropriate, as good research designs for evaluation are "a recent and still rare phenomenon" (Blane and Hewitt 1977, p. 15). Similarly, there has not been a rigorous application of the principles established by either social scientists in general or communciations scholars in particular. Despite Mendelsohn's contention that it was the neglect of these principles that proved to be the undoing of much of the public information campaigns of the past, Blane (1976) reminds us that "the potential benefits of applying social science theory and methods to an understanding of the conditions under which public information and education messages (about alcohol use) are most effective in attaining social goals have been little explored" (p. 540). So, if using what is known can be effective (Mendelsohn 1973), and if previous campaigns aimed at moderating the use of alcohol have not utilized what is known (Blane 1976, Blane and Hewitt 1977, Goodstadt 1978, McGuire 1974), then there has not been an adequate implementation of communications principles (coupled with equally rigorous evaluation) to determine whether a well-conceived public education campaign to moderate the abuses of alcohol might have more positive results. I consider in this section an integrated community organization plan as a model for the dissemination of information about how to moderate alcohol-related behaviors. The usefulness of the model depends on the extent to which mass media, existing networks of social organization, and specifically created face-to-face instructional facilities are integrated. All the knowledge necessary for the immediate mass implementation of such a campaign with a guarantee of success is not currently available. What follows is a set of guidelines based on the most promising research on how to implement a community-based strategy of health education. Such a program is likely at least to provide deeper insight into how to more successfully counteract alcohol-related health problems. The Stanford Heart Disease Prevention Program 3-Community Study, described in detail below, is an example of a successful community-based health education program.

The theoretical model for the 3-Community Study was based on the work of Cartwright (1949). He argued that for mass education to be effective, three kinds of changes must occur: changes in cognitive structures (what people know), changes in affective structures (what people want to do), and changes in action structures (what they actually do). The cognitive component (which includes attention, knowledge, information, belief, awareness, comprehension, learning, etc.) refers basically to how the focus of attention—the "attitude object"—is perceived. The affective component (including conviction, interest, desire, yielding, evaluation, etc.) pertains to a person's subjective like or dislike of the attitude object. The action, or conative, component (such as intention, behavior, adoption, etc.) refers to a person's gross behavioral tendencies toward the objective in question (McGuire 1969, Ray 1973).

Cartwright believed that the mass media are usually most effective in initiating change in cognitive and affective structures; they are not often successful, however, in initiating change in action structures.

Although the existence of these three components of a "hierarchy of effects" of any persuasive message is well established, it is of critical importance to determine which particular ordering of the three components will yield the best results.

Of the six possible permutations of the three components, three orderings have dominated the literature: (1) the "learning" hierarchy of cognitive-affective-conative (K-A-B); (2) the "dissonance attribution" hierarchy of conative-affective-cognitive (B-A-K); and (3) the "low-involvement" hierarchy of cognitive-conative-affective (K-B-A). Rather than being mutually exclusive methods for affecting persuasive change, each has its place in the process of persuasion.

Initially, most people involved with public education campaigns believed that cognitive change preceded the affective, which preceded the conative (or K-A-B), in a stairstep manner; this progression was called the "learning" hierarchy of effects. Later studies, however, showed that it occurs in some situations but not in others (Ray 1973). The learning hierarchy exists primarily when the audience is actively involved in the topic of the campaign (the attitude object is already important to them) and when there are (or people are shown that there are) distinct differences between the choices offered. In this situation, members of the audience first become aware of the object; then, since it is important to them, they may develop interest, evaluate the object, try it, and, upon favorable change in attitude toward the object, adopt a new behavior.

The dissonance attribution hierarchy (B-A-K) proposes the opposite sequence of events. When a friend suggests something new, a person may try it solely on recommendation. Affect and knowledge are altered after the behavior is performed. This sequence often occurs when the source of influence is either someone to whom the person normally turns for guidance or local peer group norms. The person, involved with the attitude object, does not see that behavioral choices are available and is forced to make a specific choice of behavior as well as affective changes to support it (usually based on the behavior). Knowledge may increase on a selective basis if the person seeks out information about his or her behavior. The primary role of the mass media in this instance, after the behavior has occurred, is to reduce dissonance within the person, which may have come about as a result of the behavior and attitude changes, or to provide information for attribution of the behavior (Bem 1972, Ray 1973).

The low-involvement hierarchy (K-B-A) was explored by Krugman (1965). He noted that most people really do not care about the content of most advertising. This low involvement means that perceptual defenses to the messages are lower. Although a single announcement will probably have no effect, repeated exposure to the same message over a period of time may lead to shifts in cognitive structure. The low-involvement hierarchy is illustrated by the National Safety Council's "If you drink, don't drive. If you drive, don't drink" campaign. People had seen or heard this message so many times that other alcohol-related campaigns were thought to convey this message even though they did not discuss the issue (Harris and Associates 1974). Excessive repetition, on the other hand, may result in the message's being blocked out entirely. When there is little objective difference between alternatives, when the audience perceives little difference between alternatives, or when the audience does not care about the magnitude of difference between alternatives, this low-involvement hierarchy is most likely to be effective.

What would be the most effective approach for an alcohol-related campaign? It would be a mistake to place all the emphasis on one particular strategy. A campaign designed to change the drinking behaviors of chronic abusers of alcohol would not use the low-involvement approach, for example, because these people have rather high involvements with the object of the campaign. A campaign designed to educate children about some of the problems of overindulgence might have more success using it.

Goodstadt (1978), reviewing various studies of the learning (K-A-B) approach to influencing drug-associated behavior, found little evidence of behavioral change resulting from attitude change. The difficulties, as he sees them, are the inability to define distinctly the behaviors to which the cognitive and affective components of attitudes refer, the inability to factor out situational variables that also have a strong influence on behavior, and difficulty in measuring the attitudes themselves. He suggests that "a more effective medium for change might be via both direct and indirect behavior influence rather than through attitude change attempts" (p. 266).

Legally required or mandated behavior change, however, especially related to drug or alcohol use among adolescents, can lead to a hardening of attitudes and clandestine behavior as an act of defiance (McGuire 1974). Without appropriate persuasive communication, mandated behavior change can do more harm than good. For example, the campaigns to require people to use seat belts (Robertson 1976) have engendered hostile responses. In response to government-mandated installation of interlock systems on belts in cars, many people disconnected the belts or the warning buzzers rather than comply with the intent of the law.

Goodstadt (1978) suggests a fourth model (V-P-B), based on the belief that values influence behavior, which he believes may help decrease the incidence of drug abuse. The emphasis on values in health education is a result of dissatisfaction with traditional methods. Familiar problems, however, have been found in most drug-related value-behavior programs: "Results from the very few programs which have been evaluated are inconclusive because of (a) their scarcity; (b) their methodological problems; and (c) conflicting or ambiguous results" (p. 271).

An adequate determination of the levels of people's knowledge, attitudes, and behaviors concerning alochol is essential to determine the most effective approach.

THE STANFORD HEART DISEASE PREVENTION PROGRAM: AN INTEGRATED MODEL OF BEHAVIOR CHANGE

Description

The Stanford Heart Disease Prevention Program 3-Community Study, as predicted by the work of Cartwright, achieved success by supplementing mass media programs with intensive interpersonal instruction. In addition, the study shows that certain reduced-risk behaviors (such as simple dietary changes and exercise) can be learned by attending to the mass media without other planned input (Maccoby and Farquhar 1975). Other behaviors (such as smoking) were shown to change only with some interpersonal training (McAlister et al. 1979). When cigarette use, plasma cholesterol, systolic blood pressure, and relative weight reductions were incorporated into the risk equation, the net difference in estimated total risk between control and treatment samples in the study was 23 to 28 percent (Farquhar et al. 1977).

The Stanford study disseminated messages aimed at five different goals throughout three communities using various media. The goals were: (1) to generate awareness about the program and its focus; (2) to increase the knowledge of the audience; (3) to motivate people to adopt the new behaviors; (4) to teach people new skills; and (5) to reinforce the new skills and behaviors people had learned as a result of the intervention so they would maintain them.

In the communities of Gilroy and Watsonville, the use of mass media was sufficient to increase levels of awareness of heart disease risk. Furthermore, on the basis of information obtained through the media, subjects in these two communities changed their health behaviors to the extent that significant decreases in plasma cholesterol and saturated fat intake were found. The use of the media alone, however, was not sufficient to aid in either significant weight loss or a decrease in cigarette consumption.

When intensive instruction was added in Watsonville, subjects were found to have higher knowledge of factors of heart disease risk as well as a much greater reduction of risk after 1 and 2 years than either the control or the mass-media-only groups. The Watsonville group also had greater reductions in cigarette consumption and weight.

Discussion

While the use of alcohol may be similar to the use of other risk-related factors, the aims of education programs about alcohol abuse are likely to be somewhat different. The goal of programs to counteract smoking, for example, is for the individual not to use or to stop using tobacco entirely. We do not, however, necessarily seek to induce a person not to drink or to stop drinking alcoholic beverages altogether; rather we wish for her or him to moderate that behavior, bringing it under more conscious control. From this perspective educating people to control alcohol abuse is, in some aspects, more similar to educating people in the efficacious methods of dietary control. An approach that combines methods found to be effective in antismoking campaigns with methods used in successful strategies for dietary change may prove valuable in affecting alcohol abuse.

One of the difficulties with any strategy for behavior change (such as moderating alcohol consumption) is that altering a person's complex behaviors necessitates learning new skills. These require individualized instruction along with some form of feedback on the individual's current level of expertise with the new skills. Feedback includes giving the individual the feeling that he or she has accomplished something and perhaps providing a reward system for successful progress. This strategy has been found to be somewhat effective in various weight loss programs (Bandura and Simon 1977, Stunkard 1975, Stunkard and Mahoney 1974). However, Stunkard and Penick (1979) point out that maintaining long-term weight loss is a goal that has as yet eluded most behavior modification programs.

For a campaign of alcohol abuse education, the first wave of messages might aim to create awareness of the problem through spot announcements (Wallack 1978); newspaper columns, television programs, etc., could be used to generate understanding of how the problem affects the receiver personally and what she or he might do to rectify the situation; personal and/or group counseling would teach new behaviors (so that the person who abuses alcohol would be able to respond positively to new cues in old alcohol-related situations) and reinforce the newly manifested behavior. Helping people to learn self-management (Bandura 1977, 1979) would be the last step of face-to-face training in achieving long-term maintenance of the changed drinking behavior.

It is important to understand the theoretical basis of attitude change in designing strategies of intervention. The next section discusses what attitudes are, how they are formed and can be changed, and what their relationship is to behavior.

ATTITUDE COMPOSITION

Rokeach (1966-1967) defined an attitude as "a relatively enduring organization of beliefs about an object or situation predisposing one to respond in some preferential manner" (p. 530). Fishbein and Ajzen (1975) add that it is the evaluative or affective nature of attitudes that distinguishes them from other concepts. Although there is some discussion involving the exact nature and measurement of attitudes (Fishbein and Ajzen 1975), there is almost universal agreement that attitudes have both a cognitive component and an evaluative component, that they are fairly well organized and long-lasting, and that there may be possible interactions between particular attitude objects and the situations that involve them. Change in an attitude indicates a change in one's predisposition to respond to a particular object in a certain situation.

The susceptibility to change of an attitude depends on the conception of how an attitude is formed. Bem (1970) and Jones and Gerard (1967) have posited that the attitude structure can be seen as a syllogism of "psycho-logic." That is, an attitude is the conclusion of a syllogism combining a belief about an object with a relevant value. The syllogism may not be perceived by the person but rather may be an implicit part of the thought process.

For example, in the syllogism, "Cigarettes cause cancer. Cancer is bad. Therefore cigarettes are bad," the conclusion ("Cigarettes are bad") is derived from a belief connecting cigarettes to cancer combined with a negative value toward cancer. The syllogism may appear to be overly simplistic, since everyone agrees that cancer is bad; in fact, the "Cancer is bad" step may be left out entirely, since it is implicit in our cognitive thought process. Yet, somehow we learned something about cancer and were persuaded by it, coming to believe that it is bad (Roberts

1975). A statement like "Frequent consumption of large amounts of alcohol is related to higher levels of high-density lipoprotein" probably has no evaluative meaning for most people. Only after people are taught or persuaded about an evaluative relationship, e.g., between high-density lipoprotein and heart disease, that an attitude about frequent consumption of sizable amounts of alcohol enters their logical structure.

Many attitudes are based on a similar combination of evaluative and nonevaluative beliefs, although the linkage is not quite so obvious. Attitudes may also be conceived of as having both a horizontal and a vertical structure. Horizontally, a person may have more than one syllogistic attitude about a particular attitude object. To extend the example above, I may also believe that "Smoking pollutes the air. Air pollution is bad. Therefore smoking is bad." The vertical structure of attitudes means that a given attitude may be a second, third, or *n*th derivative of other beliefs and attitudes. So, regarding alcohol consumption one might also think, "Having a six-pack every night after work means that I drink a lot of alcohol. Drinking is bad. Therefore having a six-pack every night is bad." To complicate matters further, higher-order beliefs are bolstered by the breadth of the horizontal beliefs around them, with more than one syllogistic chain culminating in the expressed attitude (see Figure 1).

However, as Bem (1970) suggests, the more layers there are to the vertical composition of the syllogistic structure, the more points are open to attack. Each of the underlying premises on which the individual had built her or his attitude may afford an opportunity to precipitate some change. In the chain of attitude composition, says Bem, "higher

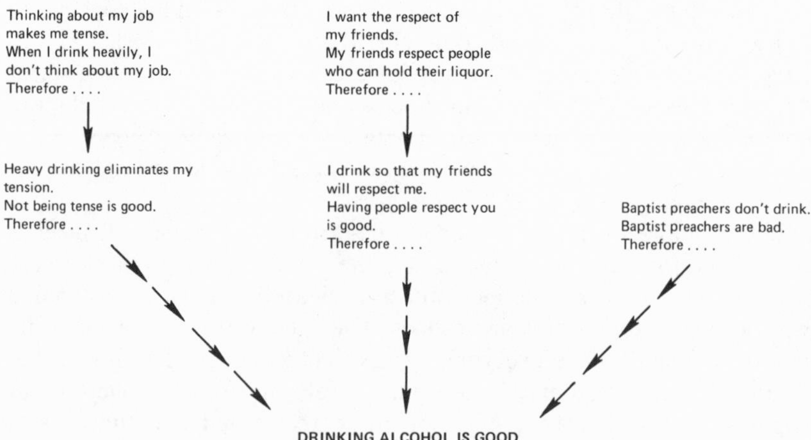

FIGURE 1 Hypothetical structure leading to a positive evaluation of drinking.

RESULTS: Have you smoked tobacco in the past month? Percent responding, "yes."

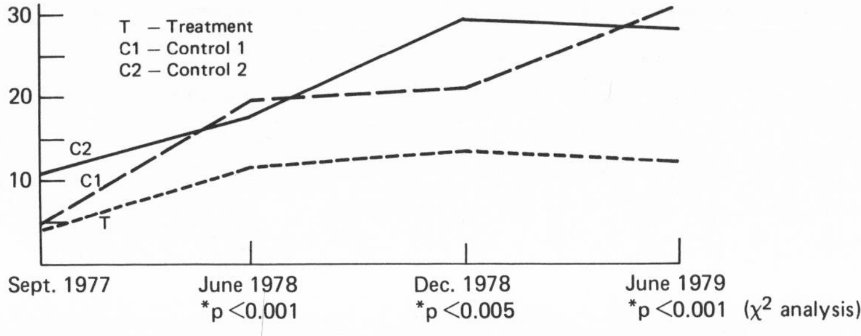

RESULTS: Have you smoked tobacco in the past week? Percent responding, "yes."

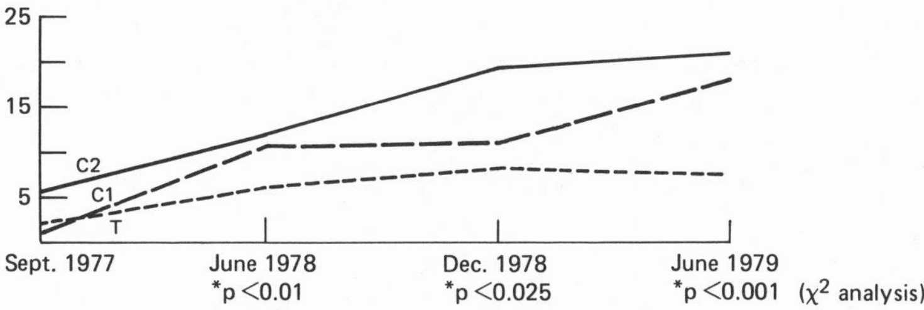

*REPLICATION with the seventh grade class (1978-79) yielded significant results in December 1978 and June 1979 ($p < 0.001$).

FIGURE 2 Results of project C.L.A.S.P. in Santa Clara County, California. *Source*: Perry (1979). Reprinted by permission.

order belief would appear to be only as strong as its weakest link" (p. 11).

As we can see in Figure 2, there are three syllogistic paths that might lead to the conclusion that drinking is good; even if one of the chains were weak and vulnerable to attack, it is supported by the others. If Figure 2 were a true depiction of a person's attitude structure about drinking, it would be possible to convince him or her that Baptist preachers are good people who may not drink for other reasons; such persuasion, however, would not change the overall attitude, since this person still wants the respect of his or her peer group and a decrease in tension at work. We can also see that this illustration depicts a "psycho-logic." That is, the syllogisms may be internally consistent to an individual but have no basis in logic to an objective observer (assuming that objective and accurate observation of all the factors of another's attitudes were possible). Conversely, because something may be objectively true does not make it subjectively true. (Indeed, there are still some people who believe the earth to be flat.)

The structure of these "psycho-logical" syllogisms helps us to conceptualize several ways to initiate a change of attitude. One might aim the persuasive appeal at a cognitive component of the attitude structure. In terms of the illustration, one might attempt to convince the person that there are many Baptist preachers who do in fact drink. One could try to modify the value premise, i.e., convince the person that Baptist preachers are good. One could try to modify the intensity of that value, i.e., show the person that although Baptist preachers may not be perfect, they are certainly not all that bad. Or one might try to circumvent the belief structure entirely and try instead to construct a new one. One might suggest new cognitions about some negative aspects of excessive drinking behavior, in the hope that the antidrinking values attached to those new beliefs would outweigh the prodrinking values of the old one.

Ross (1976) postulates that people relate to the world as amateur or intuitive psychologists. They observe events and attempt to ascribe causes to them. These causal attributions tend to influence subsequent attitudes and behaviors. This is the heart of attribution theory. This kind of causal attribution, however, often leads the attributor to a number of errors.

The most common mistakes result from the limited experience of the perceiver. A person normally has only his or her existing knowledge, attitudes, and previous attributions (which may have caused previous errors in perception) on which to base judgments. An example is a young person who sees a friend drink to intoxication, attributes this as normal ("with it") behavior, and subsequently assumes the behavior for herself or himself. Another example is seeing the hero in a Western walk into the local saloon and down shot after shot of whiskey, then beat the bad guy and win the heroine; the audience might attribute social efficacy to the behavior and try to assume it.

One focus of a public education campaign aimed at reducing alcohol abuse could be the dissemination of messages showing that the excessive drinker is not, for example, a "he-man" or that teenage peers do not consider excessive drinking to be normal or "cool" behavior.

INTERPERSONAL STRATEGIES FOR BEHAVIOR CHANGE

SOCIAL LEARNING APPROACH

In some instances, the knowledge that a problem exists is insufficient impetus to change certain health behaviors. As Cartwright (1949) indicated, it is also necessary to teach people the action structure necessary to attain the desired goal. Sometimes the psychological costs are thought to be too great to overcome. For example, many people realize that smoking cigarettes is harmful to their health, yet they continue. They may feel that the costs associated with quitting (such as "I'd put on weight," "I would crave cigarettes all the time," or "I just don't have the willpower") are greater than the benefits to be derived. This requires that they have to be taught how to quit and, having quit, how to maintain their new behaviors by responding to old behavioral cues (which had been associated with smoking) in a new manner.

An interpersonal strategy for teaching behavior change has been shown to be effective in several areas: smoking (Bernstein and McAlister 1976, Thoreson and Mahoney 1974, McAlister 1978, McAlister et al. 1979); behavior associated with cardiovascular risk (Meyer et al. 1976); and to some extent weight reduction (Harris 1969, Penick et al. 1971, Stunkard 1972). The social learning approach has its theoretical foundation in the work of Albert Bandura (1969, 1977). The theory of social learning asserts that most behavior is learned from the modeling of others and the selective reinforcement of certain behaviors directly, vicariously, or symbolically. In the modeling of new behavior, people witness stimuli, responses to those stimuli, and rewards associated with the behavior. Selective reinforcement can change the internal reward value of an undesirable set of responses from high to low while establishing new patterns of behavior connected with the high value of reward.

For example, the modeling of peers is an especially strong influence on adolescents. Peer pressure appears to be the major factor in the initiation of adolescent smoking (Maher 1977, Newman 1970a,b) as well as drug and alcohol use (Braucht et al. 1973, Gorsuch and Butler 1973, Jessor and Jessor 1971). From Bandura's (1977) analysis of social learning we can see that new behaviors tend to originate from exposures to powerful models—i.e., attractive people who appear (implicitly, at least) to be rewarded by their behavior.

Those who begin smoking early are likely to have the behavior modeled for them by popular peer group members. Furthermore, the behavior is likely to be reinforced by admiration for undertaking such a "daring" experience (Schneider and Vanmastright 1974). In this situation the interaction of influences is particularly striking. Adolescents see peers modeling a certain behavior (smoking, drinking, etc.); attribute maturity or daring to the act ("he must be 'with it' if he's doing such a thing"); and, quite frequently, fear social disgrace for not attempting to imitate the behavior.

In this case an "inoculation" strategy may be most effective in moderating behavior. This approach, outlined by McGuire (1964, 1973), is analogous to the practice of inoculation in preventive medicine. If social pressures to adopt a negative health behavior are seen as a virulent "germ," then inoculation against "infection" (actual adoption of the negative behavior) can expose people to a mild form of the argument and teach them skills for resisting pressures to adopt the unhealthful behavior—"antibodies" (Roberts and Maccoby 1973).

This concept of inoculation or the development of counterarguments can be applied to a smoking or drinking intervention. Adolescents who are likely to be pressured by peers to drink can be forewarned of these pressures and trained to give assertive counterarguments.

For example, they can be trained to reply, "If I drink to prove to you that I'm not a chicken, what I'm really showing you is that I am afraid of not doing what you want me to do. That's *really* acting like a chicken. I don't want to drink." Or, if young people are likely to see older peers showing off and acting "with it" by drinking, they can be taught to think to themselves, "If she were really 'with it' she wouldn't have to prove it by drinking."

To teach these counterarguments, a peer approach appears to work best. During early adolescence, peers take over from adults as the primary source of influence (Utech and Hoving 1969). Teaching by peers can efficiently provide traditional health education to large numbers of elementary school children (McRae and Nelson 1971). In addition, training young people to advise their peers about how to handle personal problems is also a successful approach (Alwine 1974, Hamburg and Varenhorst 1972).

Using "attractive" peers to teach counterarguments has been an effective strategy for the prevention of adolescent smoking behavior (McAlister et al. 1979). In Project C.L.A.S.P. (Counseling Leadership About Smoking Pressures), initiated in the Santa Clara County, California, school system during the school years 1977-1978 and 1978-1979, two teams of high school students taught classes in smoking prevention for a total of 9 days to all 7th graders in the treatment group. The high school students challenged the 7th graders with pressures and arguments to start smoking. They then modeled methods for handling these pressures through the use of movies, small-group discussion, slide shows, modeling, contests, and role-playing.

The results are noteworthy. The program began with a baseline analysis in September 1977. After 9 months, the treatment group had significantly fewer people who smoked "within the past month" or "within the past week" than the two control groups. Replication with the 1978-

1979 7th-grade class also yielded significant results in December 1978 and June 1979 ($p < 0.0001$).

A similar program conducted in five San Francisco Bay area high schools, using the same techniques (except the older peers came from Stanford University), also showed a significant reduction of smoking behavior for the treatment groups from September 1978 to January 1979 (Perry 1979).

A pilot study of this approach to intervention in the adoption of alcohol and drug behaviors by 7th graders was also attempted. Peer leaders described situations in which a young person was being pressured to drink alcohol and smoke marijuana. After a contest was held to see who among the students could come up with the best ways to resisting pressures, they all were given buttons that read "I'm naturally high." Peer leaders told the students that that is how college students respond to pressures to drink or to use drugs. They discussed the meaning of this phrase with the students.

Preliminary data suggest that this approach may have had some positive impact on alcohol and marijuana use: 37 percent of the students in the treatment group reported having drunk alcohol by the end of the year, while 51 percent reported doing so in the control group. Marijuana use was reported by 14 percent in the treatment group versus 25 percent in the control group (McAlister et al. 1979). However, a more extensive application along with more rigorous evaluation procedures will be necessary in order to gain a more thorough understanding of such efforts.

The social learning approach is not concerned with information per se, but rather with the particular behaviors to be changed. Initially, goals are established for the person and systems are initiated for modeling new behaviors, practicing them, and rewarding their successful practice. Ultimately, the aim is for the person to internalize the value of the behavior so that its exercise becomes self-fulfilling; at that point, the person will be better equipped to manifest the new behavior in the context of her or his environment.

Recently, Bandura has stressed cognitive aspects of learning and performance, particularly self-efficacy. People are taught incrementally that they can exert control over behavior they had previously felt powerless to affect. Each performance of the new behavior raises the level of efficacy of the person as well as the expectancy of greater efficacy in the future (Bandura 1977, 1978, 1979; Bandura and Simon 1977).

This technique has had some success in programs of weight reduction (Bandura and Simon 1977, Stunkard 1975, Stunkard and Mahoney 1974). For example, Bandura and Simon (1977) report continued weight loss after 14 weeks ($p < 0.001$). Stunkard (1975) reports significant differences in weight loss ($p < 0.001$) after 12 months. However Stunkard and Penick (1979) urge caution in being too enthusiastic about long-term findings as yet.

GROUP INFLUENCE

Cahalan (1975, p. 28) is convinced that "unless media campaigns are planned to be closely linked to people-to-people grass-roots programs to get widespread personal commitments to set an example of moderation, the campaigns will be of little consequence—especially since any campaigns for moderation in drinking have to compete against the many millions of dollars poured into advertising by the alcohol industry." He argues that only within a "social movement toward moderation" can anything be done to reduce alcohol problems. Part of that social movement would be "directly related to *setting a good parental and peer-group example* [his emphasis] at home and in other social institutions" (p. 26). In his view, we should seek to intervene to initiate (or foster) such an interpersonal social movement.

The existence of such a social support system implies the necessity of proper modes of communication. Strategically placed people diffusing the right information and modeling the correct behaviors are essential ingredients in any program of social change (Rogers and Shoemaker 1971). Part of that task can be handled by the media, which can be the original disseminators of information. Mass media messages, however, are not received equally by each person. Rather there exists a multistep flow, wherein some people get their information directly from the media, others through an opinion leader who has gotten the information from the media and/or other opinion leaders, etc. (Meyer et al. 1977).

Organizations to which people belong can function as both leaders of opinion and micro-social support systems. Similarly, local social services, to which the media can direct people, can be used to introduce new information, act as a classroom to teach appropriate techniques to modify behavior, and serve as a social support system for individuals undergoing treatment. Thus, the person who comes to one of these social service organizations would begin to have a new peer group helping to reinforce her or his commitment to a new system of drinking behaviors.

Not only would these organizations have readily available information to give to interested persons who come for help, they would also need personnel qualified in a variety of group training methods aimed at the modification of alcohol-related attitudes and behaviors. Drawing from what we know of the processes of small group dynamics, we can see that an effective program for modifying drinking behaviors is possible.

There is a respected tradition of literature on group processes. Festinger et al. (1958) documented pressure toward intragroup uniformity. If a group becomes highly cohesive, Back (1958) found, members will make an extra effort to change and reach agreement. Asch (1958) also found that people feel pressure to conform to the norm of the group, thereby sticking with the program.

Lewin (1958) found that intragroup decision making and norm-reinforcing processes facilitated change more than a simple instructor-to-group talk. The initial step could be for individuals to make a pledge within the group to change their behavior to the degree mandated by the program. "Public announcement of the decision by the individual may add to subsequent resistance, and more particularly so, by taking an irrevocable action on the basis of the decision, especially if this action involves some sacrifice on the individual's part and is mandatory" (McGuire 1974, p. 12). If everyone in the group made the statement of intent to change behavior, each individual would be provided with a "health-relevant reference group in which patients [would be] able to interact with others who are in the same boat as themselves" (Stokols 1975, p. 141). The use of this technique by SHDPP met with success (Meyer and Henderson 1974, Maccoby and Farquhar 1975).

As mentioned earlier, training people to change their behavior related to alcohol is similar to training them in dietary change, in that it seeks, not necessarily to eliminate the behavior entirely, but to teach people to control and modify their behvior in such a way that the degree of risk is reduced. Examples of successful interventions using small-group interaction processes can be found in the summary report of Lewin (1958). He reports an experiment by Bavelas, Festinger, and Zander in which the group decision method was superior to lectures in persuading women to include intestinal meats in their families' diets. The same was found by Radke and Klisurich to be true for the adoption of the use of powdered milk. Lewin, Radke, and Klisurich replicated this study by persuading mothers to adopt the use of cod-liver oil and orange juice in their infants' diets (cited in Lewin 1958).

Despite objections raised about the validity of these findings (Pelz 1958), more recent research in the medical literature supports the contention that for the behaviors we seek to modify a group treatment approach will be more effective than individual treatment. London and Schreiber (1965), for example, showed that group management was better than individual management of weight control under three experimental conditions: drug, placebo, and no medication. Somewhat similar findings were found in a program to reduce cardiovascular disease risk (Meyer and Henderson 1974).

Although negative results were obtained in a smoking program (Hunt et al. 1971), more recently Heath et al. (1979) reported a higher level of success than many previous smoking campaigns attempted in Australia. Using a group therapy approach to teach nonsmoking coping behaviors, Heath and his colleagues show a minimum 65-percent success rate, defined as complete cessation being maintained for 12 months, for each of 3 consecutive years. In their last reported intervention (1978), with 370 persons treated, Heath et al. report a success rate of 71.6 percent. They attribute their success to the use of community education for a period of 4 years, rather than previously attempted experimental exercises.

Most programs of smoking cessation are successful only with a minority of smokers. Participants most often resume smoking after a few months. Heath et al. (1979) were able to achieve a higher rate of cessation of smoking; in addition, at the end of 1 year, 72 percent were still nonsmokers. Maintenance of change is the essential element in these programs.

Bandura and Simon (1977) reported success with groups of obese people in moderating eating behaviors. Here, individuals were taught to set intermediate goals for themselves as the program progressed. As goals were attained, self-efficacy grew, which increased motivation to reach higher goals, which led to further weight loss. Since this progress took place within small groups, each person's success served to suggest to the others in the program that "If she can do it, I can do it, too."

In the Stanford Heart Disease Prevention Program 3-Community Study, programs of professionally led behavior modification to reduce the rate of smoking among high-risk individuals were implemented with a certain degree of success (a net decrease of 35 percent in number of cigarettes smoked daily after 2 years [Farquhar et al. 1977]). In the current SHDPP 5-City Project it is hoped that people can similarly be trained to teach appropriate behavior modification techniques, in order to carry on the intervention and the program of community involvement.

To apply this process to the problem of alcohol abuse, behaviors determining nonuse or nonabuse of alcohol would be specifically identified (such as how to cope at parties, business lunches, sporting events, etc.). Goals for improvement would be established and a timetable set. The instructor would model or simulate the new behaviors. For example, a party would be arranged (perhaps in a nonclinical locale for the sake of realism) and people would be shown how to cope with the social and/or internal pressures to drink. They would be encouraged to practice the new behavior and rewarded for improvement by praise, attention, a token point system, etc. The ideal behavioral outcome would be for participants to internalize the new system of behaviors so that they become self-rewarding and/or reinforced by such factors as peer approval, greater family harmony, etc. With each step each person would have a better sense of efficacy in controlling her or his own behaviors, thereby increasing the expectation of controlling them in the future. Because a great deal of alcohol abuse occurs in social situations, group counseling may be more effective than individual counseling since it gives people practice in the desired behavior in the setting in which the behavior would naturally occur.

It is necessary not only to teach change in the clinical environment but also to make the change carry over to the everyday world. The use of group learning and reinforcement is the most efficacious strategy (short of retraining everyone associated with each participant, then retraining everyone associated with them, etc.) not on an individual basis.

FACTORS OF PERSUASIVE COMMUNICATION

As mentioned before, Cartwright (1949) outlined some principles of mass persuasion that he derived from attempts to get Americans to buy U.S. savings bonds during World War II. To accomplish mass persuasion, he postulated, a program must achieve: (1) a change in the cognitive structure of people (what they know or understand), (2) changes in affective structures (what they wanted to do), and (3) changes in behavioral structure (their manifest action). Previous campaigns had been aimed primarily at the first, and in some instances the second, but none at the third, the behavioral structure. What was lacking was the specific information that could guide people in how to implement the behaviors advocated in the campaign.

The Stanford Heart Disease Prevention Program utilized the mass media in a way that is not typical: namely, to teach specific behavioral skills (Stern et al. 1976). It tried to convince people to change, taught them how to change, and how to reinforce their commitment to change so as to respond favorably to cues connected with their old behavior. Although they advise caution in the interpretation of their data, Stern et al. conclude that "a multi-media health education campaign can lead to favorable dietary changes in the general population. . . . There is also some suggestive evidence that the combination of personal counseling and mass media may produce changes more rapidly, but that, with time, individuals exposed to mass-media only tend to 'close the gap' " (p. 830).

To assess the most efficacious procedure for inducing change, Lasswell's (1948) paradigm of the flow of communication is quite helpful: Who says what to whom in which channel with what effect? These factors, the independent variables of persuasive communication, are the *source* and how it is perceived by the recipient of the communication; the characteristics of the specific *message* or messages (i.e., what is contained and how it is presented); aspects of the *channel* through which the message is transmitted; characteristics of the *receiver* of the message; and the characteristics of the *target* variable (i.e., a political belief, a specific behavior related to alcohol, etc.), and the kind of effect of the message is designed to produce.[5] We have already talked at some length about the hierarchy of effects and some of the target variables in alcohol abuse. This section focuses on source selection, message construction, channel selection, and receiver characteristics—the instruments of mass media persuasion.

SOURCE

The first independent variable is the source—the "who" in Lasswell's paradigm. Identical messages transmitted in similar fashion to the same persons may have varying degrees of persuasive impact, depending on who sends the message. Source factors are summarized by McGuire (1969) as credibility, attractiveness, and power. Despite what may appear to be obvious, these factors tend to be glossed over quite frequently.

In attempting to convince people of the correctness of a position, a source's persuasiveness depends on credibility, which in turn rests on the perception by the receiver that the source is competent, knowledgeable, and trustworthy. A number of studies have shown that differences in perceived expertise are correlated with how much opinion change a source achieves (cited in Roberts 1975). Thus, we must determine who the target population of a campaign believes to be competent and trustworthy in the area of alcohol abuse education.

There have been difficulties with studies that have tried to locate exactly what cues account for a source's credibility. Choo (1964) attributed messages about smoking and cancer to either a public health physician (high credibility) or a public relations man from the tobacco industry (low credibility). Bochner and Insko (1966) attributed messages advocating less sleep to a noted physiologist (high credibility) or a YMCA director (low credibility). Although one would like to attribute the effects to the source with higher credibility, other factors such as age, intelligence, knowledge, social status, etc., tend to confound the results. We might ask whether the testimony of an older, cured alcoholic who has had years of experience abstaining from alcohol is more effec-

[5] McGuire (1969, 1974) has done the most extensive cataloguing of the many studies of persuasion.

tive than a younger one who just kicked the habit and has a fresher knowledge of the pain involved.

Sources who are perceived as having something to gain from the recipient's behavior are less likely to be yielded to. We may well surmise that a campaign initiated by a commercial treatment facility to moderate drinking would be less effective than one initiated by the liquor industry, because the treatment facility's campaign could be seen as motivated by profit, whereas the liquor industry's campaign, if successful, could be seen as reducing its income.

One generalization that can be made from the research on source credibility is that no single source is highly credible for all messages to all audiences. Pretesting specific messages for various segments of the target population will help to determine which sources are most likely to be well received by the audience.

Here we run into an immediate problem. If the government plans to sponsor an alcohol abuse education program, should government officials be the speakers in the messages? Indeed, should the government be identified with the campaign at all? Following what could be termed the "swine flu fiasco" of a few years ago, it would seem that the public is extremely wary of certain government information programs that might attempt to influence their health behaviors (Medical Progress and the Public, n.d.). We should not assume that government affiliation with an alcohol education program will be highly credible to the public. Thus, the issue of who is a credible source and whether government affiliation has an effect on the credibility of the message must be one of the major areas of formative evaluation.

A second source factor, attractiveness, can be seen as a confluence of the receiver's feeling of similarity to, familiarity with, and likability of the source. A good deal of research indicates that the persuasive impact of a message increases linearly as a function of the perceived similarity between recipient and source. Consistency theorists (e.g., Newcomb 1953, Osgood and Tannenbaum 1955) posit that being liked by the receiver increases the source's influence.

How the misreading of source attractiveness can cause a campaign to backfire is best illustrated by the Australian "Stop a Slob from Driving" campaign. The idea was to portray drinking drivers as slobs; since people would not want to be perceived as slobs, they would reduce their driving while under the influence of alcohol. Instead, many people found "Mr. Slob" to be funny, attractive—a person they wouldn't mind being like. The result was that this particular campaign may well have done more harm than good by strengthening existing dispositions toward drinking. A pretest of this image might have revealed this flaw in the campaign.

Using celebrities may also not be the best approach. Dana Andrews,

Dick Van Dyke, Art Carney, and Don Newcombe, among others, have appeared in alcohol campaigns. Since they are reformed alcoholics they have high credibility. However, their attractiveness to the audience (especially those under 30), as well as the small degree of perceived similarity by the audience, most likely minimized the impact of their appeal.

To conclude, the sponsoring agency of an alcohol education strategy must pretest the proposed target audience to discern both the agency's credibility to the audience and the credibility, attractiveness, and power of those sources within the messages themselves.

MESSAGE

There are many aspects of message composition, including what information is included or left out, whether the message contains opposing arguments, whether the message aims at values or beliefs, whether the conclusion is explicit or left to the receivers, and how extreme a behavior change is advocated. Some general conclusions from the vast research literature in these areas (see McGuire 1969) are of particular interest for an alcohol program.

Although it was once believed that the conclusion of a message should be left to the receiver so as to involve him or her to a greater degree, there is ample research showing that including all arguments and explicitly drawing the conclusion leads to greater persuasive impact. Regardless of level of intelligence, most people seem insufficiently informed or motivated about the subjects of most campaigns to draw the intended conclusions. For communication to be persuasive, notes McGuire (1969), "it is not sufficient to lead the horse to water; one must also push his head under water to get him to drink" (p. 209).

Beyond advocating change, the message should tell the receiver how to implement the position advocated. As Cartwright (1949) tells us, "the more specifically defined the path of action to a goal (in an accepted motivational structure), the more likely it is that the structure will gain control of behavior." Furthermore, "the more specifically a path of action is located in time, the more likely it is that the structure will gain control of the behavior" (p. 443). One of the drawbacks to most campaigns is that they have not been sufficiently specific about the course of action they advocate.

Messages should be designed to advocate specific behavior regarding a specific act. For example, "If your friend is drunk at a party, take him home in his car and ask another friend to drive behind to pick you up," rather than "Friends don't let friends drive drunk." In the first instance, a specific conclusion and action is advocated; in the second, the receiver is left to devise a proper course of action. Pretesting has shown that this specific solution is more palatable to people than the confiscation of an intoxicated friend's keys. Pretesting of messages is essential.

There are also sound reasons for including more than one side of a debate. People come into contact with numerous messages advocating drinking (from friends, media, coworkers, etc.). Roberts and Maccoby (1973) show how teaching people counterarguments may be effective to help them reinforce their moderate (or non) drinking behavior. The counterargument should not be so strong as to tempt people to try the behavior the message warns them against. Also, the political realities of contemporary America suggest that we eschew an aggressive counteradvertising campaign against the liquor industry. Wallack (1978) documents how, in an alcohol campaign in California, messages had to be changed after serious objections were raised by both the liquor industry and the governor.

Many alcohol-related campaigns have relied on some degree of fear appeals to convey their messages. For example, a 1971 NHTSA print ad had a headline, "Today Your Friendly Neighbor May Kill You." The National Safety Council's "Scream Bloody Murder" campaign had such messages as "Drunk drivers add color to our highway" and "Drunk drivers bring families together . . . in hospital rooms and at funerals," etc. The extremity of these messages may have done more harm than good. Many people who have driven after having had too much to drink know that drinking and driving does not inevitably lead to a fatal accident. Research shows that extreme fear arousal leads to avoidance responses by receivers because actions recommended in the message may not be perceived to be strong enough to cope with the degree of fear aroused. According to Chu (1966), the advantage of high-fear messages is realized to a greater degree when the recommended action to cope with the problem is perceived by the receivers as highly efficacious and the danger is immediate. In other instances a moderate level of fear arousal coupled with directions for realistic coping behaviors may be the most efficacious approach. Janis and Feshbach (1953) found this to be the case in a program to teach the causes and prevention of tooth decay, as did Leventhal and Watts (1966) in a program to induce people to get chest X-rays, and the San Antonio, Texas Alcohol Safety Action Program "Fear of Arrest" campaign (Hawkins and Cooper 1976).

There are implications of this approach for an alcohol abuse education campaign. Sternhal and Craig (1974) analyzed the fear appeals used in several health and safety campaigns. They concluded that "including specific recommendations which the audience perceives to be effective in reducing physical threat" (p. 31) is an approach that may engender a substantial degree of compliance. While they feel that the threat of physical harm will increase persuasion only if the source has high credibility, other research contends that the threat of social disapproval is more effective than the threat of physical harm.

Thus, to be effective, a message that arouses fear must be accompanied with the suggestion of a specific action or behavior for the receiver to perform. It should be something that can be done easily and immediately; the person should be able to feel a sense of efficacy in performing the suggested act and feel fairly certain that it will be successful in preventing the threatened damage. In the Stanford Heart Disease Prevention Program study, a specific suggestion, such as "cut down to 2 to 3 eggs a week," produced greater responses in the heart disease program than vague, long-term suggestions (Farquhar et al. 1977). It would seem that an equally effective suggestion might be "Why not make every third drink a soft drink at the next party?"

The use of specific behavioral recommendations in the areas of smoking, diet, and exercise has been shown to be effective (Farquhar et al. 1977). The simple presentation of useful skills as alternatives to harmful behaviors will most likely lead to improved healthful acts. However, this is a very complex phenomenon; it is essential to pretest various appeals and approaches to the prevention of alcohol abuse to discover their effects; constructing messages should not be based only on communicator-perceived "common sense."

A word of caution is necessary. Several high-fear messages have deliberately disseminated untrue information in the hope that the fear aroused would be sufficient to keep people from performing some behavior. Films such as "Reefer Madness" come most easily to mind. When people, especially young people, experience firsthand that smoking marijuana will not inevitably lead to heroin addiction, prostitution, dope-crazed homicide, and an eternity of sin and degradation, they will, more than likely, not heed any messages about the negative implications of marijuana in the future. Nor will they be inclined to heed the advice of the communicator on any number of topics in the future.

Similarly, many people know that drinking and driving does not inevitably lead to a fatal accident. Thus, the high fear-arousing messages connecting this behavior with gory highway or hospital scenes may have the same ultimate effect as films like "Reefer Madness." We should be cognizant of this phenomenon in designing alcohol education programs.

CHANNEL FACTORS

The part of the communication process that has received the least research attention has been the channel. What few studies exist have focused on whether a print message is more effective than a radio or television message, whether face-to-face contact is more effective than films, etc. Reports of campaigns typically devote little, if any, attention to how the choice of particular media was arrived at, how the medium was used, and what problems were encountered in dealing with people who work in the different media. From scientific perspective this lack of attention is unfortunate, because it makes replication difficult, if not impossible.

Channels are an important link in the chain of persuasion since, in the words of a radio veteran, "You can't sell them nothin' if they don't see the spot." Typically, the credit for the success of advertising has been attributed to the artful creation of messages, jingles, etc. Little, if any, notice has been given to careful planning of the best ways to reach the target audience, both for the widest dissemination of the message and for the most effective use of the advertiser's dollar. For example, in the Stanford Heart Disease Prevention Program's 3-Community Study, radio was employed heavily for Spanish-speaking people because it was ascertained (through formative evaluation) that Hispanics in the target communities spent a great deal of time listening to Spanish-language radio.

A great deal of the effectiveness of the Stanford media campaign was in its ability to get people to think about issues that affected their health, issues about which they had probably never thought before. What change did occur after the campaign was implemented was due not only to the campaign but also to many converging factors. The campaign was most effective in setting the agenda of what was salient for people in the test area. Once people's agendas were altered, behavior change became more likely.

This agenda-setting function of the mass media has been the focus of a great deal of research attention in the past 15 years. It was best articulated by Cohen (1963), when he noted that "the press is significantly more than a purveyor of information and opinion. It may not be successful much of the time in telling people what to think, but it is stunningly successful in telling its readers what to think about" (p. 13). This function has been defined as ". . . specifying a strong positive relationship between the emphases of mass communication and the audience" (Comstock et al. 1978, p. 320).

There is ample evidence that news items that are heavily emphasized by radio, television, and newspapers become the most salient issues for the public (Becker et al. 1975; McCombs 1972, 1974; McCombs and Shaw 1972; Robinson 1972; Robinson and Zukin 1976, among others). While most research on agenda setting has focused on political information, its connection with public health information is clear. The U.S. Public Health Service reports that there has been a steady increase over the past several years of people who believe that cigarette smoking is harmful to health. Over 90 percent of those living in the three communities of the Stanford study, both smokers and nonsmokers, believed that quitting smoking would help a person live longer.

In the Stanford 3-Community Study, the information disseminated by the mass media was sufficient to induce people to change their behavior. People reduced their consumption of some high-cholesterol foods on the sole basis of what they saw on television, heard on the radio, and/or read in the local newspaper. For example, there was greater than a 40-percent reduction in the consumption of eggs over a 3-year period in the community with massive intervention and over 35 percent in the media-only community. In these communities, once the new agenda was set, the information was sufficient to effect some change in behavior in the media-only community (Farquhar et al. 1977).

Once an agenda has been set, it is possible to get people to pay attention to communications of greater depth. Thus, in the Stanford Heart Disease Prevention Program, television and radio spots got people to heed longer messages and to learn from those longer messages, e.g., advice about food purchases, cooking, etc.

Although most public education campaigns cannot mount huge, multimillion dollar campaigns due to budgetary restrictions, wise use of the various available media may make the difference between no positive outcomes and the initiation of some degree of change. The remainder of this section discusses some of the important channel factors necessary to plan a campaign.

First, different media have different capabilities. Television and radio can be used for spot announcements, public affairs programs, and entertainment programs. Mendelsohn (1973) shows that a well-planned educational program, such as "The National Driver's Test," can be entertaining as well. The Stanford Heart Disease Prevention Program aired an equally interesting program, "Heart Health Test." Spot announcements can call attention both to special programs and to various themes. Newspapers can be used to carry doctors' columns, to give people advice and information, and to increase their awareness about the various problems associated with alcohol consumption. In the Stanford program, billboards were used to reemphasize slogans associated with the campaign (Maccoby and Farquhar 1975, Maccoby et al. 1977) and used to communicate short messages (long copy can be dangerous for drivers to try to read); they can act as a tool of reinforcement.

Other media include newspaper stories, business cards, direct mail (Maccoby and Alexander 1980), pamphlets, newsletters, films, subway and bus cards, and the insertion of campaign materials in paychecks and family allowance checks (Blane and Hewitt 1977).

Typically, spots on the electronic media must be short (usually 1 minute on radio, 30 seconds on television) but can contain quite a bit of information. Since people can read at their own pace, newspaper and magazine ads can contain somewhat more information. As noted, billboards, bus and subway cards, and business cards (called "out-of-home" media) must be very short, with little information.

A problem usually encountered involves public service announcements. Many radio and television stations take public service announcements (PSAs) and run them at 3 a.m., when commercial time is the cheapest and often unsold. If there is no money to buy spots at more convenient times, however, all is not lost. All stations are licensed by the Federal Communications Commission (FCC) and are mandated to operate in the "public interest, convenience and necessity." Leaving aside a discussion of whether they fulfill that mandate, they reapply for a license every 3 years. At that time the FCC becomes very interested in how the station has fulfilled its public service requirements. When planning the media campaign, planners can remind the station operators of this fact. They could then promise to write the FCC a letter commending the station for its assistance in promoting the public health and welfare for inclusion in that station's file (as planners of the Stanford study did with much success). This tactic frequently engenders a positive response.

Another important channel factor is coordination (Maccoby and Alexander 1980). Confusion arises when people are exposed to a wide variety of messages from both local and national sources; the sources compete for the attention of the audience and can give contradictory information (Blane and Hewitt 1977). The messages from each medium must be keyed to the others (see Wallack 1978, 1979).

Another important channel factor is targeting the audience that is intended to be reached. The Stanford Heart Disease Prevention Program found that it could not simply translate English-language spots into Spanish to reach California's large Chicano population. Rather, different messages were developed based on the popular "radio-novella" format (Maccoby and Alexander 1980).

Probably the most important factor involving the channel is determining who uses which medium and when. For example, teenagers listen to radio more frequently than they view television (Comstock et al. 1978). Furthermore, FM radio has grown in the last 15 years so that, according to ratings released in August 1979, for the first time it is listened to more than AM radio (Media and Marketing Decisions 1979). In addition, magazines and newspapers that have similar editorial formats also tend to have similar constituencies, and different times of different days attract different audiences to radio and television.

Different markets respond differently as well. For example, television stations in New York City (the Number 1 market) cover a tristate region (New York, New Jersey, and Connecticut), whereas television stations in Los Angeles (the Number 2 market) reach people in only a part of one state. This may be an important consideration if campaigns are mounted on a state-by-state basis.

Yet another channel factor is audience fragmentation. There are, for example, more than 70 radio stations that cover the greater Los Angeles market. There are 14 major formats (disco, soft rock, progressive rock, top 40, jazz, black, country, middle-of-the-road, beautiful music, Spanish, classical, talk, news, and religious), and each station targets its appeal to a different audience. Those who create messages should try to adjust the content and style of radio spots to appeal to the tastes of particular audiences by conforming to the format of the particular station. Anyone who has heard a rock-format spot on a classical music station or a spot featuring Johnny Cash on a black-oriented jazz station can attest to how intrusive the contrast of sounds can be. The incongruity often creates hostility to the message, greatly limiting its ability to persuade. Since adolescents listen to radio more often than they watch television, the potentials as well as the hazards of radio should not be overlooked.

It is essential to recognize the importance of these channel factors in planning a good campaign. Research is needed on the audiences available to particular channels, as is continuous monitoring of the existing knowledge, beliefs, attitudes, risk-related behavior, and media use of those audiences. We should account for the differing level of need and the differences in ethnic group, socioeconomic status, sex (public education messages about alcohol use tend to have a distinctly masculine orientation, according to Blane and Hewitt [1977]), and geographic location (for example, the level of alcohol abuse is much higher in Fresno than it is in Los Angeles).

RECEIVER FACTORS

Source, message, and channel can be considered a prelude to what occurs after a person has received a message. Different people respond to the same message differently, often because of personality differences. Studies have shown that some people are more persuadable than others, while certain personal characteristics (i.e., demographic background, emotional stability, ability, motivation, and other personality factors) also play a role in the relative efficacy of a message from one person to another. Although there is quite a literature (see McGuire 1969) on this topic, the discussion in this section is confined to a few pertinent points.

Some factors mentioned before operate as a function of receiver characteristics. For example, in an alcohol campaign the decision whether to mention opposing arguments depends a great deal on the level of intelligence of the receiver, his or her existing predisposition about alcohol, and, most important, the likelihood that she or he will subsequently be exposed to counterarguments. High-fear appeals to those who are prone to anxiety or who have feelings of vulnerability to illness or death should be avoided. On the other hand, they may effectively reach those who feel complacent or invulnerable to the dangers discussed in a campaign. These audiences may be separated by ascertaining patterns of media use through pretesting in target communities.

Self-esteem is also an important factor. McGuire (1969) states that self-esteem creates different reactions to persuasive messages as a function of the relative complexity of the situation. With simple messages, minimal self-esteem is usually sufficient to increase resistance to conflicting information. In more complex situations, however, levels of self-esteem appear to separate levels of effects, probably because when more information is to be processed, a person with lower self-esteem may avoid the message as a means of coping with dissonant information, whereas a person with high self-esteem may not feel as threatened by dissonant information. The optimal persuasion situation would occur with those receivers who have enough self-esteem to actually receive the message and process the information—but not so much that they feel they cannot be told anything new. Unfortunately, this balanced degree of self-esteem and high risk for alcohol abuse are not well correlated.

Age is an important demographic factor in some instances. McGuire (1974) notes that children in junior high school are at an age at which they appear to be most easily persuaded by information contained in a drug (or alcohol) education program. As he notes, this may be fortunate since this age-group is exposed to the greatest amount of initial drug-related information.

In determining the target audience, factors such as self-esteem, intelligence, and persuadability are important, as, of course, are demographic data. Campaign planners should seek to determine the locus of problems of alcohol abuse and the groups who have these problems in common.

CONCLUSION

Attitudes are complex and require complex strategies for effective change. A message should not try to be all things to all people; it should be targeted to a specific audience, which involves defining who that audience is, what those people are like, what it is we want to change, what the best strategy is to change it, and what is the most efficacious method of disseminating that information. Campaign planners must be willing, in the face of negative evaluative information, to change strategy if necessity warrants. What is most important is change, not the immediate viability of the models.

This approach, by acknowledging that the audience is an equal partner in the communication process, necessitates taking a developmental approach when planning a campaign in order to successfully predict those factors that facilitate acceptance or rejection of messages. Mendelsohn (1973, p. 52) suggests the following conditions for communication programs to be successfully planned and executed: (1) they are planned around the assumption that most of the public to which they will be addressed will be either only mildly interested or not at all interested in what is being communicated; (2) middle-range goals that can be reasonably achieved as a consequence of exposure are set as specific objectives; and (3) careful consideration is given to delineating specific targets in terms of their demographic and psychological attributes, their life-styles, value and belief systems, and mass media habits.

This kind of approach to message design based on principles derived from research was the strategy in the Stanford Heart Disease Prevention Program's 3-Community Study. Butler-Paisley (1975) maintains that future public information campaigns would have a greater likelihood of success if program planners were more cognizant of the social science literature. Program planners and developers should be less intuitive and more empirical in deciding on their creative strategies and more willing to commit the time, money, and resources to weigh alternative strategies.

PROJECTS OF THE NATIONAL INSTITUTE ON ALCOHOL ABUSE AND ALCOHOLISM

It may be useful to look at current efforts of the National Institute on Alcohol Abuse and Alcoholism (NIAAA) to set up programs of preventive education in light of the foregoing analyses. NIAAA is currently conducting the Prevention Model Replication Project, in which three alcohol abuse prevention prototypes have been selected for replication. Each of these prototypes (CASPAR—Somerville, Massachusetts; King County, Washington; and University of Massachusetts, Amherst, Massachusetts) takes a somewhat different approach to the dissemination of information about alcohol abuse prevention. These prototypes, which contain elements indicating some promise for future success, are briefly described.

CASPAR ALCOHOL EDUCATION PROGRAM

The premise of this program is that specially trained teachers can teach their students about alcohol-related issues rather than alcoholism, modify their attitudes about drinking, and help them develop strong decision-making skills about drinking. It is anticipated that these factors will lead to a reduction of alcohol abuse by teenagers. Peer leaders are used as part of the curriculum to supplement teachers' instruction inside and outside the classroom.

The CASPAR program includes a number of evaluations throughout the course of the intervention, which are used on the assumption that all contingencies cannot be foreseen. Adjustments to the intervention are made on the receipt and evaluation of a variety of surveys.

These studies (1979) have substantiated not only that teacher training is necessary, but also that a properly implemented curriculum is essential for effective alcohol education. The proponents of CASPAR report some change in knowledge and attitudes as well as a gradual change toward moderation of alcohol abuse. While this program shows promise on a number of dimensions, the limited data available at this time preclude an analysis of greater depth.

ALCOHOL EDUCATION CURRICULUM PROJECT

This prototype is structured on the belief that specially trained teachers, without supplemental peer leaders, can influence attitudes toward drinking that will lead to more responsible decisions by students about their use of alcohol. These teachers are used to teach the curriculum to other teachers. Kits of materials designed for differing grade levels are used.

Evaluations of the curricula on levels of knowledge are made through the use of questionnaires supplemented by self-reported alcohol use by junior and senior high school students. Post hoc analyses have shown a slight influence on teachers' attitudes and that a relationship exists between reported drinking habits and the effect of the curriculum on students. No detailed data were presented to support these contentions, however, and no change in drinking behavior was reported.

UNIVERSITY OF MASSACHUSETTS DEMONSTRATION ALCOHOL EDUCATION PROJECT

This project, aimed at college students, attempts to promote a campus environment that is conducive to responsible alcohol use. The premises of the program are:

(1) Students' drinking behavior influences and is influenced by the environment of the campus.

(2) This environment can be changed by extended efforts conducted simultaneously throughout the social system.

(3) The use of alcohol by students can be influenced by using a combination of mass media to promote awareness (setting the agenda) and small-group interactive approaches to examine alcohol-related attitudes and behavior.

The goals of the program are to increase the knowledge of those in the university community about responsible alcohol use, to help individuals understand and moderate personal and social alcohol-related behaviors, and to promote early identification of those with alcohol-related problems. The program uses a comprehensive set of specific approaches including media campaigns, peer-leader training, seminars, in-service training of service staffs, and attention to institutional policies to facilitate early identification and treatment of those with alcohol problems. Evaluation data include mailed attitude surveys, random monitoring of dormitory behavior, interviews with dormitory directors, and data on alcohol-related arrests, on-campus sales, medical incidents, and property damage.

DISCUSSION

Summary reports (1979) of these projects indicate that there has been some change in attitudes about alcohol use, but the reports present no specific data. The ratio of alcohol-related behavior to total behavior (neither is defined) in dormitories (Amherst, Massachusetts, project) evidently decreased from spring 1977 to fall 1977, but insufficient data are supplied with which to make a proper evaluation of this finding.

It would appear that this program holds a great deal of promise in principle, since a well-integrated mass-media and group-counseling approach is used. The major problems at this time appear to be that the questionable reliability and validity of the data hinder inferences about alcohol-related behavior change; the program has not been implemented for a sufficient amount of time for a meaningful assessment of behavior change; and the absence of a control group design makes it difficult to interpret the significance of the data that are reported.

Those who wish to mount campaigns with the ultimate goal of changing behavior must be willing to invest more than 6 months before deciding the success or failure of their efforts and to stay cognizant of the quasi-experimental nature of their efforts—and hence the need for use of quasi-experimental design features as development of controls. As the Stanford Heart Disease Prevention Program's 3-Community Study showed, unlearning old behaviors (which may have taken years to develop) by replacing them with new ones is a process that will, in all likelihood, take a long period of time and require careful design in order to demonstrate effects.

In brief, on the basis of general principles, the two projects from Massachusetts show the most promise. For example, student peers make credible, trustworthy sources to disseminate messages and to lead group discussions. The University of Massachusetts program uses small-group approaches led by trained peers to provide positive social situations for discussing alcohol and behavior. This combines the use of trustworthy, credible peers with the beneficial effects of the open participation of people in small, interactive groups. The CASPAR program also makes use of peers to disseminate information and lead group discussions.

They differ greatly from the formal classroom situation (used in King County), which does not lend itself to frank discussions, peer assistance, or meaningful problem solving. Teachers may not be seen by students to be trustworthy sources of information, since an adversary relationship frequently exists between students and faculty. Furthermore, the reliance on a self-reports of alcohol use administered by teachers may be seen by students as a tactic to catch them engaging in a proscribed activity. This may heighten the lack of trust, which will not help to modify the problem.

The reliance of the University of Massachusetts program on FM radio indicates the recognition by program planners that the use of television for information and entertainment is not ubiquitous. Rather, college students tend to rely more on radio than television (Comstock et al. 1978) than does the average person. In a college community that is not within a major metropolitan area (such as Amherst, Massachusetts), student participation in as well as the use of college radio may be quite substantial. This may be true of similar campus communities; it can easily be ascertained through a preintervention survey.

There is insufficient evidence in NIAAA's reports of how closely these programs adhere to all the principles described in this paper. That would require minute descriptions of, for example, the procedures by which the usefulness of sources used was determined vis-à-vis their availability to program planners; the subjects' media use vis-à-vis the funds, facilities, and expertise available; and the usefulness of the programs versus the nature and extent of the target population's alcohol use.

Rather than serve as a scorecard for NIAAA or other programs, this paper should serve as a guide to those implementing these and future programs in the creation of present and future interventions, taking local factors into consideration.

STEPS FOR CONSTRUCTING A CAMPAIGN

Ideally, the goal of a health intervention campaign is to change behavior in the direction of a healthier life. While laudable, such an undertaking is beyond the scope of the mandate, knowledge, and resources of this panel, even if we could predict a very high rate of success. We have to set our sights somewhat lower. Summarized below is a set of general guidelines for the implementation of a campaign to change health behavior specific to alcohol abuse.

First, clear and specific objectives must be set. What do we want to produce? To train adolescents in the judicious use of alcohol? To reduce the number of drunken drivers on the road? To train parents so that fewer of them will abuse their spouses and/or children?

These are just three behaviors associated with alcohol abuse. Each is different and involves somewhat different populations, patterns of mass media use, messages, sources, etc. Each target behavior has a distinct impact on the costs of the project.

The campaign's objectives determine the costs. To simply try to make people aware of certain slogans in a given community over a given length of time, a certain budget level may suffice (Wallack 1979). If, however, the goal is to teach chronic abusers of alcohol to diminish their level of problem behaviors 10 percent, the cost factor will be much higher. The more comprehensive the goals, the higher the cost.

The next step is to decide how the knowledge, attitudes, and/or behaviors associated with drinking should be changed and how to measure those changes. For example, do we propose to influence attitudes of others toward drunk drivers? To influence the drinking driver to stay off the road? Or to seek long-term help? Do we want to suggest that people think more negatively about intoxication? Will we advocate less frequent drinking to intoxication? Do we wish to eliminate intoxication per se, or to neutralize certain of its effects?

Each goal requires a different strategy as well as different baseline measures and different outcome measures.

Establishing interim criteria for success or failure is the next step. They will indicate whether and when we must "go back to the drawing board," and how drastically we must redesign. The criteria should entail a causal model of anticipated campaign effects, including the difficulties of each step.

Where do we begin to test our program? Each locality has unique qualities: the population is somewhat different and the particular levels of alcohol consumption (as well as associated problems) may differ.

In the field of marketing there are certain cities—such as Columbus, Ohio, Phoenix, Arizona, and Portland, Oregon—that manufacturers routinely use to test new products. They are usually chosen because they have demographic cross-sections that very closely match the U.S. population as a whole. In addition, they are isolated media markets, i.e., the local mass media have no direct competition from the media of other markets.[6] This allows for greater experimental control of the target population's media input.

In these larger cities, however, media and other costs are high. The Stanford Heart Disease Prevention Program 3-Community Study was conducted in towns with populations of less than 15,000 (Tracy, Gilroy, and Watsonville, California). Operating costs are much lower in these areas and their relative isolation made for easier experimental manipulation and monitoring.

Each of the foregoing steps requires the expertise of people from various disciplines. As Mendelsohn (1973) recommends, research-oriented behavioral and social scientists, evaluation planners, and those experienced in working with the media and community organizations should collaborate in the implementation of the intervention.

Involvement of these kind of experts is especially important in constructing the baseline evaluation. Questions should be constructed with

[6] Such as Newark, New Jersey, which has competition from New York newspapers, radio and television stations.

an eye to their relevance to the outcome goals. The baseline survey should seek to determine certain social and demographic characteristics of the target population. The credibility and usefulness of various sources, messages, and media should be ascertained. Will the people assemble in groups? What kinds of groups will they (not) attend? How regularly? Will they participate on their own? How likely are their family, friends, and/or work associates to be supportive of their participation in the program? What types of media do they generally rely on for information? For entertainment?

This information need not be obtained solely by obtrusive means. For example, in several localities the names of those convicted of driving under the influene of alcohol can be obtained through police records. This minimizes a major cost since part of a target population could be readily identified. One can carry out the baseline survey with these people without telling them why they have been singled out. Although these people may have a greater vested interest in drinking-driving countermeasures, stigmatizing them as alcohol abusers should be avoided since their cooperation in the program is essential.

The next step, once the baseline data have been gathered and analyzed, is to plan a small-scale, rough draft of the campaign. This can be done fairly inexpensively. This prefield test is a good way to find out how effective, convincing, and credible the proposed messages and sources are. For example, in the Stanford Heart Disease Prevention Program, the "Heart Health Test" was tried out in a shopping center. This generated a lot of excitement among patrons; the study elicited responses from a good cross-section of the community because many people wanted to participate. Proper planning can make this kind of intervention fun for the community and useful for program planners.

Evaluate the results of the prefield test. Which messages were found to be credible? Which were not? Were people interested in the goals of the campaign? If not, what could be changed to generate more interest or more public commitment?

At every step in the process, evaluations should incorporate new information that may change the premises on which decisions are based. Any campaign plan should be based on the best evidence from previous experience. Each new campaign should generate new data to help refine the techniques and understanding of how to affect human behavior.

Once the evaluations have been made, planners should compare the outcomes with the initial objectives and continue to make modifications in strategy as the evidence warrants.

Concurrently, there should be an assessment of the social organizations in the local community to see how helpful they can or want to be to the goals of the campign. Usually, there must be some training of the personnel in the various techniques of the modification of the target behavior. If no such organizations exist, they may have to be organized by the program planners.

Once the messages, sources, strategy, etc. have been well designed and tested, the campaign can get under way.

CONCLUSION

It was once believed that, of the masses of people who consume a lot of television, radio, newspapers, etc. every day, many, perhaps most of them, were impervious to any educational information carried by these media. It was slowly discovered that people *can* be educated to change their behavior if the right information is properly presented. What this required, primarily, was a lowering of goals, careful preplanning, judicious use of mass media (frequently in conjunction with community-based counseling programs), and adequate concurrent, as well as post hoc, evaluation procedures.

The Stanford Heart Disease Prevention Program was shown to to be effective on a limited basis. A similar program is now in the formative stage of testing to see whether such a program will work in larger communities in California and in Finland and Australia. The results will produce more evidence on how to educate people to be more aware of their well-being and how to act to lower their risk of a variety of disorders. This paper has argued that an equivalent strategy has never been attempted to combat alcohol abuse but, given proper application and enough time, it may be a viable approach.

This paper does not advocate an all-or-nothing approach. Regulation of alcohol consumption and sale could be more effective if coupled with a properly planned program of public education. If laws are passed regulating people's alcohol behaviors but are not accepted by the public, enforcement will be much more difficult. If, however, the law is supplemented with community-based education programs founded on sound principles derived in the social sciences, then the changes for compliance and the reduction of the abusive behaviors will be greater. A concerted effort to affect such behaviors as alcohol abuse is a complex, long-term, expensive task. The evidence argues against beginning an intervention before investing considerable time, money, expertise, and patience in the project. Alcohol abuse campaigns that are formulated with little thought for implementation increase public cynicism and apathy; such campaigns may act to inoculate many people against well-planned programs in the future.

However, all the evidence reviewed in this paper suggests that wise planning of an alcohol abuse education campaign may overcome whatever damage was done by previous campaigns.

REFERENCES

Alwine, G. (1974) If you need love, come to US—An overview of a peer counseling program in senior high school. *Journal of Scholastic Health* 44:463-464.

Asch, S. E. (1958) Interpersonal influence: Effects of group pressure upon the modification and distortion of judgement. In E. Maccoby et al., eds., *Readings in Social Psychology*, 3rd ed. New York: Holt, Rinehart and Winston.

Back, K. W. (1958) Influence through social communication. In E. Maccoby et al., eds., *Readings in Social Psychology*, 3rd ed. New York: Holt, Rinehart and Winston.

Bandura, A. (1969) *Principles in Behavior Modification*. New York: Holt, Rinehart and Winston.

Bandura, A. (1977) *Social Learning Theory*. Englewood Cliffs, N.J.: Prentice-Hall.

Bandura, A. (1978) The self system in reciprocal determinism. *American Psychologist* 33(4):344-358.

Bandura, A. (1979) Self-efficacy: An integrative construct. Stanford, Calif.: Stanford University.

Bandura, A., and Simon, K. M. (1977) The role of proximal intentions in self-regulation of refractory behavior. *Cognitive Therapy and Research* 1:177-193.

Becker, L. B., McCombs, M. E., and McLeod, J. M. (1975) The development of political cognitions. In S. H. Chaffee, ed., *Political Communication*. Beverly Hills, Calif.: Sage Publications.

Bem, D. J. (1970) *Beliefs, Attitudes, and Human Affairs*. Belmont, Calif.: Brooks/Cole Publishing Co.

Bem, D. J. (1972) Self perception theory. In L. Berkowitz, ed., *Advances in Experimental Social Psychology*, Vol. 6. New York: Academic Press.

Bernstein, D. A., and McAlister, A. (1976) The modification of smoking behavior: Progress and problems. *Addictive Behavior* 1:89-102.

Blane, H. T. (1976) Education and prevention of alcoholism. Pp. 519-578 in B. Kissin and H. Begleiter, eds., *Social Aspects of Alcoholism*. New York: Plenum.

Blane, H. T., and Hewitt, L. E. (1977) Mass Media, Public Education and Alcohol: A State-of-the-Art Review. Final report prepared for the National Institute on Alcohol Abuse and Alcoholism.

Bochner, S., and Insko, C. A. (1966) Communicator discrepancy, source credibility and opinion change. *Journal of Personality and Social Psychology* 4:614-621.

Braucht, G. N., Brakarsh, D., Follingstad, D., and Berry, K. L. (1973) Deviant drug use in adolescence: A review of psychological correlates. *Psychological Bulletin* 79:92-106.

Butler-Paisley, M. (1975) Public Communications Programs for Cancer Control. Institute for Communication Research, Stanford, Calif.

Cahalan, D. (1975) Implications of American drinking practices, attitudes for prevention, and treatment of alcoholism. Paper presented at the Conference on Behavioral Approaches to Alcohol and Drug Dependencies, Seattle, Wash.

Cartwright, D. (1949) Some principles of mass persuasion: Selected findings of research on the sale of U.S. war bonds. Pp. 426-447 in W. Schramm and D. F. Roberts, eds. (1971), *The Process and Effects of Mass Communication*. Urbana, Ill.: University of Illinois Press.

Choo, T. (1964) Communicator credibility and communication discrepancy as determinants of opinion change. *Journal of Social Psychology* 64:1-20.

Chu, G. C. (1966) Fear arousal, efficacy and imminency. *Journal of Personality and Social Psychology* 4:517-524.

Cohen, B. C. (1963) *The Press, the Public and Foreign Policy*. Princeton, N.J.: Princeton University Press.

Comstock, G., Chaffee, S., Katzman, N., McCombs, M., and Roberts, D. (1978) *Television and Human Behavior*. New York: Columbia University Press.

Douglas, J. D. (1976) The effects of mass media and education program on problems of drinking driving. Unpublished paper, Department of Sociology, University of California at San Diego.

Driessen, G. J., and Bryk, J. A. (1972) Alcohol countermeasures: Solid rock and shifting sands. Paper presented at the Vermont Symposium on Alcohol, Drugs, and Driving, Warren, Vt.

Farquhar, J. W., Maccoby, N., Wood, P. D., Alexander, J. K., Breitrose, H., Brown, B. W., Jr., Haskell, W. L., McAlister, A. L., Meyer, A. J., Nash, J. D., and Stern, M. P. (1977) Community education for cardiovascular health. *The Lancet* (June): 1192-1195.

Festinger, L., Riecken, H. W., and Schachter, S. (1958) When prophecy fails. In E. Maccoby et al., eds., *Readings in Social Psychology*. New York: Henry Holt and Co.

Fishbein, M., and Ajzen, I. (1975) *Belief, Attitude, Intention, and Behavior*. Menlo Park, Calif: Addison-Wesley Publishing Co.

Flynne, L. P., and Haskins, J. B. (1968) *Verbal Confidence Levels as Predictors of Driving Performance: An Exploratory Methodological Study for Drinking/Driving Communications Pretesting*. Syracuse, N.Y.: Newhouse Communications Center.

Gillies, M. (1975) *The Social Policy and Alcohol Use Survey: A Preliminary Report*. Toronto, Ont.: Addiction Research Foundation.

Gillies, M., Goodstadt, M. S., and Smart, R. G. (1976a) *Social Policy and Alcohol Use Survey: 4. The Public's View on the Prevention of Alcoholism*. Toronto, Ont.: Addiction Research Foundation.

Gillies, M., Goodstadt, M. S., and Smart, R. G. (1976b) *Social Policy and Alcohol Use Survey: 5. Public Concern About Alcohol Abuse*. Toronto, Ont.: Addiction Research Foundation.

Gillies, M., Goodstadt, M. S., and Smart, R. G. (1976c) *Social Policy and Alcohol Use Survey: 6. Errata in Alcohol Consumption Data*. Toronto, Ont.: Addiction Research Foundation.

Goodstadt, M. S. (1978) Alcohol and drug education: Models and outcomes. *Health Education Monographs* 6(3):263-279.

Gorsuch, G. L., and Butler, M. C. (1973) Initial drug use: A review of predisposing social psychological factors. *Psychological Bulletin* 79:92-106.

Grey Advertising, Inc. (1975a) *Communications Strategies on Alcohol and Highway Safety. Volume I: Adults 18-55*. U.S. Department of Transportation Publication No. DOT HS-801 400. Springfield, Va.: U.S. National Technical Information Service.

Grey Advertising, Inc. (1975b) *Communications Strategies on Alcohol and Highway Safety. Volume II: High School Youth*. Final Report. U.S. Department of Transportation Publication No. DOT HS-801 400. Springfield, Va.: U.S. National Technical Information Service.

Hamburg, V. A., and Varenhorst, B. B. (1972) Peer counseling in the secondary schools: A community mental health project for youth. *American Journal of Orthopsychiatry* 42:566-581.

Harris, L., and Associates, Inc. (1974) *Public Awareness of the NIAAA Advertising*

Campaign and Public Attitudes Toward Drinking and Alcohol Abuse: Phase Four: Winter 1974 and Overall Summary. New York: Louis Harris and Associates, Inc.

Harris, M. B. (1969) Self-directed program for weight control: A pilot study. *Journal of Abnormal Psychology* 74:263-270.

Haskins, J. B. (1969) Effects of safety communication campaigns: A review of research evidence. *Journal of Safety Research* 1:58-66.

Hawkins, T. E., and Cooper, E. J. (1976) *San Antonio Alcohol Safety Action Project Analytic Study No. 7: Analysis of Public Information and Education 1975*. San Antonio, Tex.: Southwest Research Institute.

Heath, E. A., Harris, U. J., and Radford, A. J. (1979) A total health approach to smoking cessation. Paper presented at the Anzserch National Conference, Perth, Australia.

Hunt, W. A., Barnett, L. W., and Brasch, L. G. (1971) Relapse rates in addiction programs. *Journal of Clinical Psychology* 27:455.

Hyman, H. H., and Sheatsley, P. B. (1947) Some reasons why public information campaigns fail. *Public Opinion Quarterly* 11:413-423.

Janis, I. L., and Feshbach, S. (1953) Effects of fear-arousing communications. *Journal of Abnormal and Social Psychology* 48:78-92.

Jessor, R., and Jessor, S. (1971) *Problem Behavior and Psychological Development*. New York: Academic Press.

Jones, E. E., and Gerard, H. B. (1967) *Foundations of Social Psychology*. New York: Wiley.

Kinder, B. N. (1975) Attitudes toward alcohol and drug abuse. II. Experimental data, mass media research, and methodological considerations. *International Journal of Addictions* 10:1035-1054.

Klapper, J. T. (1960) *The Effects of Mass Communications*. Glencoe, Ill.: Free Press.

Krugman, H. E. (1965) The impact of television advertising: Learning without involvement. *Public Opinion Quarterly* 29.

Lasswell, H. D. (1948) The structure and function of communication in society. Pp. 84-99 in W. Schramm and D. F. Roberts, eds. (1971), *The Process and Effects of Mass Communication*. Urbana, Ill.: University of Illinois Press.

Lazarsfeld, P. F., Berelson, B., and Gaudet, H. (1948) *The People's Choice*, 2nd. ed. New York: Duell, Sloane and Pearce.

Leventhal, H., and Watts, J. (1966) Sources of resistance to fear-arousing communications on smoking and lung cancer. *Journal of Personality* 34:155-175.

Lewin, K. (1958) Group decision and social change. In E. Maccoby et al., eds., *Readings in Social Psychology*. New York: Holt, Rinehart and Winston.

Lipset, S. M. (1953) Opinion formation in a crisis situation. *Public Opinion Quarterly* 17:20-46.

London, A. M., and Schreibe, E. D. (1965) A controlled study of effects of group discussions and an anorexiant in outpatient treatment of obesity with attention to psychological aspects of dieting. *Annals of Internal Medicine* 65(1):80.

Maccoby, N. (1979) Promoting positive health-related behavior in adults. Paper presented at the Fourth Vermont Conference on Primary Prevention of Psychopathology.

Maccoby, N., and Alexander, J. (1980) Use of media in life-style programs. In P. O. Davidson and S. M. Davidson, eds., *Behavioral Medicine: Changing Health Lifestyles*. New York: Brunner/Mazel.

Maccoby, N., and Farquhar, J. W. (1975) Communication for health: Unselling heart disease. *Journal of Communication* 25:114-126.

Maccoby, N., Farquhar, J. W., Wood, P. D., and Alexander, J. (1977) Reducing the risk of cardiovascular disease: Effects of a community based campaign on knowledge and behavior. *Journal of Community Health* 3:100-114.

Maher, R. G. (1977) The antecedents and influences of smoking behavior in high school students. Unpublished thesis, Harvard College.

McAlister, A. L. (1978) Mass communication of cessation counseling: Combining television and self-help groups. Paper presented at the International Conference on Smoking Cessation, New York.

McAlister, A. L., Perry, C., and Maccoby, N. (1979) Adolescent smoking: Onset and prevention. *Pediatrics* 63:650-658.

McCombs, M. E. (1972) Mass communication in political campaigns: Information, gratification and persuasion. Pp. 176-187 in F. G. Kline and P. J. Tichenor, eds., *Current Perspectives in Mass Communications Research*. Beverly Hills, Calif.: Sage Publications.

McCombs, M. E. (1974) A comparison of intra-personal and interpersonal agendas of public issues. Paper presented at the convention of the International Communication Association, New Orleans, La.

McCombs, M. E., and Shaw, D. L. (1972) The agenda-setting function of mass media. *Public Opinion Quarterly* 36: 176-187.

McGuire, W. J. (1964) Inducing resistance to persuasion: Some contemporary approaches. In L. Berkowitz, ed., *Advances in Experimental Social Psychology*, Vol. 1. New York: Academic Press.

McGuire, W. J. (1968) Personality and susceptibility to social influence. In E. F. Borgatta and W. W. Lambert, eds., *Handbook of Personality Theory and Research*. Chicago: Rand McNally.

McGuire, W. J. (1969) The nature of attitudes and attitude change. In G. Lindzey and E. Aronson, eds., *The Handbook of Social Psychology*, Vol. 3. Menlo Park, Calif.: Addison-Wesley Publishing Co.

McGuire, W. J. (1973) Persuasion, resistance and attitude change. In I. Pool and W. Schramm, eds., *The Handbook of Communication*. Chicago: Rand McNally.

McGuire, W. J. (1974) Communication persuasion models for drug education. In M. Goodstadt, ed., *Research on Methods and Programs of Drug Education*. Toronto: Addiction Research Foundation of Ontario.

McRae, C. F., and Nelson, D. M. (1971) Youth to youth communication on smoking and health. *Journal of Scholastic Health* 41:445.

Media and Marketing Decisions (1979) *Update: FM's Earful*. P. 42, August.

Medical Progress and the Public (n.d.) *Swine Flu*. Stanford, Calif.: School of Medicine, Stanford University.

Mendelsohn, H. (1973) Some reasons why information campaigns can succeed. *Public Opinion Quarterly* 37:50-61.

Meyer, A. J., and Henderson, J. B. (1974) Multiple risk factor reduction in the prevention of cardiovascular disease. *Preventive Medicine* 3:225-236.

Meyer, A. J., Maccoby, N., and Farquhar, J. W. (1977) The role of opinion leadership in a cardiovascular health campaign. Paper presented at the 27th annual conference of the International Communication Association, Berlin.

Meyer, A. J., McAlister, A., Nash, J., Maccoby, N., and Farquhar, J. W. (1976) Maintenance of cardiovascular risk reduction: Results in high risk subjects. Paper presented at the 49th scientific session of the American Heart Association.

Newcomb, T. M. (1953) An approach to the study of communicative acts. *Psychological Review* 60:393-404.

Newcomb, T. M. (1958) Attitude development as a function of reference groups. In E. Maccoby et al., eds., *Readings in Social Psychology*. New York: Holt, Rinehart and Winston.

Newman, I. M. (1970a) Peer pressure hypothesis for adolescent cigarette smoking. *Scholastic Health Review* 1(2):15.

Newman, I. M. (1970b) Status configurations and cigarette smoking in a junior high school. *Journal of Scholastic Health* 40(1):23.

Osgood, C. E., and Tannenbaum, P. H. (1955) The principles of congruity in the prediction of attitude change. *Psychological Review* 62:42-55.

Pelz, E. B. (1958) Some factors in group decision. In E. Maccoby et al., eds., *Social Psychology*. New York: Holt, Rinehart, and Winston.

Penick, S. B., Filion, R., Fox, S., and Stunkard, A. J. (1971) Behavior modification in the treatment of obesity. *Psychosomatic Medicine* 33:49-55.

Perry, C. (1979) Project C.L.A.S.P.: Counseling Leadership About Smoking Pressures. Stanford, Calif.: Stanford Heart Disease Prevention Program.

Ray, M. L. (1973) Marketing communication and the hierarchy of effects. In P. Clarke, ed., *New Models for Mass Communication Research*. Beverly Hills, Calif.: Sage Publications.

Roberts, D. F. (1975) Attitude change research and the motivation of health practices. In *Applying Behavioral Science to Cardiovascular Risk*. Washington, D.C.: American Heart Association.

Roberts, D. F., and Maccoby, N. (1973) Information processing and persuasion: Counterarguing behavior. In P. Clarke, ed., *New Models for Mass Communication Research*. Beverly Hills, Calif.: Sage Publications.

Robertson, L. S. (1976) Whose behavior in what health marketplace? Paper presented at the Steinhart Symposium on Consumer Behavior in the Health Marketplace. Lincoln, Neb.

Robinson, J. P. (1972) Mass communication and information diffusion. In F. G. Kline and P. J. Tichenor, eds., *Current Perspectives in Mass Communication Research*. Beverly Hills, Calif.: Sage Publications.

Robinson, M. J., and Zukin, C. (1976) Television and the Wallace vote. *Journal of Communication* 26:79-83.

Rogers, E. M., and Shoemaker, F. F. (1971) *Communication of Innovations: A Cross-Cultural Approach*. New York: Free Press.

Rokeach, M. (1966-1967) Attitude change and behavior change. *Public Opinion Quarterly* 30:529-550.

Ross, M. (1976) The self perception of intrinsic motivation. In J.H. Harvey, W. J. Ickes and R. F. Kidd, eds., *New Directions in Attribution Research*. Hillsdale, N.J.: Earlbaum.

Schneider, F. W., and Vanmastrigt, L. A. (1974) Adolescent-preadolescent differences in beliefs and attitudes about cigarette smoking. *Journal of Psychology* 87:71.

Schramm, W., and Carter, R. F. (1959) Effectiveness of a political telethon. *Public Opinion Quarterly* 23:121-126.

Sears, D. O. and Freedman, J. L. (1967) Selective exposure to informa- tion: A critical review. *Public Opinion Quarterly* 31:194-213.

Star, S. A., and Hughes, M. M. (1950) Report of an educational campaign: The Cincinnati plan for the United Nations. *American Journal of Sociology* 55:385-400.

Stern M. P., Farquhar, J. W., Maccoby, N., and Russell, S. H. (1976) Results of a two-year health education campaign on dietary behavior. *Circulation* 54:826-833.

Sternhal, B., and Craig, C. S. (1974) Fear appeal: Revisited and revised. *Journal of Consumer Research* 1(3):22-34.

Stokols, D. (1975) The reduction of cardiovascular risk: An application of social learning perspectives. In *Applying Behavioral Science to Cardiovascular Risk*. Washington, D.C.: American Heart Association.

Stunkard, A. J. (1972) New therapies for eating disorders: Behavior modification of obesity and anorexia nervosa. *Archives of General Psychiatry* 26:391-398.

Stunkard, A. J. (1975) From explanation to action in psychosomatic medicine: The case of obesity. *Psychosomatic Medicine* 37(3):195-236.

Stunkard, A. J., and Mahoney, M. J. (1974) Behavioral treatment of eating disorders. In H. Leitenberg, ed., *Handbook of Behavior Modification*. New York: Appleton-Century-Crofts.

Stunkard, A. J., and Penick, S. B. (1979) Behavior modification in the treatment of obesity. *Archives of General Psychiatry* 36:801-806.

Swinehart, J. W. (1976) Some lessons from the "Feeling Good" television series. Paper presented at the Steinhart Symposium on Consumer Behavior in the Health Marketplace, Lincoln, Neb.

Thoreson, G. E., and Mahoney, M. J. (1974) *Behavioral Self-Control*. New York: Holt, Rinehart and Winston.

Utech, D. A., and Hoving, K. L. (1969) Parents and peers as competing influences on the decisions of children of differing ages. *Journal of Social Psychology* 78:267-274.

Wallack, L. M. (1978) Evaluating a large-scale prevention demonstration program: Procedures, problems, new directions. Paper presented at the Minnesota Prevention Evaluation Conference, Breezy Point, Minn.

Wallack, L. M. (1979) *The California Prevention Demonstration Program Evaluation: Description, Methods, Findings*. Berkeley, Calif.: Social Research Group.

Reducing the Costs of Drinking and Driving

DAVID S. REED

INTRODUCTION

Public concern over the dangers of drunken driving is almost as old as the automobile. Indeed, few authors on the subject can resist citing the "motor wagons" editorial in the *Quarterly Journal of Inebriety* in 1904. Despite the long history of concern and the many attempts to control the problem, drunken driving is still perceived as a major highway safety problem.

Paradoxically, the widespread familiarity in our society with drinking, driving, and their combination may have hindered the development of effective countermeasures. An individual who is very familiar with the elements of a problem may let preconceived notions interfere with the gathering, assimilating, and applying of information. The first section of this paper examines the costs generated by the drinking-driving problem, what part of these costs are potentially preventable, and how the magnitude and distribution of costs might be viewed in assessing the priority of the problem. The next section taps the extensive experience worldwide with programs to reduce drunken driving. We find some

David S. Reed, a doctoral candidate in the program in public policy at the John F. Kennedy School of Government, Harvard University, is currently an intern at the Federal Communications Commission.

This work was completed during my enrollment at the John F. Kennedy School of Government, Harvard University. I would particularly like to thank Mark Moore and Dean Gerstein, who supervised its preparation.

promising avenues for future action and—equally important—some unpromising avenues that still have vocal advocates. The following section examines efforts to reduce the risk associated with a given amount of drunken driving. Because there have been few such efforts, this section offers few conclusions; it does raise questions that seem to warrant further investigation.

The final section examines the manner in which the federal government has designed and managed programs of drinking-driving countermeasures. It concludes that changes are necessary if we are to learn from experience and improve our ability to reduce the costs of drunken driving.

COSTS AND THEIR IMPLICATIONS FOR THE ROLE OF GOVERNMENT

THE NEED FOR EVALUATION

Any attempt to ameliorate drinking-driving problems will consume scarce government resources and may impose monetary and nonmonetary costs (such as some restriction of civil liberties) on individuals. We must therefore determine the costs generated by the problem in order to compare them with the costs imposed by possible solutions. This section examines the magnitude and distribution of costs resulting from drunken driving. It also provides information to help determine the priority of reducing this problem by governmental efforts.

This question of costs is of more than academic interest. The comptroller general of the U.S. (1979, p. 3) estimates that "Federal, State, and local governments spent over $100 million in 1976 for their drinking-driver countermeasure activities." This level of resources, and, as we discuss below, the large human and economic costs potentially at stake suggest the importance of determining the appropriate role of government in efforts to reduce the costs of drunken driving.

PREVENTABLE ACCIDENTS

Information presented to policy makers about drunken driving (U.S. Department of Transportation 1968, Noble 1978a, Comptroller General of the U.S. 1979) has typically expressed the importance of the problem in terms of the costs associated with it. For example, Noble (1978a, p. 61) reports that "approximately one-third of the . . . injuries and one-half of the fatalities [from traffic accidents] are alcohol related." The term "alcohol-related" refers to any accident in which a driver, or some-

TABLE 1 Percentage of Accident-Involved Drivers and Control Group Drivers Found to Be Within Various BAC Ranges, by Worst Consequence of Accident and Place and Time of Study

BAC (%)	Fatal,[a] Vermont, 1967–1968, Accident	Control	Injury, Huntsville, Ala., 1974–1975, Accident	Control	Injury, Grand Rapids, Mich., 1962–1963, Accident	Control	Property Damage, Grand Rapids, Mich., 1962–1963, Accident	Control
<0.01	64%	84%	73.98%	88.18%	81.83%	89.01%	83.87%	89.01%
0.01–0.049	5%	9%	4.86%	4.29%	6.55%	7.76%	6.89%	7.76%
0.05–0.999			7.97%	4.68%	4.11%	2.46%	3.54%	2.46%
0.10–0.149	31%	7%	4.84%	2.11%	7.51%	0.76%	5.70%	0.76%
⩾0.015			8.35%	0.74%				

[a] Vermont data adjusted per Appendix B.
Source: Appendix A.
Note: BAC levels less than 0.01% are effectively equal to zero. The minimum BAC at which it is illegal to drive in most states is 0.10%.

times any person involved in an accident, had a positive blood alcohol content (BAC).[1]

Clearly, a better indication of the importance of solving a problem than calculating the associated costs would be to calculate the costs that would be eliminated if it were solved. To determine the maximum preventable costs of drunken driving, I first examine the BAC levels of drivers involved in various types of accidents and those of control groups of drivers selected at random from times and places similar to those at which the accidents being controlled for occurred. Table 1 shows these data, which I have selected from several studies (also see Appendix A). As Table 1 shows, the BAC levels of drivers involved in more serious accidents are generally higher than those of drivers involved in less serious accidents.

Once the distribution of BAC levels among accident-involved drivers and the control groups is known, it is possible to compute how many accidents would be avoided if all driving was done at the risk level associated with the lowest BAC level; that is, accidents that would be prevented by a "perfect" drinking-driving countermeasure. Appendix C demonstrates these computations, and the results are shown in Table 2.

Since the figures in Table 2 are based on the association between

[1] Blood alcohol content (BAC) and the equivalent term blood alcohol concentration refer to the standard measure of the concentration of alcohol present in a person's body at a given time (not including any alcohol that has been drunk but not yet absorbed).

TABLE 2 Expected Reduction in Motor Vehicle Traffic Accidents If All Drivers Had a Zero BAC

Type of Accident, Place and Time	Expected Reduction
Fatal,	
Vermont, 1967–1968	23.7%
Injury,	
Huntsville, 1974–1975	15.8%
Injury,	
Grand Rapids, 1962–1963	8.2%
Property damage,	
Grand Rapids, 1962–1963	5.7%

Source: Appendix C.

BAC and accident involvement, we must ask whether any other factors, correlated with both BAC while driving and accident risk, are confounding our analysis. There appear to be two confounding factors biasing the results in opposite directions. First, there is evidence that greater frequency of drinking is positively associated with more frequent drunken driving and negatively associated with accident risk at any given BAC (Borkenstein et al. 1974, Hurst 1973). As Appendix C shows, the risk of accident as computed from the data on Grand Rapids and Vermont is actually lower for the 0.10-0.049 BAC range than for the <0.01 range. The correlations with drinking frequency bias Table 2 toward underestimating the accident reduction, since those who currently drive drunk, if sober, would have a *lower* accident risk than those who currently drive sober.

The second bias results from a disputed but probable positive association between "problem drinking" or "alcoholism" and both drunken driving and accident risk while sober (Smart 1969, Noble 1978b, pp. 238-240). This bias would result in drunken drivers, as a group, having higher accident risk than the general driving population even if they were prevented from driving with positive BAC levels.

Available information is not sufficient to quantify either of these biases, therefore I present the expected reductions in Table 2 as they are and hope that the biases are small or cancel each other out.

A final caution is that my figures for maximum achievable accident reduction are based on samples at specific times and places. Their validity for the nation as a whole is not ensured. It is heartening that the Huntsville and Grand Rapids studies of injury accidents, despite their spatial and temporal separation, yield reduction estimates within 8 percentage points of each other. The Vermont data on distribution across

BAC ranges of drivers killed is in agreement with several other studies of driver fatalities (Jones and Joscelyn 1978, p. 12, Noble 1978b, pp. 233-240).[2] I have encountered no studies of property-damage-only crashes other than the Grand Rapids study.

MAGNITUDE OF PREVENTABLE COSTS

The figures on accident reduction can be roughly converted to savings in terms of lives, injuries, and dollars of property damage. In 1977, motor vehicle traffic accidents resulted in 49,500 deaths, 1.9 million disabling injuries, and $15.5 billion of property damage in the United States (National Safety Council 1976). (The $15.5 billion figure is based on $520 per accident-involved vehicle [Jones and Joscelyn 1978].) Multiplying these figures by the percentages of accident reductions in Table 2 yields maximum achievable savings in 1977 of 11,700 deaths, 156,000 to 300,000 disabling injuries, and $963 million in property damage (based on the total reduction in accident-involved vehicles).

Of course, like any estimates based on sketchy information, these figures must be interpreted with some care. The figures in Table 2 refer to reductions in the number of accidents, not directly to reductions in the consequences of accidents. For instance, alcohol-related accidents more frequently involve only a single car and driver than do accidents in general (Borkenstein et al. 1974, p. 105, Noble 1978b, p. 235). Therefore, the reduction in the number of people killed, the number of people injured, and the number of cars damaged will be less than the reduction in number of accidents. It is also likely, however, that the average severity of injury and damage in the accidents prevented will be greater than the severity of injury and damage in accidents in general; therefore only in the case of fatalities do the above figures clearly overstate potential savings from drinking-driving countermeasures.

It is beyond the scope of this paper to develop estimates of preventable costs for all other problems that may compete with drinking-driving for government resources. Table 3, however, provides some perspective. For example, we see from Table 3 that 22 percent of all accidental deaths resulted from alcohol-related traffic accidents and that 10 percent of accidental deaths would have been prevented if all drivers had zero BAC.[3]

[2] I could not produce accident reduction estimates from these other studies because they lacked control groups.

[3] Table assumes that 50 percent of all traffic accident deaths are alcohol related (Comptroller General of the U.S. 1979, Noble 1978b), and that 23.7 percent of all traffic accident deaths would be prevented if all drivers had zero BAC (see Table 2).

TABLE 3 Deaths from Alcohol-Related Motor Vehicle Traffic Accidents, and Deaths Prevented If All Drivers Had Zero BAC as Percentages of Various Categories of Deaths, 1977

Cause of Death	Related to Drinking-Driving	Prevented If All Drivers Had Zero BAC
Alcohol-related motor vehicle traffic accidents	100%	47%
All motor vehicle traffic accidents	50%	24%
All accidents	22%	10%
All causes except cardiovascular diseases and malignancies	4%	2%
All causes	1%	0%

Source: Bureau of the Census (1978, pp. 75, 78).

In considering the data in Table 3, one should be aware that deaths caused by traffic accidents, particularly alcohol-related traffic accidents, occur typically among a younger age-group than deaths from other major causes. For example, Perrine et al. (1971, p. 46) report that of drivers killed in accidents, 47 percent were under 30 years of age, and of those with positive BAC levels, 49 percent were under 30. Many think that deaths of younger persons should be counted more heavily in making policy decisions, because each such death results in more years of life lost.

Another problem of weighting involves deaths of drinking drivers themselves versus deaths of innocent victims of drinking-driving accidents. I estimate that 61 percent to 78 percent of persons killed in drinking-driving traffic accidents are drivers with positive BAC levels (see Appendix B). One may argue that since people can choose how much to drink before driving, it may not be the proper role of government to protect them from the consequences of a decision freely made.

Alternatively, one might argue that people may drive after excessive drinking due to a momentary lapse of judgment, an episode of severe stress, or a habitual behavior pattern that is difficult to control (i.e., an "addiction"). One study reports that 75 percent of drivers admit to driving after drinking at least occasionally (U.S. Department of Transportation 1968, p. 61).[4] Viewed in this way, drunken driving looks less

[4] This is not to say that 75 percent of drivers drive with the high BAC levels associated with greatly increased accident risk (see Table C-1 in Appendix C). However, there is evidence that less frequent drinkers, and less frequent drinking drivers, show elevated accident risk at lower BAC levels than does the drinking-driving population in general (Hurst 1973).

like a risk some choose to take and more like a random risk to which many people will be exposed whether or not they would choose to in a moment of calm reflection. Furthermore, the death or injury of someone who is driving drunk may impose costs on friends, loved ones, those dependent for financial support, and the economy in general.

In setting policy, the proper weight to give to deaths of drunk drivers cannot be determined by empirical or logical analysis, although these may inform the decision. I feel that such deaths should be weighted equally to others for purposes of setting policy, but that we should be wary of providing perverse incentives to potential drunk drivers—a topic I address later in this paper.

In conclusion, it seems that drinking-driving countermeasures can be legitimate and useful government actions, but that even if such countermeasures were perfectly successful, the savings in lives, injuries, and property loss would be less than widely quoted figures would lead one to believe. Which countermeasures show the best prospect of success is the topic of the rest of this paper.

EXPOSURE REDUCTION

OVERVIEW

The strategy of drinking-driving countermeasure that occurs first to most people is exposure reduction: reducing the amount of drunken driving that takes place and thereby reducing accident costs. This section examines several potential ways of achieving exposure reduction:

- General deterrence: countermeasures that seek to prevent drivers in general from combining driving with drinking in excess of legally prescribed limits (0.10 percent BAC in most states).
- Recidivism reduction (specific deterrence): countermeasures that seek to specifically compel those people who have already been arrested for driving while intoxicated (DWI) not to drive drunk again.
- Third-party intervention: countermeasures that seek to influence those around potential drunk drivers (servers of alcohol, fellow party guests, or bar patrons, etc.) to prevent them from driving while intoxicated.
- Altering the legal minimum drinking age.
- Screening the driving population for those most likely to drive drunk.
- Installing devices in vehicles to automatically detect drunk drivers.
- Providing alternative transportation for potential drunk drivers.

GENERAL DETERRENCE

Risk of Punishment

The most effective programs of general deterrence seem to have been those that raised drivers' perceived risk of arrest and punishment for drunk driving.

The classic program of this type is the British Road Safety Act of 1967. The act "defined driving with a BAC of .08% or higher per se as an offense," and "gave police the power to require pre-arrest roadside tests [for breath alcohol] from drivers who had been involved in traffic accidents or moving violations or where there was reasonable cause to suspect them of drinking and driving. Drivers who refused to provide a breath test were considered guilty" (Cameron 1978, p. 22). Passage of the act was accompanied by a great deal of publicity and public awareness of its provisions (Jones and Joscelyn 1978, pp. 66-67).

The immediate impact of the act was positive and dramatic. For the 3 months following passage of the act, casualties from traffic accidents were reduced 16 percent from the same period the previous year, and fatalities were reduced 23 percent (Ross 1973, p. 20). The percentage of fatally injured drivers with BAC levels of greater than or equal to 0.08 percent dropped from 27 percent before passage to 17 percent the following year (Comptroller General of the U.S. 1979, p. 27). A careful quasi-experimental study by Ross (1973) attributed these reductions substantially to the act and the wide publicity it received. He also noted (p. 75):

> Unfortunately, there are many signs that the initial effect of the legislation is diminishing. Although there are problems in speculating on what would have happened in the absence of the legislation, the significant change in the slope of the casualty rate curves . . . suggests that the savings achieved ought to be regarded as temporary. This conclusion is bolstered by the fact that blood analysis of fatalities shows that the initial drop in the percentage with an illegal alcohol concentration, from 25 percent in 1967 to 15 percent in 1968, was progressively diminished, and the percentage has now returned to its former level.

The trend noted by Ross appears to have continued. By 1975 the percentage of drivers killed in England and Wales in road accidents with a BAC level of 0.08 percent or more had reached 36 percent, substantially above its pre-1967 level (Comptroller General of the U.S. 1979, p. 27).

The explanation offered by Ross (1973, pp. 75-78), which has achieved wide acceptance, is that the well-publicized passage of the act convinced many drivers that the risk of arrest when driving drunk had become much higher than it used to be be. Potential drunk drivers were deterred by the threat of arrest and punishment. As time went on, however, drivers realized that enforcement was rather slack and that risk of arrest was not all that high. This realization caused an "evaporation" of the act's deterrent effect.[5]

In September 1975, the Cheshire County Police seemed to recapture the Road Safety Act's deterrent effect through a publicized policy of administering breath tests to all drivers pulled over for violations or involved in accidents during "drinking hours." The resulting accident reduction "evaporated" a month after the policy came to a publicized end (Ross 1977).

How applicable is Britain's experience to that of the United States? Ross (1973, p. 1) asserts that "in broad culture and narrower legal structure Britain is the closest of European countries to the United States." Comparative statistics indicate that Britain is very similar to the United States in number of traffic fatalities per vehicle mile (Borkenstein et al. 1974, p. 20) and has a slightly smaller representation of alcohol-influenced drivers among driver fatalities (Organisation for Economic Co-operation and Development 1978, p. 25). More convincing is the fact that the Road Safety Act experience has been partially replicated in countries other than Britain. As Robertson (1977, p. 6) describes:

> A law providing for breath testing and penalties for blood alcohol above .08% by weight or refusing the test came into force in Canada in 1969 but with less widely publicized predictions of increased chances of arrest than in Britain. Death rates were about 8 percent less than expected in the subsequent two years but, again, the effect of the law was temporary. After 2 years of the law, death rates in Canada returned to levels that would have been expected without the law.

In 1978, France amended its drunk-driving law to allow police to

[5] There are alternatives to the "evaporation" theory regarding the British Road Safety Act. In support of alternative or supplementary explanations, we should note that since the date of Ross's study, incidence of illegally high BAC levels among drivers killed in Britain has continued to climb, until it now substantially exceeds the incidence before passage of the act. Mere evaporation of the deterrent effect cannot account for this. Other factor(s) must be strengthening the association between BAC level and driver fatalities over time. It is not certain how much of what Ross observed was due to "evaporation" and how much was the result of the aforementioned unidentified factor(s).

The "evaporation" theory, however, is still attractive as a partial explanation. It is well grounded in theory and predicts the outcomes of several other programs patterned after the British Road Safety Act.

administer breath tests to drivers even in the absence of an accident or traffic violation. Dorozynski and Volnay (1978, p. 44) describe the initial effects (presented here in translation):

> A certain number of [breath testing] operations were organized immediately, [and] announced in the press, in several French provinces and in Paris. Their results were entirely unexpected: less than 0.5% of the "alcooltests" were positive, and in Paris, zero.
>
> Taking account of what is known of the accuracy of the "alcooltest," and of its sensitivity, this roughly indicates that the "French at the wheel" had stopped drinking.

While the information from France so far is too sketchy to tell us anything, it does not appear inconsistent with Ross' observations in Britain.

Unfortunately, the only agreed-upon success in the United States similar to the British Road Safety Act was a 1-year countermeasure program at Lackland Air Force Base, Texas, in 1959. While the program achieved "a statistically significant decrease of 50% to 60% in the number and rate of accidents, driver injuries, and other injuries during the operational period" (Cameron 1978, p. 23), it is not clear how applicable this experience would be to a civilian environment.

Despite the lack of documented successes in the United States for countermeasures to deter drunken driving by increasing the risk of arrest and punishment, this approach appears to have won favor among state highway safety officials. A recent survey by the U.S. General Accounting Office (Comptroller General of the U.S. 1979) asked officials from the 50 states, the District of Columbia, and Puerto Rico to choose and rank the three most important current or past efforts in their state to combat drunken driving. "Instituting or increasing the use of special police patrols for the drinking driver" was ranked number one by 25 percent of respondents, more than any of the other nine responses. Another 27 percent of respondents ranked it second.

Of six states examined in depth in the GAO report (California, Georgia, Louisiana, Minnesota, New York, Washington state), five were operating some police patrols targeted at drinking drivers, four reported increased numbers of driving while intoxicated (DWI) arrests, and two reported evaluations of the patrol's effect on accident rates. In King County, Washington, a "drinking-driver emphasis patrol" was operated as part of the federally funded Alcohol Safety Action Project (ASAP). Evaluation failed to find a change in accident fatalities or injuries resulting from the patrol, although DWI arrests had increased. In Hennepin County, Minnesota, emphasis patrols were also operated as part of ASAP. "Alcohol involvement in fatal crashes was reduced from 63 percent in 1972 to 38 percent in 1976" (Comptroller General of the U.S. 1979, p. 22). Even in the absence of a control group, this finding is very suggestive of success.

From present evidence, then, it appears that when drivers' perceived risk of arrest and punishment for drunk driving is sufficiently increased, drunk driving is deterred and accidents are reduced. In Britain, fatalities from traffic accidents decreased by 23 percent in response to the Road Safety Act, and similar legislation in Canada brought about a temporary reduction of 8 percent. In order to sustain a high perceived risk of arrest and punishment, the risk must be made high and kept high.

Targeting patrols by day of week, time, and geographic location; legislative and technical progress toward making breath tests for alcohol easier to administer; and simplifying the process of making a DWI arrest and providing police with motivation to make such arrests are all ways to increase the risk of arrest. Using such methods, ASAPs were able to double and triple the number of DWI arrests, although it is unclear how much of this increase resulted merely from charging drivers with DWI rather than a specific moving violation (Zimring 1978 pp. 151-152).

What remains unknown is just what levels of risk are necessary to achieve various degrees of deterrence and what it would cost to bring about such increases in risk. These questions appear to require empirical study.

Severity of Punishment

If increasing risk of punishment can deter drunken driving, then what about increasing severity of punishment? It seems at first glance easier and less expensive to hand out stiffer penalties to convicted drinking drivers than to beef up enforcement. The archetypes of the severe punishment approach are the Scandinavian countries. Imprisonment or fines exceeding one-tenth of the convicted person's after-tax income, combined with license suspensions exceeding 1 year, are common punishments for first DWI offense in these countries.

The Scandinavian drunk-driving laws are widely reputed to be effective deterrents, and there is anecdotal evidence in accord with this reputation, but during a 3-month study in Scandinavia, Ross (1975) was unable to find any scientific evidence that the laws had deterrent effects. He then performed time series analyses of drunk driving and traffic casualty measures in Sweden, Norway, and Finland. In no case did these measures change systematically with changes in drunk-driving laws so as to indicate a causal relationship. Of course, the failure to demonstrate the laws' efficacy does not prove them ineffective. The Scandinavian laws are quite old and have changed only gradually over the years, hampering time series analysis.

Since there is no scientific evidence of the Scandinavian laws' effectiveness, and since cultural traditions—such as drinking in large quantities when drinking—may be adequate to account for the anecdotal reports of differences with American drinking and driving behavior, we must look elsewhere to try to assess the potential of deterrence through severe punishment.

A less well-known test of the severe punishment approach occurred in the United States in Chicago. For 7 months during 1970 and 1971, "magistrates in Chicago's traffic courts were directed by the supervising judge to sentence persons convicted of driving while intoxicated (DWI) to seven days in jail and to recommend to the Secretary of State's office that such drivers' licenses be suspended for one year. The policy was publicly announced and widely publicized" (Robertson et al. 1973, p. 57). While motor vehicle fatalities decreased during this period, evaluators concluded that the decrease was not statistically significant compared with variations during the preceding 5 years and noted a similar decline during the period in another, similar city without such a program.

The Chicago program's failure could be attributed to a frequently identified pitfall of this approach—judges, juries, and even police and prosecuters are thought to be reluctant to subject DWI offenders to severe punishment. Up to 75 percent of drivers admit to driving after drinking at least occasionally (U.S. Department of Transportation 1968, p. 61), so many people in our society do not view driving after drinking as deviant behavior. If the general feeling of the public is "there but for the grace of God go I," it is doubtful that severe penalties will be applied often even if they are authorized by law.

In fact, what DWI offenders do is unusual. In order to achieve a high enough BAC level to be considered intoxicated in most states, a person weighing 140 pounds would have to consume four to five drinks within an hour (more if he or she were drinking on a full stomach), a "drink" being 12 ounces of beer, 3 ounces of wine, or 1 ounce of 80- to 90-proof spirits. A typical BAC level for a person brought to trial for DWI is 0.15 percent and would be the result of consuming six to seven drinks. It is probably safe to say that most drivers have never driven after consuming this much alcohol.

Severe penalties for drunk driving may nevertheless result in fewer arrests, fewer convictions, and more plea bargaining. In the Chicago experiment it is unknown how frequently judges actually complied with the supervising judge's instructions to give 7-day jail sentences for DWI convictions. It is known that DWI arrest levels did not change significantly, but there was a decrease in the number of convictions, accounted for by a decrease in the conviction rate for defendants who did not undergo a test for alcohol after arrest. "[T]his change appears to be a result of changes in plea bargaining or reluctance of judges or juries to convict and sentence to seven days in jail those drivers for whom objective evidence of impairment was not available" (Robertson et al. 1973, p. 66).

The difficulty of convicting any but the most blatantly impaired may be the reason that a Norwegian government committee, in the wake of Ross' Scandinavian study, has recommended abolishing the mandatory prison term for DWI convictions with a BAC level of 0.05-0.12 percent, and reducing the mandatory prison sentence for those convicted with BAC levels exceeding 0.12 percent from 21 days to 7 days (Comptroller General of the U.S. 1979, pp. 31-32).

Even if judges and juries could be educated to adopt severe sentences for DWIs, there is a question of how severe a penalty we as a society are willing to levy for this crime. A 7-day jail sentence may not be severe enough to achieve significantly more deterrence than the risk of license suspension. We simply do not know at present what combinations of risk and penalty will achieve deterrence. I speculate, however, that people's behavior is relatively insensitive to changes in the seriousness of an adverse outcome that is viewed as very unlikely to occur. After all, those who currently drive drunk are not deterred by the small risk of a very severe penalty—accidental death. While information is sparse, the most commonly accepted figure for the risk of arrest when driving while illegally intoxicated in the United States is 1 in 2,000 (Jones and Joscelyn 1978, p. 53) and in Britain, in the wake of the Road Safety Act, it was estimated to be 1 in 1,000 (Comptroller General of the U.S. 1979, p. 27).

In addition to the dubious potential for effectiveness, we should remember that the severe punishment approach would not be as costless as it may seem. As the severity of penalties increases, the length of trials and number of appeals are likely to increase (assuming the penalties are not circumvented by plea bargaining), thus further crowding the overburdened court system. If penalties are to include imprisonment, then the cost of a large addition to the country's prison population is part of the calculation.

Public Information and Education

The third approach to achieving general deterrence is public information and education. In fact, a recent report by the U.S. General Accounting Office (Comptroller General of the U.S. 1979, p. i) makes this claim:

> Before any significant reduction in alcohol-related traffic accidents will occur, a long-term continuous educational commitment must be made. Governments, educational institutions, and the general public need to work together to change attitudes about drinking and driving.

The report concludes that "the Secretary of Transportation should take the lead in a massive effort to start changing social attitudes about drinking and driving" (front cover).

Is such a "massive effort" necessary? And would it be productive? Public education, taken in its broadest sense, would include programs to arrest drinking drivers and to subject them to punishment and/or other treatment. If the argument is thus trivialized, it is considerably strengthened but of less interest. Public education in a narrower sense means communication with the public through symbols rather than actions.

Furthermore, not all efforts at public information and education are intended to achieve general deterrence—that is, not all are targeted at affecting the behavior of potential drinking drivers. For instance, the efforts included in the U.S. Department of Transportation's Alcohol Safety Action Project (ASAP) program were intended to build political support for countermeasure programs. The objectives, as stated by the U.S. Department of Transportation (Hawkins et al. 1976, p. 15), were:

- Make the problem of alcohol-related crashes a higher priority among community concerns.
- Make key officials and professional groups (police, judiciary, etc.) aware that two-thirds of the drunk driving fatalities involve problem drinkers, rather than social drinkers. [This is a weakly supported assertion.]
- Create support for the hypothesis that this relatively small segment of the driving population that abuses alcohol can be effectively controlled.
- Inform key officials, professionals, and the public about modern countermeasure methods.

The propriety of a federal agency's funding advertising campaigns to try to influence public priorities and to advance controversial scientific theories is open to question. This is not to underrate the importance of information and technology transfer to state and local governments.

Another use of public information and education is to try to influence third parties to intervene in potential drunk-driving situations (e.g., driving an intoxicated friend home from a party). This use is discussed later.

There are three avenues for using public information and education to achieve general deterrence. The first is to inform potential drunk drivers of the risks they face—accident and arrest—if they drive while drunk. The potential effectiveness of this avenue is dubious, since it appears that the public is quite familiar with these risks. A national opinion survey in the United States in 1973 found that "86 percent of the respondents rated drinking-driving as 'a very serious problem' in the United States and 78 percent felt that police and courts should be 'tougher than they are' " (Hawkins et al. 1976, p. 74). A recent study in Canada (Wilde et al. 1975) asked members of the public to rank 40 traffic safety measures in order of effectiveness and in order of desirability. "Double penalties for drunk driving" rated second in effectiveness (behind "make the wearing of seat belts mandatory") and fourth in desirability. This information leads me to concur with Cameron (1978, p. 53) that:

> While the public might not always be aware of the specific details of the laws against drinking and driving or understand the more technical aspects of how alcohol affects the body, the majority of people are aware of the basic message, that drinking and driving is dangerous and is against the law. Thus, the drinking and driving behavior observed in the U.S. today is, generally speaking, characteristic of a fairly well informed and educated public.

A campaign of public information and education that merely repeats what is generally known or fills in small details seems unlikely to cause much change in drinking-driving behavior.

The second avenue is to use information and education to try to alter attitudes rather than provide information—that is, to alter norms and standards of behavior of people who drink and drive so as to make drunk driving less likely. As Maloff et al. (1980) point out, norms concerning substance use are set and reinforced by a person's entire social environment, including family and peer group. An advertising campaign advocating conflicting norms is likely to be too weak a stimulus to have much effect and may be rejected out of hand as an attack on the groups an individual identifies with.

An example of the apparent failure of a public information and education campaign to effect a desired attitude change is described by Hochheimer (in this volume) with regard to the Australian "Stop a Slob from Driving" campaign:

> The idea was to portray drinking drivers as slobs; since people would not want to be perceived as slobs, they would reduce their alcohol-related driving. Instead,

many people found "Mr. Slob" to be funny, attractive, a person they wouldn't mind being like.

This incident represents more than a random backfiring of an advertising campaign. The campaign was undertaken because many Australians perceived drunk driving as an acceptable, even attractive thing to do. Thus, any character presented as a drunk driver already had an advantage with the campaign's target audience. The potential for reducing drunk driving by altering attitudes of potential drunk drivers through such campaigns seems quite limited.

A final avenue is to use public information and education to provide potential drunk drivers with information that will make it easier for them to avoid driving while dangerously or illegally drunk. Such information might include simple rules of thumb for determining how many drinks a person of given body weight can drink on a full and an empty stomach before reaching 0.10 percent BAC (e.g., simple sobriety tests one might give oneself to decide whether or not to drive, or socially and economically acceptable alternatives to driving when one realizes one has had too much to drink). Of course, such a campaign would hinge on the existence of such rules of thumb, tests, and alternatives.

Attempts to achieve general deterrence of drunk driving through public information and education have generally employed the first two avenues, describing the risks of drunken driving and trying to form attitudes against it. While there have been many such campaigns, a relatively small number have been subjected to scientific evaluation of their impact on drinking-driving behavior (Jones and Joscelyn 1978, Organisation for Economic Co-operation and Development 1978, Wilde 1971). Of these, the one that came nearest to success was a campaign conducted in Edmonton, Canada, during December 1972 (Organisation for Economic Co-operation and Development 1978, p. 88): "In this campaign evaluation a control city (Calgary) was used and initial results suggested a trend towards reduced blood alcohol levels among Edmonton drivers apparently as a result of the campaign." There is no evidence that the Edmonton campaign reduced the number of traffic accidents.

Conclusion

In conclusion, general deterrence of drunk driving does seem possible if a high perceived risk of arrest can be sustained. Severe punishment does not appear as promising as increased arrest risk for achieving general deterrence. Public information and education campaigns that provide information useful to those who wish to avoid driving while dangerously or illegally drunk, without radically changing their drinking or driving behavior, may also be useful.

REDUCTION OF RECIDIVISM

Maximum Potential Savings

The potential reduction in traffic accidents obtainable by reducing DWI recidivism is sharply limited by the small number of persons with previous DWI arrests among drivers involved in accidents while driving with a positive BAC level. Sterling-Smith (1976, pp. 111, 135) reports that of drivers responsible for fatal accidents in the Boston area during the early 1970s, 4 percent had a previous DWI arrest and 39 percent had a BAC level exceeding 0.05 percent at the time of the accident. Thus, even if all drivers responsible for fatal accidents with a previous DWI arrest had a BAC exceeding 0.05 percent at the time of their accident, only 10 percent (i.e., 4 percent/39 percent) of drunk drivers responsible for fatal crashes would have had a previous DWI arrest.

Hurst (1973) has suggested that more experienced drinking drivers, when driving with a positive BAC level, may drive in a slow, cautious, but erratic manner that reduces their chances of serious accident compared with other drivers at the same BAC level. This manner of driving would probably increase the chance of less serious accidents as well as arrest. Since drunk drivers with previous DWI arrests are more likely to be experienced drunk drivers than those without, we would expect drunk drivers with previous DWI arrests to be more heavily represented among drivers in less serious accidents than in fatal accidents.

This expectation is borne out in data reported by Waller (U.S. Department of Transportation 1968, p. 67), that of male drivers in crashes involving alcohol or hit-and-run, 20 percent had one or more previous arrest for DWI. Based on the Sterling-Smith and Waller findings, I estimate that drivers with previous DWI arrests comprise 10 percent of all alcohol-influenced drivers in fatal accidents, 15 percent in injury accidents, and 20 percent in property-damage-only accidents.

Multiplying these estimates by my figures from Table 2 for the percentage of accidents that would be prevented if all drunk driving was eliminated, and the resulting cost savings in 1977, we arrive at estimates of the savings if all drunk driving by persons with previous DWI arrests was eliminated—that is, the savings from a "perfect" countermeasure reducing recidivism. The results are shown in Table 4. Of course, if the risk of arrest for drunk driving increased, so would the percentage of accident-involved drunk drivers with previous DWI arrests. Thus, increased risk of arrest would raise the potential savings from reducing DWI recidivism.

TABLE 4 Expected Reduction in Motor Vehicle Traffic Accidents and Costs as a Result of Preventing All Drunk Driving by Persons with Previous Arrests for Driving While Intoxicated

Type of Accident, Place and Time	Expected Reduction in United States, 1977	
Fatal, Vermont, 1967–1968	2.4%	1,188 lives
Injury, Huntsville, 1974–1975	2.4%	45,600 injuries
Injury, Grand Rapids, 1962–1963	1.2%	22,800 injuries
Property damage, Grand Rapids, 1962–1963	1.1%	$171 million

Effectiveness of Treatments

We are still left with the question of what is the best way to treat those arrested for DWI. Possible treatments fall into two categories. The first is punitive, involving treatments such as fines, imprisonment, license suspension and revocation, and license restriction (e.g., to allow driving only to and from work). Many "punitive" treatments are also prophylactic, in that they temporarily or permanently restrict the subject's opportunity to drive drunk again. The second is educational and therapeutic, including such treatments as drinking-driver schools, group therapy, and treatment for general alcohol abuse.

In general, educational and therapeutic treatment is more expensive than punitive treatment (an exception being long-term imprisonment) but not necessarily more effective in preventing accidents. A review of relevant literature (Preusser et al. 1976, Jones and Joscelyn 1978, pp. 56-63, Organisation for Economic Co-operation and Development 1978, pp. 91-96, Nichols et al. 1978, Comptroller General of the U.S. 1979, pp. 4-34, U.S. Department of Transportation 1979a,b) yields the following observations:

- No credible evaluation has shown that any educational or therapeutic treatment reduces future accidents of a person arrested for DWI more than traditional punitive treatment does. However, at least two studies of intensive treatment for general alcohol abuse showed a decline in the future incidence of accidents and rearrests combined (Organisation for Economic Co-operation and Development 1978, pp. 92, 95).
- Some evaluations have reported a decrease in the DWI rearrest

rate due to educational or therapeutic treatment, but such reports appear to vary inversely with the scientific rigor of the evaluation. Even when an improvement is reported, it is small. Aggregate data from the ASAP program indicate that for DWI arrestees classified as "social drinkers," the group found to be most responsive to educational and therapeutic treatments, such treatments are associated with a small (less than 15 percent) increase in the number of persons not rearrested for DWI within 3 years (Nichols et al. 1978, p. 180).

- While DWI arrestees classified as "problem drinkers" (including "alcoholics") have been shown to be less responsive than "social drinkers" to educational and therapeutic treatment in general, they appear to be more responsive to more personalized and interactive treatment than to more formal treatment, and more responsive to programs classified as therapeutic than those classified as educational. These differences are not apparent when treating "social drinkers" (Nichols et al. 1978, pp. 183-185).

A poor record of past performance does not preclude future success, but the burden of proof seems to rest with advocates of a particular educational or therapeutic treatment program to show reason to believe that such a program will reduce recidivism more sharply than the cheaper punitive approach.

Although an analysis of the treatment of general alcohol abuse is outside the scope of this paper, it should be noted that court referral of DWI offenders has become an important case-finding mechanism for alcoholism treatment programs. Persons thus referred tend to be younger, lighter drinkers, and to have suffered less disruption of their lives from alcohol abuse than others entering alcoholism treatment (Chatham and Batt 1979).

THIRD-PARTY INTERVENTION

The potential of this approach obviously hinges on what fraction of drunk drivers involved in accidents were drinking and driving in the presence of others before their accidents. Unfortunately, this information does not seem to be available, although data concerning the predriving drinking venue of DWI arrestees may be available.[6] It seems reasonable to assume, however, that a large fraction of drunk drivers,

[6] In the process of its "DUI Project," described in the text, the California Department of Alcoholic Beverage Control collected data on where persons had been drinking who were arrested for driving under the influence of alcohol.

perhaps a majority, were drinking in the presence of other persons before driving.

Servers and fellow guests or patrons can take various steps to reduce drunk driving, such as:

- Make it less convenient or less socially acceptable for a guest or patron to drink to intoxication.
- Suggest that intoxicated guests or patrons wait to sober up before driving or have a friend or taxi take them home.
- Physically restrain or report to police an intoxicated guest or patron who insists on driving.

All of these steps impose some costs on the third party: reduction in profit or apparent hospitality, the expense or inconvenience of arranging alternative transportation or lodging for the inebriate, or the unpleasantness of telling a person that he or she is for the moment incompetent to drive. The problem, then, is how to convince third parties to bear these costs.

As mentioned above, public information and education campaigns have been used to try to increase third-party intervention. These campaigns face the same difficulties as those attempting general deterrence; no truthful information that could be provided is likely to have much impact on a third party's perception of the risk inherent in drunk driving by others, since present perceptions appear to be fairly accurate. Moreover, a media campaign may not have sufficient persuasive force to alter social behavior that is reinforced by groups important to the individual. As for its use in general deterrence, however, the use of public information and education for third-party intervention is unproven rather than discredited.

The other way to convince third parties to intervene in potential drunk-driving situations is to impose legal liability on them. Such liability may be imposed by statute (referred to as a dramshop law) or by court interpretation of common law. Mosher (1979, pp. 6-8) describes the present situation:

> Today there are eighteen state dram shop acts. A typical statute provides that a commercial seller of alcoholic beverages will be found liable for injuries caused by his or her patrons if the server sold or gave alcoholic beverages to the patron in violation of the law. A violation occurs if the patron is a minor, habitual drunkard, or someone "already" or "obviously" intoxicated when served. . . . Because most of these laws were passed before Prohibition, many are now outdated and not relevant to current social problems. . . . The habitual drunkard provisions are now largely ignored because of the difficulty and potential con-

> stitutional problems involved in identifying those who can be considered "habitual drunkards." . . . Courts in ten states without dram shop acts, five states with such acts, and the District of Columbia have imposed civil liability on commercial servers of alcoholic beverages by court decision . . . based on the server's alleged negligence.

There is variation among states as to whether the drunk driver can sue a third party who contributed to his intoxication (Mosher 1979, p. 7). In general, liability is imposed on "social hosts," those not in the business of serving alcohol, only when an underaged person is served, not when an "obviously intoxicated" person is served. Court decisions in Iowa and California that social hosts were liable if they served "obviously intoxicated" guests were quickly followed by legislation in each state overturning the decision (Mosher 1979, pp. 12-14). In another case, a California court decided that a person who merely encourages another to drink without furnishing alcohol is not liable (p. 19).

One would hope that third-party liability would cause those liable, primarily commercial servers, to take steps to prevent drunk driving by patrons. Unfortunately, the criteria by which liability is judged do not encourage servers to take precautions. A server is considered liable if he or she served an underaged or "obviously intoxicated" person, and if that person subsequently did damage. To take reasonable precautions to avoid serving such persons or to prevent those served from doing damage is not considered a valid defense (Mosher 1979, pp. 22-23). If a single patron or guest leaves and does damage and if a court decides post hoc that he or she was "obviously intoxicated" when served, then liability is determined. Since there is no standard of practice that, if adhered to, will absolve the server of responsibility, servers tend to view dramshop liability as a random risk, and insure against it rather than altering their serving practices (Mosher 1979, pp. 23-25).

If there were accepted standards of practice for servers of alcohol, and if following these practices absolved the server of liability even if a patron "slipped through," drove drunk, and had an accident, then presumably servers would follow these practices to protect themselves and avoid high insurance costs. (Insurance premiums increased 500 percent to 1,000 percent for commercial servers in California in the wake of a court decision imposing liability on them. Insurance rates for relatively small chains of on-premise sales reached $100,000 per year [Mosher 1979, p. 24].)

In addition to liability for the actions of intoxicated patrons, the mere serving of an intoxicated person by a commercial server is in violation of the alcoholic beverage control (ABC) laws of most states. Unfortunately, state ABC departments have traditionally concentrated much more effort on ensuring the operation of an orderly industry with a low level of competition than on preventing alcohol-related problems such as drunk driving (Matlins 1976).

An exception is the "DUI (Driving Under the Influence) Project" of the California Department of Alcoholic Beverage Control (Mosher and Wallack 1979). In this experimental program drivers arrested for drunk driving were asked where they had been drinking. On-premise retailers who were twice identified in this way were offered free training for their personnel by the Department of ABC on methods of detecting intoxicated patrons and curtailing their consumption. It seems that the concern over dramshop liability was largely responsible for the good cooperation the Department of ABC received from retailers during this project.

There was no evaluation of the project's impact on drunken driving or accidents. Nonetheless, it is significant as an example of involvement of a state ABC department in drunk-driving countermeasures, and as an instance of a government agency developing and recommending standards of practice for servers of alcohol to avoid intoxication and subsequent drunk driving by patrons.

MINIMUM DRINKING AGE

If people are prevented from drinking, then they are also prevented from driving drunk. A return to prohibition would be politically unfeasible even if it were desirable, but persons under a given age are routinely prohibited from purchasing or consuming alcohol. Throughout the United States, the minimum drinking age is set within the range from 18 to 21 years. It is probably unrealistic to consider setting a minimum drinking age outside this range, but the question remains of what value within the range is optimal.

It is clear from several studies that when the drinking age is lowered from 21 or 20 to 18, the number of accidents involving 18-, 19-, and 20-year-old drivers increases (Organisation for Economic Co-operation and Development 1978, pp. 96-98, Comptroller General of the U.S. 1979, pp. 43-45, Haddon 1979, pp. 56-57, Scotch 1979, pp. 2-4). Various studies have found the percentage increase to range from undetectable to 26 percent.

The fact that prohibiting 18- to 20-year-olds from drinking reduces their accident involvement does not in itself make a convincing argument for setting the drinking age at 21. After all, prohibiting persons of any age-group from drinking would probably reduce their accident involvement. On what basis can we decide that persons who are old enough to drive, vote, and enter into contracts may not have the same access to alcohol as all other adults?

A possible justification is that accident risk rises more quickly with

BAC level for drivers under 21 than for any other age-group (Jones and Joscelyn 1978, pp. 31-33). It is unclear, however, to what degree this is merely because younger drivers tend to be less experienced at drinking and drinking and driving. Those who drink more often are at less risk at any given BAC level. To the degree that the association between age and sensitivity to BAC is due to lack of drinking experience, raising the drinking age will merely transfer high accident risk from drivers under 20 to those recently turned 21.

SCREENING

A strategy that has received little attention is screening the population of drivers for those most likely to drive drunk and targeting countermeasures to them as individuals. An example of this strategy is a program that had been planned by federal and District of Columbia officials in which every driver applying for a renewal of his or her driver's license (and, presumably, an initial license) would be given a written test to determine whether she or he is, or is likely to become, a problem drinker. Those drivers so identified would be required to attend a program on alcohol education before they would be issued unrestricted drivers' licenses. One consultant involved in the project compared the measure to the current requirement that diabetics, epileptics, and other persons whose driving could be impaired by recurrent illness must certify that they are under proper medical care and control before being issued licenses.

As a pilot study for the contemplated project, drivers renewing their licenses in the District of Columbia during 1976 were asked to voluntarily, and anonymously, take part of a widely used test for current or prospective alcohol abuse. The test took less than 15 minutes to complete and consisted of a series of yes/no and true/false questions. Many considered the questions, some of which dealt with the driver's income, relationship with spouse, and arrest and drinking-driving history, to be too personal and inappropriate for a motor vehicle licensing agency to ask.

Despite the fact that the test had been given on a voluntary and anonymous basis, press coverage, citizens' complaints, and protests by the American Civil Liberties Union led the mayor to suspend the project for further study and eventually to order the program aborted and all collected data destroyed. (The above information was culled from the *Washington Post* for August 5, 1976, August 7, 1976, August 31, 1976, December 22, 1976.)

This experience does not bode well for countermeasures involving mass screenings. There seem to be two basic problems.

- The screening device. The test or procedure used for screening must use only information that is considered proper for licensing authorities to examine. Information in this class might include convictions for drunken driving and other alcohol-related offenses, and perhaps whether the driver has ever been treated for alcohol abuse. The screening device must produce a low level of false positive errors in order not to inconvenience or stigmatize persons without drinking problems.
- The treatment of persons identified by the screening. Treatment is largely dependent on what the screening device tests for. If potential drunk drivers are identified, then some sort of alcohol education or other preventive measure may work. If actual drinking drivers are identified, then the problem is the same as that of reducing recidivism (specific deterrence).

A program in North Carolina screened driver license records to identify "habitual offenders" based on past traffic convictions (Li and Waller 1976). Many of the "habitual offenders" were drinking drivers. Despite a law authorizing 5-year license revocations for habitual offenders, 62 percent of the referrals from the screening program were never acted on by prosecuters. The screening program was widely viewed as superfluous, because there was opportunity to impose similar sanctions on "habitual offenders" when they were brought to court for a particular offense. This may be a generic weakness of screening programs based on public record information.

DETECTION DEVICES IN VEHICLES

The suggestion has been made that cards be equipped with devices that will detect an intoxicated driver and either prevent the car from starting or make it very conspicuous on the road, for example, by automatic flashing headlights. Such a device could be installed in all cars or in only those driven by persons who seem likely to drive after drinking (e.g., persons with previous DWI convictions).

Of course, any such device could be bypassed: a driver alone or with the help of a mechanic could remove the device, render it inoperative, or even install a switch to reactivate it if the car was subject to periodic inspection.

Of the two types of detectors that have been developed, either are to some degree capable of being circumvented by having a sober companion take the "test" for the drinking driver. One type of detector is a breath alcohol analyzer; the other is a skill tester, such as one requiring the driver to punch into a keyboard a series of random numbers that are briefly displayed on a screen.

Although it is clear that any of the detection devices so far suggested can be defeated, they may still be of use since they require the driver to admit to himself or herself and to anyone else whose aid has been enlisted, that he or she is too drunk to drive. It is not known how much exposure would be reduced if potential drunk drivers and those around them were given unambiguous and immediate evidence that they were incapacitated.

More advanced detectors than are currently available, such as those that would continuously "sniff" the air around the driver's head for alcohol, or that would continuously monitor the driver's behavior for such signs of intoxication as oversteering, would eliminate the problem of someone's starting a drunk's car for him or her, although they would still be susceptible to bypass.

The widespread installation of detection devices in vehicles may meet hostile public reaction, since even those who never wish to drive drunk are likely to oppose the inconvenience as well as the added expense. The inconvenience and expense would be more easily justified if detectors were installed only in the cars of persons with previous DWI convictions. One would expect such persons to have a greater likelihood of driving drunk in the future than do drivers in general and, indeed, empirical evidence suggests that this expectation is correct (Jones and Joscelyn 1978, p. 37).

ALTERNATIVE TRANSPORTATION

To have an intoxicated person ride public transportation is safer for himself or herself and, of course, for others. It therefore seems promising to provide public transportation as an alternative to drinking and driving at times and places with a high concentration of drinking. For example, it is reported that the alcoholic beverage industry supplies free mass transit in Toronto on New Year's Eve. A large part of Washington, D.C., is closed to traffic on the night of the Fourth of July, and expanded bus and subway service is provided free of charge (although this is probably targeted as much at relieving traffic congestion as preventing drunk driving).

There do not seem to be any evaluations of alternative transportation programs for drunken-driving countermeasures, therefore little can be said regarding this strategy's effectiveness and efficiency.

FINDINGS AND RECOMMENDATIONS

While our knowledge regarding the ways in which the amount of drunk driving can be reduced is far from complete, it does lead to some conclusions that may be useful to the policy maker.

General Deterrence

There is good evidence that if drivers in general perceive a high risk of being arrested if they drive drunk they will be deterred from doing so, and alcohol-related traffic accidents will decrease substantially. It is currently unknown how high the risk of arrest must be maintained to achieve deterrence or how much it would cost to maintain such risk. Carefully designed and evaluated experimental countermeasure programs should be executed in which steps are taken to increase the risk of arrest for drunk drivers, the risk of arrest is measured, and the impact on drinking-driving behavior and accidents is assessed.

There is no convincing evidence that severe punishments deter drunk driving in the absence of high risk of arrest.

A massive effort to start changing social attitudes about drinking and driving, as urged by a recent report by the General Accounting Office to be the major thrust of federal drinking-driving countermeasure efforts, would be a poor allocation of resources. Public information and education campaigns explaining the risks of drunk driving are unlikely to affect behavior, since the public at large has a fairly accurate perception of these risks already. Campaigns aimed at altering individual norms and attitudes concerning drinking and driving are unlikely to have much impact, since such norms are set and constantly reinforced by social groups with which individuals identify. If, however, information is available that would help individuals avoid driving when dangerously or illegally drunk without radically changing their values and social behavior, then disseminating such information would be a useful countermeasure.

Reducing Recidivism (Specific Deterrence)

Since only a small number of drivers involved in alcohol-related accidents have previous arrests for drunk driving, the potential savings from reducing recidivism is sharply limited. Even if all persons with drunk-driving arrests were prevented from ever combining drinking and driving again, the reduction in motor vehicle traffic accidents would be only about 1-3 percent, although absolute cost savings would not be trivial. Increasing the risk of arrest for drunk driving would increase the potential savings from reducing recidivism.

Educational and therapeutic treatments for arrested drunk drivers, such as the drinking driver schools operating in many states, rarely cause a reduction in recidivism compared with traditional punitive measures such as fines. The burden of proof should rest with the advocates of a particular educational or therapeutic treatment program to show that such a program will reduce recidivism more sharply than the punitive approach, which has lower costs.

Court referrals of drunk-driving offenders deemed to be problem drinkers or alcoholics has become a significant case-finding mechanism for programs of alcohol abuse treatment.

Third-Party Intervention

Servers of alcohol and other persons can intervene to prevent an individual from becoming intoxicated and driving. Such actions impose costs on the intervener, so finding ways to motivate intervention is important.

Public information and education campaigns are likely to be no more effective in increasing third-party intervention than in bringing about general deterrence.

Dramshop laws and court decisions to the same effect impose liability on commercial servers of alcohol for damage done by patrons who were underage or served while "obviously intoxicated." Under current practice, servers are held liable regardless of whether they took reasonable precautions to avoid drunk driving by patrons. For this reason, servers tend to insure themselves against dramshop liability (and engage in political action against it) rather than altering their serving practices to reduce the risk. Standards of practice for servers that would reduce the risk of drunk driving should be developed and disseminated. Courts and legislatures should be encouraged to absolve servers who follow these standards of liability for damage done by patrons who drive drunk despite the server's efforts.

Drinking Age

The minimum drinking age in all states is currently between 18 and 21, and it does not seem feasible to adopt a drinking age outside this range. When the drinking age is lowered from 21 to 18, the accident involvement of 18-, 19-, and 20-year-old drivers increases. The same effect would probably occur to any age-group that was prohibited from drinking, so the case is not clear for setting the drinking age at 21. Drivers under 21 are more sensitive in their accident risk to BAC than are other drivers, but this may be merely because they are inexperienced drinkers. If so, then raising the drinking age would merely transfer increased accident risk from drivers under 21 to those just over 21. There is at present no compelling case for denying persons between the ages of 18 and 21 the same access to alcohol as all other persons who have attained majority.

Screening

It may be possible to screen the driving population for those most likely to drive drunk and expose these people to preventive treatments, but this approach has yet to be shown workable. An attempt in Washington, D.C., failed, largely because the screening test was viewed as an invasion of privacy. Although it may be better developed in the future, the screening approach seems to have little potential in the near future.

Detection Devices in Vehicles

It is technologically possible to install in a car a device that will prevent an intoxicated person from starting the car. Such devices may be circumvented, but they may still have significant effectiveness in reducing drunk driving. The cost and inconvenience of having such devices on all cars may arouse public opposition and indeed may not be warranted by their effectiveness as a countermeasure. The use of such devices could be tested on the cars of a sample of drivers convicted of drunk driving. Since such persons are presumably more likely to drive drunk in the future than is the average driver, the cost and inconvenience of the detection devices are more likely to be warranted in this application.

Alternative Transportation

Providing public transportation at times and in places in which drinking and drinking and driving are concentrated may in some cases be an effective and efficient countermeasure. No evaluation of this approach is currently available, but a test may be useful and not very expensive in an area with an existing mass transit system.

REDUCING RISK

PERVERSE INCENTIVES

Reducing risk refers to lowering the expected cost, in term of deaths, injuries, and property damage, of each unit of drunk driving. A possible objection to such a strategy is that as drunk driving becomes safer, people will do more of it.

The question involves estimating the "risk elasticity of drunken driving." If a 1-percent decrease in the risk of drunk driving led to a 1-percent increase in the amount of drunk driving (i.e., if elasticity equals one), then our efforts to reduce risk would seem vain. The total costs resulting from drunk driving would remain the same. Conceivably, the elasticity may be greater than one, which is to say that a decrease in risk would lead to a greater increase in exposure, so that total costs from drunk driving would increase.

I would like to suggest that the elasticity is less than one; that when the adverse consequences of an act are both remote in probability and so serious that they are painful to contemplate, as is the possibility of a serious accident resulting from drunk driving, that a person will tend to evaluate the risk at less than its expected cost and will be insensitive to small changes in the expected cost. This would explain why drivers seemed no more effectively deterred from drunk driving by the threat of a jail sentence than of a license suspension in the Chicago experiment (see above). The probability of being subjected to either penalty was quite small, and both penalties were so serious that one would be tempted to think "that couldn't happen to me."

If this speculation is accurate, then changes in the risk of drunken driving brought about by risk-reducing measures would not have a large impact on the amount of drunk driving and would result in a reduction of total costs resulting from drunk driving (net of the cost of bringing about the risk reduction). This speculation could be tested experimentally if the amount of drunk driving in an area was measured before and after a quick and significant reduction in the risk of drunken driving.

GENERALLY APPLIED RISK REDUCTION

Some risk reduction measures are applied to drivers in general. They may be differentially more (or less) effective in lowering the risk of drivers with elevated BAC levels, but implementing the measures does not require knowing which drivers are likely to be impaired.

Passive restraint systems, for instance (such as air bags or automatic seat belts), would protect vehicle occupants regardless of alcohol involvement in a crash, but they would be differentially effective in protecting drunk drivers involved in accidents since they are less likely to use conventional seat belts than are drivers involved in accidents in general (Sterling-Smith 1976, p. 160). The same is true of other attempts to make vehicles in general more crashworthy. Other changes in the driving environment would reduce the probability of accident or the probable severity of accidents that do occur for all drivers while having a differentially greater effect on drinking drivers. For example, the ability to divide attention between tasks has been found to be one of the driving-related skills degraded first and most severely as BAC increases (U.S. Department of Transportation 1968, pp. 42-52). Therefore, speed governors in cars, redesigned road markings, and other changes in the driving environment that reduce the driver's need to frequently shift attention would probably result in greater risk reduction among drinking drivers than drivers in general.

Some generally applied risk reduction measures would benefit only those persons with elevated BAC levels. Haddon and Baker (1978, p. 16) point out that "injured people are rarely tested for alcohol in emergency rooms although symptoms, diagnoses, response to drugs and anesthesia, and even prognosis may be influenced by alcohol." Informal consultation with physicians experienced in emergency room situations suggests that physicians do not routinely test accident victims for alcohol, relying on cues such as breath odor, drunken behavior, and unaccounted-for unconsciousness to warn them of possible alcohol involvement. William Haddon, M.D., of the Insurance Institute for Highway Safety, has suggested that medical personnel may be reluctant to test for alcohol in part because doing so might tend to involve them in litigation concerning the accident in which the patient was injured.

It is certainly not clear *prima facie* whether the incremental improvement in medical care resulting from more frequent tests for alcohol would be worth the expense of such testing. It might be found that testing some subgroups of accident victims, such as those injured in nighttime traffic accidents, is cost-effective.

All measures to reduce the risks of drunken driving (as opposed to amount of drunken driving)—even measures that offer no protection to persons with zero BAC levels—should be evaluated and assigned priority in relation to all proposed measures to reduce the risk of driving in general. In a world of limited resources there is no defense for spending a dollar to reduce alcohol-related accidents if the same expenditure would be more effective if applied to reducing motor vehicle traffic accidents in general. Similarly, if the lives of drinking drivers should be weighted no less heavily than those of innocent victims when selecting countermeasures, then a marginal dollar should be devoted to alcohol-specific risk reduction if this produces the greatest expected savings in accident costs (lives, injuries, and property damage).

It is important, therefore, that government agencies choosing among motor vehicle traffic safety measures evaluate their benefits in a way that recognizes drunk drivers are a minority on the road and that they have differential sensitivity to some safety measures. This will promote the efficient mix of alcohol-specific and nonalcohol-specific risk reduction.

SPECIFICALLY APPLIED RISK REDUCTION

Some measures to reduce the risk of drunken driving may be uneconomical or impractical to apply to drivers in general, but cost-effective and practical when applied specifically to drivers with a higher than normal probability of driving with elevated BAC levels.

Some modifications to improve a vehicle's crashworthiness or ease of driving, as mentioned in the previous section, may be cost-effective only when applied to the vehicles of persons with previous drunk-driving arrests, persons requesting such safety devices for their cars, or persons willing to buy such protection on the free market.

In the case of specifically applied risk reduction the problem of political acceptability may be greatest. For instance, it is known that drowsiness, one of the obvious effects of drinking, impairs driving ability, yet public information and education campaigns from government and private sources consistently omit such suggestions as taking caffeine, driving with the window open, or playing the radio when driving after drinking. (Although it is frequently and accurately pointed out that coffee does not reverse the intoxicating effects of alcohol.) Presumably, such suggestions are omitted because they could be perceived as encouraging drunk driving by lowering its expected cost. There does not appear to be empirical evidence as to whether a driver with an elevated BAC level has less risk of accident with or without antidrowsiness measures, and the answer is not clear *a priori* (e.g., is a more awake drunk driver also more reckless?). Such questions are not even asked when the problem of perverse incentives is viewed as a moral issue rather than an issue of effectiveness.

FINDINGS AND RECOMMENDATIONS

Reducing the risk associated with drunken driving would tend to increase the amount of drunken driving that is done. I speculate that the "risk elasticity of drunken driving" is less than one, meaning that a decrease in risk is likely to bring about a smaller increase in exposure, resulting in a net reduction in accidents.

Risk-reducing measures that affect drinking drivers by the same mechanism that they affect drivers in general are best considered in the context of general traffic safety. Those concerned with alcohol-related problems should check that the procedures used by traffic safety authorities to select countermeasures accurately take into account differential impacts on drinking drivers. (For example, drinking drivers use seat belts less often than drivers in general.)

To the extent that general traffic safety measures are successful, they will reduce the accident reduction attributable to drinking-driving countermeasures. The extent of the drinking-driver problem should be periodically reviewed to determine if extensive countermeasures are still warranted.

Research should be performed to determine whether it would be cost-effective for emergency medical personnel to test traffic accident victims for alcohol.

Alcohol affects particular skills important to driving. It may be possible to redesign portions of the driving environment to decrease the impact on accident risk of degradation of these skills.

Cars driven by persons likely to drive drunk, such as those with previous DWI convictions, could be held to higher standards of crashworthiness than cars in general. This might, however, tend to make drivers of such cars overconfident and be counterproductive, and public reaction may be hostile.

GOVERNMENT ACTION TOWARD DRINKING AND DRIVING

In spite of the large reduction in deaths, injuries, and property damage that could be achieved by effective drinking-driving countermeasures, we have no technology to bring about these savings. I have demonstrated above that a consideration of possible countermeasures raises more questions than it answers. In the case of risk-reducing countermeasures, there is unfortunately little experience to draw on. There have been many applications of exposure-reducing countermeasures, yet despite this experience we have not developed any dependable and effective techniques. If our ability to prevent losses from drunken driving is ever to improve, we must begin to learn from experience.

INSTITUTIONAL BIAS AGAINST LEARNING

Cameron (1978, p. 11) has observed that:

> Only a small proportion of drinking-driving programs in the U.S. have ever been subjected to a scientific evaluation of their effectiveness in reducing alcohol-traffic problems. In fact, much of what is known about the effectiveness of some types of drinking-driving countermeasures is based primarily on data from other countries.

An example of evaluation problems is the Alcohol Safety Action Project (ASAP) program. The program consisted of 35 local ASAPs, each approximately 3½ years in duration, funded by the U.S. Department of Transportation between 1969 and 1975 at a cost to the federal government exceeding $88 million (U.S. Department of Transportation 1979a,b). Each ASAP attempted to integrate and improve enforcement, prosecution, screening, and treatment countermeasures so as to reduce drunken driving and associated accidents in its geographic area.

The design of the ASAP program did not allow for evaluation of its effectiveness as a drinking-driving countermeasure. Zimring's critique (1978) and the Department of Transportation's reports on the Program (U.S. Department of Transportation 1979a,b) reveal several important flaws:

- Noncomparable sites. The original eight grantees included four cities, one "twin city" site, two counties, and the state of Wisconsin.
- Inadequate control sites. Control sites for comparison with the ASAP sites were not selected until the projects had ended operation. This delay restricted potential control sites to the relatively few localities that happened to have kept sufficient accident records for comparison with ASAPs.
- Proliferation instead of replication. The program expanded to 35 projects before the initial 8 had been completed and analyzed, therefore preventing evaluation problems from being detected in time to be corrected in the later projects.
- Overaggregated evaluation. Evaluation focused on the change in accident occurrence at each ASAP site during the entire operational period at that site, even though each project used several types of countermeasures and many changes occurred in each project's countermeasure programs during its operational period. At best, each evaluation could have determined the efficacy of focusing money and attention on drinking and driving; it did not have the potential to determine the efficacy of particular countermeasures.

None of the above is to denigrate the skillfully prepared final reports on ASAP (U.S. Department of Transportation 1979a,b), but these were obviously post hoc efforts to scavenge findings from a program not well designed to supply them. Exceptions to the generally poor experimental design associated with ASAP include the Short Term Rehabilitation study (Nichols et al. 1978) and a well-designed experimental evaluation of a "drunk driver school" at the Nassau County, New York, ASAP (Preusser et al. 1976).

The lack of adequate evaluation is not confined to the efforts of the

U.S. Department of Transportation. During the early 1970s the National Institute on Alcohol Abuse and Alcoholism (NIAAA) funded 18 problem drinking driver programs (PDDPs) to operate in conjunction with ASAPs. Stanford Research Institute was contracted to evaluate them and released its final report in 1975 (Eagleston et al. 1975).

Although the 131-page document reflects considerable work and expense, it is of little use in determining whether PDDP, or any components of it, brought about improvement in the persons treated. The paper states (p. II-6):

> It is apparent that PDDP treatment and rehabilitation does affect [sic] a positive change on client drinking patterns and behavior as measured in various ways at intake and six months after intake. This is accomplished at a relatively low cost per client—$153 on the average. As a result we recommend that these programs be continued.

Even assuming that the measures at intake and 6 months later are entirely adequate, the report's data do not adequately support its conclusion. First, without a control group, we do not know from the report whether the programs' clients would have shown just as much or more improvement in the absence of treatment. There is evidence that drinking problems are often transitory and exhibit spontaneous remission in the absence of treatment (Cahalan and Roizen 1974). Second, the report does not account for or even mention distortion of the data through selection by potential clients and by program personnel of program entrants. Only 77.2 percent of clients making "initial contact" with a PDDP actually entered treatment (Eagleston et al. 1975, pp. V-5, V-6). Even if there had been a control group, superior progress among those in treatment might be a result of selection rather than of treatment.

What can account for the persistent pattern of neglect of the most basic precepts of experimental design when planning and evaluating government drinking-driving programs? The answer, I believe, is that the persons and organizations with influence over these programs view them not as research, but as service provision. The programs are treated as if they were applying well-developed technology to solving a problem (I use the term "service" here in a broad sense to include arresting drunk drivers, repairing roads, etc).

From the perspective of service provision, the program traits we have noted are predictable and expected and in many cases represent good management. If the purpose of a program is to provide service, then why spend money establishing control groups or collecting baseline data? Why wait until one project is evaluated before starting the next? These efforts would reduce the amount of service one could provide. Of course, even a service provider sees the need for project evaluation, but the evaluation required for such purposes is a management and performance audit rather than a scientific investigation. If the technology has been assumed sound, then why question it?

The political pressure on government agencies is no doubt to *do* something about drunken driving rather than to learn something about it. There is much evidence that agencies are quite responsive to this pressure. Remember that ASAP stood for Alcohol Safety *Action* Project. The ASAPs, touted as "demonstration projects" to *show* what could be done and how to do it, resulted in a set of handbooks for state and local governments on how to run an ASAP-type program, but with no assurance that this is a good way to reduce drinking-driving accidents (Hawkins et al. 1976, U.S. Department of Transportation 1979b). The problem drinking driver program "evaluation" is probably all a service provider could want in terms of reviewing the organization, management, and operation of these programs. It has little to offer those interested in whether PDDPs and programs like them are effective because it was never meant to address that question.

RECOMMENDATION

Unfortunately, we do not know how to prevent drinking-driving accidents and their related costs in a dependable and cost-efficient manner. It is therefore counterproductive to run government programs relating to drunken driving as if the primary purpose was service provision.

Persons concerned with research and experimental design should work with those concerned with service provision and program management from the earliest planning and budgeting stages of drinking-driving countermeasure programs. Federal agencies sponsoring such programs should view the development and rigorous testing of the countermeasure being used as a goal at least coequal to the application of the countermeasure in an attempt to reduce drinking-driving accident costs. Programs should not be termed "demonstration projects" unless they are indeed demonstrating a countermeasure that the agency has good reason to believe will be effective.

Such recommendations will doubtlessly evoke protests that resources are being diverted from saving lives to performing research. But if our goal is immediate saving of lives then we are investing our resources poorly. The U.S. Department of Transportation (1979a,b) estimates that the ASAP program saved 506 lives at a cost of $156,306 each. I think a fairer estimate is that 425 lives were saved at a cost on the order of $300,000 each.[7] Using either set of estimates, it appears that more lives could have been saved by using the ASAP money elsewhere. As the U.S. Department of Transportation (1979a,b) points out, "The National Highway Safety Needs Report . . . ranks 37 countermeasures [by] cost per fatality forestalled. . . . The ASAP cost [$156,306] would rank between number 14—motorcycle lights-on practice—and number 15—impact absorbing and roadway safety devices."

Given our present level of technology for preventing drinking-driving accidents, additional expenditures seem warranted only if they promise to produce findings that will help us improve the technology and save more lives in the future, as well as contributing to current traffic safety.

APPENDIX A: REVIEW OF EPIDEMIOLOGICAL STUDIES

To my knowledge, the five investigations reviewed here comprise all controlled epidemiological studies of the risk of automobile-accident involvement (as opposed to causation) associated with various BAC levels for drivers in the United States. I examine the studies here solely to determine the relationship between BAC and accidents, although the studies are rich in other information.

Each study is briefly analyzed. The results of my analysis are summarized in Table A-1. The "risk" figure is a simple Bayesian calculation of the risk of having an accident per unit of driving within a BAC interval, as explained in Appendix C.

The variation among results of the studies judged methodologically adequate seems explicable by differences in location, time, and accident severity. In rural areas such as Vermont, the small number of accidents involving traffic congestion would make accidents due to gross driver error (such as a drinking driver might make) a more important part of the total accident picture, thus increasing the risk figures for high BAC intervals. Similarly, the lower road congestion and average driving speed of the early 1960s would lead to a more important role for alcohol in

[7] The DOT figure of 506 lives saved is based on the sum of accident reduction at all ASAP sites that exhibited a statistically significant decrease in accidents. I think it is more informative to subtract from this estimate the sum of accident increase at the two ASAP sites that exhibited a statistically significant increase in accidents, in order to balance gratuitous accident reductions at the sites exhibiting reductions: thus my figure of 425 lives saved. To estimate ASAP costs, I deducted the portion of federal ASAP funds devoted to evaluation and then doubled the balance to roughly account for other federal funds, state and local expenditures, and expenditures by persons arrested for drunk driving. The resulting estimated cost for ASAP countermeasures is $141 million.

TABLE A–1 Epidemiological Studies of BAC and Risk of Accident Involvement

Description of Study	Relative Risk Within BAC Intervals				
	<0.01	0.01–0.049	0.05–0.099	0.10–0.149	≥0.15
Evanston, Ill., 1935–1938					
Injury accidents	1.00	3.82	2.90	12.73	53.92
Comparatively poor methodology					
Grand Rapids, Mich., 1962–1963				5.70	18.85
All accidents	1.00	0.91	1.52		
Methodologically adequate				8.82	
Injury accidents	1.00	0.92	1.86		
Ill-matched control group				10.77	
Damage-only accidents	1.00	0.94	1.53		
Ill-matched control group				7.95	
Huntsville, Ala., 1974–1975					
Injury accidents	1.00	1.34	2.02	2.73	13.42
Methodologically adequate					
New York, N.Y., 1959–1960					
Driver-fatal accidents	1.00	0.82	1.27	3.21	39.73
Inadequate sample size					
Vermont, 1967–1968					
Driver-fatal accidents	1.00	1.02	3.27	21.07	53.43
Methodologically adequate					

accidents than it had in the mid-1970s. This might account for the Grand Rapids study showing greater alcohol-related risk than the Hunstville study.

The tables on the following pages show the derivation of relative risk figures from each study's data. Table columns labeled "Ratio" show the ratio of percent of all accident group observations within the BAC interval to percent of all control group observations within that interval. The column labeled "Risk" shows the values in the "Ratio" column multiplied by the constant necessary to make risk in the lowest BAC interval equal to one.

EVANSTON

Richard L. Holcomb (1938) Alcohol in relation to traffic accidents. *Journal of the American Medical Association* 111(12):1076-1085.

Case group was some accidents resulting in hospitalization, in and near Evanston, Ill., between Feburary 1, 1935, and February 1, 1938. Method of selection is not explained in report.

Control group was vehicles stopped at eight locations in Evanston "which approximated the area in which the accidents of the first part of the study occurred." Samples were taken during the following periods during a week beginning Saturday, April 23, 1938:

Saturday	0000 to 2400
Sunday	0000 to 2400
Monday	0000 to 0600 and 1800 to 2400
Tuesday	0600 to 1800
Wednesday	0000 to 0600 and 1800 to 2400
Thursday	0000 to 1800
Friday	0000 to 0600 and 1800 to 2400
Saturday	0000 to 0200

Of the control group 1.35 percent refused the BAC test. It is not reported what fraction of the case group refused, but the report claims "rather good cooperation was obtained from the drivers examined. A representative cross section of all drivers involved in injury accidents was thus obtained" (p. 1077).

TABLE A–2 Evanston: Survey Results and Relative Risk

BAC	# Control	% Control	# Accident	% Accident	Ratio	Risk
<0.01	1,538.78	87.93	144.10	53.37	0.61	1.00
0.01–0.049	100.98	5.77	36.26	13.43	2.33	3.82
0.05–0.099	77.35	4.42	21.14	7.83	1.77	2.90
0.10–0.149	26.08	1.49	31.21	11.56	7.76	12.73
≥0.15	7.35	0.42	37.29	13.81	32.88	53.92

The selection of neither the case nor control group is explained in detail, and neither appears to have been selected with as much rigor as in more recent studies. Thus, the reliability of this study's results is questionable.

GRAND RAPIDS

R. F. Borkenstein, Crowther, R. F., Shumate, R. P., Ziel, W. B., and Zylman, R. (1974) The Role of the Drinking Driver in Traffic Accidents,

2nd ed., *Blutalkohol* XI (Supplement 1):8-131. (1st ed. unpublished, Department of Police Administration, Indiana University, 1964.)

Case group was all accidents in Grand Rapids, Michigan, between July 1, 1962, and June 30, 1963.

Control group was four vehicles stopped at random at each of 2,000 time-place sites selected at random from sites of all reported accidents between May 1, 1959, and April 30, 1962. Control group sampled over same period as accident group.

Of control group 2.18 percent and 4.25 percent of accident group refused the BAC test (p. 39). Refusal rates were generally higher in demographic groups more prone to drunken driving (pp. 52-57). The probable result is a significant but small bias toward underestimating the relative risk of driving at higher BAC levels.

TABLE A–3 Grand Rapids: Survey Results and Relative Risk

BAC	# Control	% Control	# Accident	% Accident	Ratio	Risk
<0.01	6,756	89.01	4,992	83.41	0.94	1.00
0.01–0.049	589	7.76	406	6.71	0.86	0.91
0.05–0.099	187	2.46	210	3.51	1.43	1.52
0.10–0.149	44	0.58	186	3.11	5.36	5.70
≥0.15	14	0.18	191	3.19	17.72	18.85

In order to promote driver cooperation, the study was heavily publicized while observations were being made. Although Borkenstein claims that this publicity was insignificant as a drinking-driving countermeasure and therefore as a source of bias (p. 52), I suspect that it caused a small decrease in the frequency of drunken driving by moderate drinkers (who would have positive BAC levels below about 0.10). I cannot predict with any confidence its impact on the "risk" associated with various BAC levels. The study also presents the distributions across BAC categories of only those accidents that resulted in injury, including the 15 accidents that resulted in death, and of only those accidents that resulted in property damage. The study does not, however, show the distribution across BAC intervals of control group observations corresponding to these two severities of accidents. Thus, the risk figures presented in Tables A-4 and A-5 are based on case group observations for the severity of accident indicated and control group observations for all accidents.

TABLE A–4 Injury Accidents: Survey Results and Relative Risk

BAC	# Control	% Control	# Accident	% Accident	Ratio	Risk
<0.01	6,756	89.01	1,162	81.83	0.92	1.00
0.01–0.049	589	7.76	93	6.55	0.84	0.92
0.05–0.079	132	1.74	27	1.90	1.09	1.19
0.08–0.109	76	1.00	47	3.31	3.31	3.61
⩾0.11	37	0.49	91	6.41	13.08	14.62

I have interpolated these results into my standard BAC categories by assuming uniform density of observations within the interval 0.08-0.109. (There are insufficient data to show ⩾0.15 as a separate interval.)

<0.01	6,756	89.01	1,162	81.83	0.92	1.00
0.01–0.049	589	7.76	93	6.55	0.84	0.92
0.05–0.099	187	2.46	58.33	4.11	1.67	1.86
⩾0.10	58	0.76	106.67	7.51	9.88	10.77

TABLE A–5 Damage-Only Accidents: Survey Results and Relative Risk

BAC	# Control	% Control	# Accident	% Accident	Ratio	Risk
<0.01	6,756	89.01	3,833	83.87	0.94	1.00
0.01–0.049	589	7.76	315	6.89	0.89	0.94
0.05–0.079	132	1.74	105	2.30	1.32	1.40
0.08–0.109	76	1.00	85	1.85	1.85	1.96
⩾0.11	37	0.49	232	5.08	10.37	10.99

I have interpolated these results into my standard categories as above:

<0.01	6,756	89.01	3,833	83.87	0.94	1.00
0.01–0.049	589	7.76	315	6.89	0.89	0.94
0.05–0.099	187	2.46	161.67	3.54	1.44	1.53
⩾0.10	58	0.76	260.33	5.70	7.50	7.95

HUNTSVILLE

R. Farris, Malone, T. B., and Lilliefors, H. (1976) *A Comparison of Alcohol Involvement in Exposed and Injured Drivers, Phases I and II.* National Highway Traffic Safety Administration technical report DOT HS-801 826. Washington, D.C.: U.S. Department of Transportation.

Case group was automobile drivers in injury-causing accidents in

Huntsville, Alabama, between July 1974 and July 1975, including the six fatal accidents during that period.

Control group was an unreported number of drivers (averaging 2.61) stopped at same place, time of day, day of week, and direction of travel as each of the last 314 (of a total of 615) observations in the case group. Control observations were taken within 30 days of the case observation being matched.

Two percent of control group and three percent of case group refused the BAC test.

The subgroup of case observations to which controls were matched did not show a significantly or systematically different distribution of BAC levels than the earlier, unmatched case data. I have therefore pooled all case data.

TABLE A–6 Huntsville: Survey Results and Relative Risk

BAC	# Control	% Control	# Accident	% Accident	Ratio	Risk
<0.029	728	90.32	459	76.63	0.85	1.00
0.03–0.059	25	3.10	23	3.84	1.24	1.46
0.06–0.099	30	3.72	38	6.34	1.70	2.00
0.10–0.149	17	2.11	29	4.84	2.29	2.69
⩾0.15	6	0.74	50	8.35	11.28	13.27

I have interpolated these results into my standard BAC categories by allowing each of the report's categories to "steal" enough observations from the next lower category to preserve its (the higher category's) original observation density as it expands to become one of my categories.

<0.01	710.76	88.18	443.14	73.98	0.84	1.00
0.01–0.049	34.55	4.29	29.12	4.86	1.13	1.34
0.05–0.099	37.69	4.68	47.74	7.97	1.70	2.02
0.10–0.149	17	2.11	29	4.84	2.29	2.73
⩾0.15	6	0.74	50	8.35	11.28	13.42

NEW YORK

James R. McCarroll and William Haddon, A controlled study of fatal automobile accidents in New York City. *Journal of Chronic Diseases* XV:811-826.

Case group was fatally injured drivers in New York City, exclusive of Staten Island, between June 1, 1959, and October 24, 1960.

Control group was six cars going in the same direction at same place, time of day, and day of week as each accident in case group. "All site visits were made in 1960, within a few weeks of the calender week of occurrence" (p. 812).

Of the control group 0.39 percent refused the BAC test. BAC test was performed by the municipal medical examiner on each member of case group.

TABLE A–7 New York: Survey Results and Relative Risk

BAC	# Control	% Control	# Accident	% Accident	Ratio	Risk
0	195	77.38	14	41.18	0.53	1.00
<0.02	14	5.56	0	0.00	0.00	0.00
0.02–0.099	34	13.49	3	8.82	0.65	1.23
0.10–0.249	9	3.57	2	5.88	1.65	3.12
0.25–0.399	0	0.00	15	44.12	—	—

I have interpolated these results into my standard BAC categories by assuming uniform density of observations within each of the study's BAC categories.

<0.01	202.00	80.31	14.00	41.12	0.51	1.00
0.01–0.049	19.48	7.75	1.10	3.23	0.42	0.82
0.05–0.099	21.09	8.39	1.86	5.46	0.65	1.27
0.10–0.149	2.96	1.18	0.66	1.94	1.64	3.21
≥0.15	5.99	2.38	16.43	48.25	20.27	39.73

The predominant flaw of this study is the small sample size, which allows us to put little confidence in the results.

VERMONT

M. W. Perrine, Waller, J. A., and Harris, L. S. (1971) *Alcohol and Highway Safety: Behavioral and Medical Aspects*. National Highway Traffic Safety Administration technical report DOT-HS-800-599. Washington, D.C.: U.S. Department of Transportation.

Case group was fatally injured drivers in Vermont from July 1, 1967, to April 30, 1968.

Control group was pooled data of: (1) six drivers at same place, time of day, and day of week as each observation in case group, either within a few weeks or one year after accident being matched; and (2) six drivers at same place, time of day, and day of week as each observation from a group of crashes occurring in Vermont in 1966 and resulting in an injury warranting hospitalization, each observation of which was chosen as a close match to one observation in the case group.

The report's account of control group selection is unclear and self-contradicting. The report justifies pooling the two samples by showing that there is a significant difference in the distribution within either sample of only 3 of 22 variables. I suspect that the control group shows fewer high BAC levels than were actually present among drivers at times and places similar to those of case group crashes.

Of control group 1.3 percent refused the BAC test. BAC test was performed on all members of the case group.

The lowest BAC category used in the report is <0.02. I have interpolated into my standard BAC categories by assuming that the density of observations in the interval 0.01 ≤BAC <0.02 is the same as the density in the interval 0.02 ≤ BAC ≤0.04. All other categories in the report are comparable to mine.

TABLE A–8 Vermont: Survey Results and Relative Risk

BAC	# Control	% Control	# Accident	% Accident	Ratio	Risk
<0.01	942.10	83.74	47.62	44.92	0.54	1.00
0.01–0.049	104.90	9.32	5.38	5.08	0.55	1.02
0.05–0.099	54	4.80	9	8.49	1.77	3.27
0.10–0.149	13	1.16	14	13.21	11.39	21.07
≥0.15	11	0.98	30	28.30	28.88	53.43

The "risk" figures for the higher BAC categories are probably biased upward somewhat due to flaws in the control group.

APPENDIX B: ADJUSTING VERMONT DATA TO REFLECT ALL FATAL ACCIDENTS

The Vermont Study (see Appendix A) reports the BAC levels only for fatally injured drivers, but for purposes of computing the possible savings from drinking-driving countermeasures we need to know the BAC distribution of all drivers involved in accidents in which *anyone* was killed.

To adjust the Vermont data to reflect all fatal accidents, I use data presented by Sterling-Smith (1976, p. 135). Sterling-Smith studies the drivers judged to have been "most responsible" for each of 267 fatal traffic accidents in the Boston area from 1971 to 1974. He presents a breakdown of this group of drivers by BAC range and whether the fatality was the driver or another person. Table B-1 shows these results.

TABLE B–1 Number of Drivers Responsible for Fatal Accidents, Boston, 1971–1974

BAC (%)	Driver Killed	Other Killed
<0.01	29	116
0.01–0.04	6	13
⩾0.05	68	35
TOTAL	103	164

Source: Sterling-Smith (1976, p. 135).

Table B-1 cannot be directly applied to the Vermont data. Since Boston is much more urbanized than Vermont, there is a greater density of vehicles and pedestrians for a driver to collide with, and we would expect a lower ratio of driver fatalities to all fatalities than in Vermont. The data bear this out. According to Table B-1 driver fatalities comprise 39 percent of fatal accidents in the Boston area, whereas Perrine et al. (1971, p. 42) report that the ratio in Vermont in the early 1970s was 59 percent.

To adjust the Boston data to those of Vermont, I reduced the number of nondriver fatalities ("other killed") until driver fatalities comprised 59 percent of total fatalities. I multiply the number of nondriver fatalities in each BAC range by 72/164 to arrive at the figures in Table B-2.

TABLE B–2 Adjusted Number of Drivers Responsible for Fatal Accidents

BAC (%)	Driver Killed	Other Killed	Total Killed
<0.01	29	51	80
0.01–0.04	6	6	12
⩾0.05	68	15	83
TOTAL	103	72	175

By dividing the "total killed" figure by the "driver killed" figure for each BAC range in Table B-2, I obtain for each range an estimate of the ratio of all fatal accidents to driver-fatal accidents in Vermont. Let us call the ratio for the ith BAC range R_i. The percentage of driver fatalities in each BAC range, as listed in Appendix A, is D_i. Then it is clear that the percentage of all fatalities in the ith BAC range is

$$T_i = \frac{R_i D_i}{\sum R_i D_i} 100\%.$$

Table B-3 shows T_i for each BAC range. The control group figures need no adjustment, so they are merely reproduced in Table B-3 as they appeared in Appendix A (with the three highest BAC ranges summed).

TABLE B–3 Percentage of Drivers Involved in Fatal Accidents, and of Control Group Drivers, Estimated to Be Within Three BAC Ranges

BAC (%)	Accident (%)	Control (%)
<0.01	64	84
0.01–0.04	5	9
≥0.05	31	7

APPENDIX C: RELATIVE RISK AND ACCIDENT REDUCTION

CONTROLLED EPIDEMIOLOGICAL STUDIES

Among the most valuable studies of drinking-driving problems are controlled epidemiological studies, such as those reviewed in Appendix A of this paper. In these studies, a random (or exhaustive) sample of drivers involved in accidents in a set geographic area and time period is made. The BAC of each driver in the sample is measured at the time of the accident. These drivers constitute the "accident group."

In addition, researchers match a "control site" to each observation in the accident group. The control site is ideally the same place, at the same time of day and day of the week, under similar weather conditions, and otherwise nearly identical to the circumstances under which the accident being matched occurred. Several drivers passing each control site are randomly selected and their BAC levels measured. These observations constitute the "control group."

Drivers in both the accident group and control group are classified into BAC ranges, and the results are presented in Table C-1.

I use the results of a controlled epidemiological study, as presented in Table C-1, to estimate the results of an imaginary experiment that is impractical to actually perform. The imaginary experiment is to randomly sample "units of accident exposure" and record driver BAC and whether an accident occurs for each unit of exposure selected. A unit of accident exposure would be a period of driving during which a "standard" driver would have some particular expected number of accidents.

TABLE C–1 Results of a Controlled Epidemiological Study

BAC Range	Number of Accident Group Drivers	Number of Control Group Drivers
0	a_0	c_0
1	a_1	c_1
n	a_n	c_n

For example, driving for 1 hour (or 1 mile) at night in the rain on a narrow twisting road would constitute more units of exposure to accident than would 1 hour (or 1 mile) of driving on a sunny day on a well-designed expressway.

If we could perform the imaginary experiment, then we could answer two important questions: (1) What are the probabilities of an accident occurring when one unit of accident exposure is driven by a driver at various BAC levels? (2) What would be the reduction in number of accidents if all driving was done by drivers in the lowest BAC range?

ESTIMATING RELATIVE RISK[1]

What is the probability that a random unit of exposure results in an accident, given that the driver is in some particular BAC range? To answer this question, I invoke Bayes's theorem:

$$P(A|i) = \frac{(A \cap i)}{P\,(i)}, \tag{1}$$

where

$P(A|i)$ is the probability that a given unit of exposure results in an accident, given that the driver is in the ith BAC range.

$(P \cap i)$ is the probability that a randomly chosen unit of driving results in an accident, and that the driver is in the ith BAC range.

$P(i)$ is the probability that a randomly selected unit of exposure is driven by a driver in the ith BAC range.

[1] The Bayesian analysis used in this and the following section has been applied to drunk driving before, notably by Hurst (1970, 1973).

Also from Bayes's theorem:

$$P\,(A \cap i) = P(i|A)\,P(A). \tag{2}$$

Substituting (2) into (1):

$$P(A|i) = \frac{P(i|A)\,P(A)}{P(i)}. \tag{3}$$

Referring back to Table C-1, we see that $P(i|A) = a_i/\Sigma a_i$, since the accident group is a random sample of units of exposure resulting in accidents. We can use $c_i/\Sigma c_i$ as a proxy for $P(i)$, because the probability of a driving trip being included in the control group is primarily determined by the likelihood that a similar trip resulted in an accident. Thus, the control group approximates a random sample of units of exposure.[2] We can thus rewrite (3) as:

[2] The probability of any given trip being included in the control group can be expressed as:

$$P(T \,\varepsilon\, C) = N\,E\,D\,(S/N - R),$$

where

$P(T \,\varepsilon\, C)$ is the probability that trip T becomes an element of the control group.

N is the number of trips (including trip T) to which trip T could be matched as a control (i.e., the number of trips passing sites for which the site of T might be a control site).

E is the mean exposure to accident of each of the N trips.

D is the risk multiplier due to the fraction of drivers among the N trips who have elevated BAC levels.

S is the number of trips sampled at each control site.

R is the probability that trip T ends in an accident before the car reaches the site where it would have been sampled for the control group.

I make the simplifying assumptions that D approaches one and R approaches zero. We are left with:

$$P(T \,\varepsilon\, c) = ES.$$

S is a constant, so the probability that trip T is included in the control group is approximately equal to the expected value of the exposure to accident of trip T.

We can thus rewrite (3) as:

$$P(A|i) = \frac{(a_i/\Sigma a_i)\ P(A)}{(c_i/\Sigma c_i)} \ . \tag{4}$$

Note that controlled epidemiological studies do not provide sufficient information to determine $P(A|i)$, the probability that a randomly selected unit of exposure will result in an accident. We cannot determine the absolute value of $P(A|i)$. We can, however, determine its value for all values of i relative to its value for $i = 0$ (i.e., for the lowest BAC range):

$$\frac{P(A|i)}{P(A|0)} = \frac{a_i\ c_0}{c_i\ a_0} \ . \tag{5}$$

I will refer to $P(A|i)/P(A|0)$ as the relative risk of driving in BAC range i. Thus, the relative risk of the lowest BAC range is always one. Table C-2 shows the relative risk of driving in several BAC categories, as derived from the controlled epidemiological studies in Table 1 of this paper.

TABLE C–2 Relative Risk of Accident Involvement per Unit of Driving in Various BAC Ranges, as Computed from Several Studies

BAC (%)	Fatal, Vermont, 1967–1968	Injury, Huntsville, Ala., 1974–1975	Injury, Grand Rapids, Mich., 1962–1963	Property Damage, Grand Rapids, Mich., 1962–1963
<0.01	1.0	1.0	1.0	1.0
0.01–0.049	0.7	1.4	0.9	0.9
0.05–0.099	} 5.8 (0.05–0.15+)	2.0	1.8	1.5
0.10–0.149		2.7	} 10.8 (0.10–0.15+)	} 8.0 (0.10–0.15+)
≥0.15		13.4		

PREDICTING ACCIDENT REDUCTION

The total number of accidents of a given severity that one would expect to occur in a given area during a given period may be expressed as:

$$\Sigma B_i = \Sigma P(A|i)E_i \ , \tag{6}$$

where B_i is the expected number of accidents involving drivers in BAC range i; E_i is the units of exposure to accident driven by drivers in BAC range i.

Note that as long as P (A|0) remains constant, relative risk for BAC range i is a linear function of $P(A|i)$. Also, c_i approximates a linear function of E_i (see note 2). Thus, we can rewrite (6) as:

$$\Sigma\hat{B}_i = \frac{P(A|i)}{P(A|0)}\, c_i\,, \tag{7}$$

where $\hat{B}_i$ is a linear function of B_i.

I will refer to $\hat{B}_i$ as the relative number of accidents involving drivers in BAC range i. If all drunken driving was eliminated, without changing any individual's driving patterns (mileage, route, etc.), then we would compute a new $\hat{B}_i$ for each BAC range merely by setting relative risk for every range equal to one. Thus, if a "perfect" drinking-driving countermeasure was to go into effect, the total expected relative number of accidents would be:

$$\Sigma\hat{B}_i = \Sigma c_i. \tag{8}$$

In Table C-3, I present $\hat{B}_i$ for each BAC range, and $\Sigma\hat{B}_i$, for each of the studies in Table C-2, computed both with and without a "perfect" drinking-driving countermeasure in effect. The percentage change in B_i when the perfect countermeasure is figured in predicts the percentage change in the absolute expected number of accidents if such a countermeasure was implemented.

TABLE C–3 Relative Number of Accidents Involving Drivers in Various BAC Ranges, Without and With a Perfect Drunk-Driving Countermeasure, for Four Studies

BAC (%)	Fatal, Vermont, 1967–1968 (Without/With)		Injury, Huntsville, Ala., 1974–1975 (Without/With)		Injury, Grand Rapids, Mich., 1962–1963 (Without/With)		Property Damage, Grand Rapids, Mich., 1962–1963 (Without/With)	
<0.01	84	84	88.2	88.2	89.0	89.0	89.0	89.0
0.01–0.049	6	9	6.0	4.3	7.0	7.8	7.0	7.8
0.05–0.099			9.4	4.7	4.5	2.5	3.8	2.5
0.10–0.149	41	7	5.7	2.1	8.6	0.8	6.4	0.8
≥0.15			9.4	0.7				
SUM	131	100	118.7	100.0	109.1	100.1	106.2	100.1

REFERENCES

Borkenstein, R. F., Crowther, R. F., Shumate, R. P., Ziel, W. B., and Zylman, R. (1974) *The Role of the Drinking Driver in Traffic Accidents* (The Grand Rapids Study), 2nd ed. *Blutalkohol* 11 (Supplement 1).

Bureau of the Census (1978) *Statistical Abstract of the United States 1978*. Washington, D.C.: U.S. Department of Commerce.

Cahalan, D., and Roizen, R. (1974) Changes in drinking problems in a national sample of men. Paper presented at Session XI-E, Social Research in Alcohol and Drug Use, St. Francis Hotel, San Francisco.

Cameron, T. (1978) *The Impact of Drinking-Driving Countermeasures: A Review and Evaluation*. Berkeley, Calif.: Social Research Group, University of California.

Chatham, L. R., and Batt, T. (1979) Results of NIAAA funded court referral programs. Paper presented at the Second National DWI Conference, Rochester, Minn.

Comptroller General of the U.S. (1979) *The Drinking-Driver Problem—What Can Be Done About It?* Washington, D.C.: U.S. General Accounting Office.

Dorozynski, A., and Volnay, M. (1978) *Pratique de L'Alcool et de L'Alcooltest*. Paris: Bordas.

Eagleston, J., Fridlund, G., Hentzel, S., Kelley, A., Lotridge, J., Mothershead, A., Paltenghi, B., Pyszka, R., and Resnic, A. (1975) *Continued Monitoring and Evaluation of NIAAA-Funded Problem Drinking Driver Programs*, Final Report. Washington, D.C.: National Institute of Alcohol Abuse and Alcoholism.

Haddon, W. (1979) Options for prevention of motor vehicle injury. Insurance Institute for Highway Safety, Washington, D.C.

Haddon, W., and Baker, S. P. (1978) *Injury Control*. Washington, D.C.: Insurance Institute for Highway Safety.

Hawkins, T. E. (1976) *Summary of ASAP Results for Application to State and Local Programs*. Washington, D.C.: National Highway Traffic Safety Administration, U.S. Department of Transportation.

Hurst, P. M. (1970) Estimating the effectiveness of blood alcohol limits. *Behavioral Research in Highway Safety* 1(Summer):87-99.

Hurst, P. M. (1973) Epidemiological aspects of alcohol in driver crashes and citations. *Journal of Safety Research* 5(3):130-148.

Jones, R. K., and Joscelyn, K. B. (1978) *Alcohol and Highway Safety 1978: A Review of the State of Knowledge*, Summary Volume. Ann Arbor, Mich.: University of Michigan Highway Safety Research Institute.

Li, L. K., and Waller, P. F. (1976) Evaluation of the North Carolina Habitual Offender Law. Highway Safety Research Center, University of North Carolina, Chapel Hill.

Maloff, D., Becker, H. S., Fonaroff, A., and Rodin, J. (1980) "Informal social controls and their influence on substance use." Pp. 5-38 in D. Maloff and Peter K. Levison, eds., *Issues in Controlled Substance Use*. Washington, D.C.: National Academy of Sciences.

Matlins, S. M. (1976) *A Study in the Actual Effects of Alcoholic Beverage Control Laws*. Rockville, Md.: National Institute on Alcohol Abuse and Alcoholism.

Mosher, J. F. (1979) Dram shop liability and the prevention of alcohol-related problems. Berkeley, Calif.: Social Research Group, University of California. (Forthcoming in *Journal of Studies on Alcohol*.)

Mosher, J. F., and Wallack, L. M. (1979) *The DUI Project*. Sacramento, Calif.: Department of Alcoholic Beverage Control, State of California.

National Safety Council (1976) *Accident Facts*. Chicago, Ill.: National Safety Council.

Nichols, J. L., Weinstein, E. B., Ellingstad, V. S., and Struckman- Johnson, D. L. (1978) The specific deterrent effect of ASAP education and rehabilitation programs. *Journal of Safety Research* X(4):177-187.

Noble, E. P., ed. (1978a) *Third Special Report to the U.S. Congress on Alcohol and Health*. Rockville, Md.: National Institute on Alcohol Abuse and Alcoholism, U.S. Department of Health, Education, and Welfare.

Noble, E. P., ed. (1978b) *Third Special Report to the U.S. Congress on Alcohol and Health*. Technical Support Document. Rockville, Md.: National Institute on Alcohol Abuse and Alcoholism, U.S. Department of Health, Education, and Welfare.

Organisation for Economic Co-operation and Development (1978) *New Research on the Role of Alcohol and Drugs in Road Accidents*. Paris: Organisation for Economic Co-operation and Development.

Perrine, M. W., Waller, J. A., and Harris, L. S. (1971) *Alcohol and Highway Safety: Behavioral and Medical Aspects*, Final Report. Washington, D.C.: National Highway Traffic Safety Administration, U.S. Department of Transportation.

Preusser, D. F., Ulmer, R. G., and Adams, J. R. (1976) Driver record evaluation of a drinking driver rehabilitation program. *Journal of Safety Research* VIII(No. 3, September):98-105.

Robertson, L. S. (1977) *Evaluation of Community Programs*. Washington, D.C.: Insurance Institute for Highway Safety.

Robertson, L. S., Rich, R. F., and Ross, H. L. (1973) Jail sentences for driving while intoxicated in Chicago: A judicial policy that failed. *Law and Society Review* (Fall):55-67.

Ross, H. L. (1973) Law, science, and accidents: The British Road Safety Act of 1967. *Journal of Legal Studies* II(1):1-78.

Ross, H. L. (1975) The Scandanavian myth: The effectiveness of drinking-and-driving legislation in Sweden and Norway. *Journal of Legal Studies* IV(2):285-310.

Ross, H. L. (1977) Deterrence regained: The Cheshire constabulary's "breathalyzer blitz." *Journal of Legal Studies* 6(January):241-249.

Scotch, N. A. (1979) Proposal for a Project to Study the Raising of the Drinking Age in Massachusetts. Department of Socio-Medical Sciences, Boston University School of Medicine.

Smart, R. G. (1969) Are alcoholics' accidents due solely to heavy drinking? Waller, J. A. Impaired driving and alcoholism: Personality or pharmacologic effect? Smart, R. G. Dr. Smart's reply. *Journal of Safety Research* I(4):170-178.

Sterling-Smith, R. S. (1976) *Psychosocial Identification of Drivers Responsible for Fatal Vehicular Accidents in Boston*. Washington, D.C.: National Highway Traffic Safety Administration, U.S. Department of Transportation.

U.S. Department of Transportation (1968) *Alcohol and Highway Safety Report*. Washington, D.C.: U.S. Department of Transportation.

U.S. Department of Transportation (1979a) *Alcohol Safety Action Projects Evaluation Methodology and Overall Program Impact*, Vol. 3. National Highway Traffic Safety Administration (DOT-HS-803-896). Washington, D.C.: U.S. Department of Transportation.

U.S. Department of Transportation (1979b) *Results of National Alcohol Safety Action Projects*. National Highway Traffic Safety Administration (DOT-HS-804-033). Washington, D.C.: U.S. Department of Transportation.

Wilde, G. J. S. (1971) *Road Safety Campaigns: Design and Evaluation*. Paris: Organisation for Economic Co-operation and Development.

Wilde, G. S., O'Neill, B., and Cannon, D. (1975) *A Psychometric Investigation of Drivers' Concern for Road Safety and Their Opinions of Various Measures for Accident Prevention.* Kingston, Ontario: Queen's University.

Zimring, F. E., (1978) Policy experiments in general deterrence: 1970-1975. Pp. 140-173 in Alfred Blumstein et al., eds., *Deterrence and Incapacitation: Estimating the Effectiveness of Criminal Sanctions on Crime*. Washington, D.C.: National Research Council.

The Role of Nonalcohol Agencies in Federal Regulation of Drinking Behavior and Consequences

JAMES F. MOSHER *and* JOSEPH R. MOTTL

INTRODUCTION

This paper outlines and discusses an important finding—that significant aspects of the federal response to alcohol problems are formulated by federal agencies not usually associated with alcohol policy. We provide a survey of federal agencies with various types of jurisdiction over alcohol distribution and alcohol-related problems and examine their potential role in the federal effort to address these issues effectively. In addition to identifying potential new actors, the paper also suggests potential new prevention strategies. We hope that we have provided a constructive, preliminary analysis of possible policy initiatives.

Research for this report was completed March 1980, and any changes in regulations or policies since that time are not included. Preparation entailed contacting numerous federal employees who, in virtually every instance, were both extremely courteous and helpful, often in the face of many competing demands. Their assistance was invaluable.

FEDERAL AUTHORITY TO REGULATE THE ALCOHOL MARKET

Since the end of Prohibition, federal responsibility for regulating the distribution of alcoholic beverages and the prevention of alcohol-related

James F. Mosher and Joseph R. Mottl are at the Social Research Group, School of Public Health, University of California, Berkeley.

Preparation of this report was partly supported by a National Research Center grant to the Social Research Group, AA-03524.

problems has been generally viewed as limited—a secondary, advisory responsibility that gives deference to the various states' primary role. The Bureau of Alcohol, Tobacco, and Firearms (BATF) and the National Institute of Alcohol Abuse and Alcoholism (NIAAA) are the primary federal agencies affecting the national alcohol policy. Both of these agencies defer to state authority in important ways. Although many BATF regulations are mandatory, they may not contradict any of the more restrictive state policies. In fact, many BATF provisions take effect only if particular states have similar or identical provisions. NIAAA has emphasized treatment and rehabilitation programs that, although based on a perceived need for national action, are usually voluntary and require state or private cooperation.

The basis for this limited view of federal jurisdiction rests in large part on the 21st Amendment, which ended Prohibition in 1933. That amendment gave alcohol a unique position in interstate commerce by limiting federal preemptory and interstate commerce powers.[1] It provides that alcohol may not be imported into any state contrary to the laws of that state. Thus, unlike any other legal commodity, the federal government in most instances cannot regulate alcohol in a state unless federal regulations are at least as strict as the state's provisions.

However, despite this grant of special power to the states—to restrict the importation of alcohol and its production and distribution within its boundaries—federal power remains substantial in four ways. First, the federal government may enforce stricter regulations than those found in the states. For example, BATF regulations prohibiting moonshining would stand and federal agents would still have authority to enforce them, even if particular states chose to legalize moonshine. In effect, states and the federal government have concurrent jurisdiction over the alcohol market, with federal regulations at least as strict as comparable state provisions.

Second, the states, as a practical matter, have chosen for the most part not to regulate the alcohol industry other than at the retail level, a decision that the federal government has encouraged (Mosher 1978c).

[1] The 21st Amendment makes alcohol a unique commodity vis-à-vis federal powers over interstate commerce. See *State Board of Equalization of California* v. *Young's Market*, 299 U.S. 59 (1936). However, the U.S. Supreme Court, in more recent cases, has made clear that the state powers under the 21st Amendment are not absolute. See, e.g., *Hostetter* v. *Idlewild Bon Voyage Liquor Corporation*, 377 U.S. 324 (1964) (in which the Court held that the state could not regulate sales by a retailer who sold exclusively to persons as they entered an airplane to leave the state and travel overseas). For a review of Supreme Court action regarding the 21st Amendment, see the majority opinion in *California* v. *LaRue*, 409 U.S. 109 (1972). For a discussion of state and federal powers over the alcohol market, see Mosher (1979b).

Only those states with a significant local alcohol industry have taken an active role with respect to regulating production, usually in a manner that fosters development of the industry (Bunce 1979).

Third, a significant portion of the alcohol regulations and alcohol-related problems are tangential to the constitutional provision. For example, the 21st Amendment does not prohibit federal taxation or national import-export jurisdiction, yet these powers may have a powerful influence on the structure of the alcohol market. Safety regulations relating to auto travel or house construction may affect the extent of alcohol-related problems, despite their being unrelated to state alcohol jurisdiction.

Finally, there are significant exceptions to the 21st Amendment. Federal reserves and lands, including Indian reservations, are not necessarily considered to be within a state's boundary for the purpose of the constitutional amendment.[2] A state may not block access to and may not tax alcohol or regulate alcohol use or distribution on federal property unless agreed to by the federal government.

SCOPE OF THE PAPER

Much of this federal authority, which rests largely outside BATF and NIAAA, has been ignored in the formation of federal alcohol policy and in most alcohol literature. Two areas in particular appear to have been neglected—the regulation of alcohol availability and the regulation of drinking contexts. As to availability, there has been a recent emphasis in the literature on studying alcohol beverage control laws and pricing policies as a means to deter alcohol problems (Bruun et al. 1975, Douglass and Freedman 1977, Medicine in the Public Interest 1976, Mosher 1979a, Room and Mosher, 1979). This strategy has been viewed as problematic because of the reluctance of state legislatures, alcohol beverage control (ABC), and the BATF to accept a preventive role (Medicine in the Public Interest 1976, Mosher 1978). Federal agencies and departments that have broad taxing and availability powers (many of which are acting at cross purposes to NIAAA priorities) have been ignored.

Regulation of drinking contexts has been more generally ignored. The advent of prevention of alcohol problems as a specialized field of study has created a broad range of literature linking alcohol to a host of social

[2] See *Collins* v. *Yosemite Park and Curry Co.*, 304 U.S. 518 (1938); *Yellow Cab Transit Co.* v. *Johnson*, 48 F. Supp. 594, affirmed, 137 F. 2d 274, affirmed 321 U.S. 383 (1942). For discussion of federal authority on native American reservations, see Mosher (1975).

problems—accidents, crimes, and diseases (Aarens et al. 1977). Alcohol has been demonstrated to be a causal factor in these problems, and prevention strategies have emphasized deterring the drinking behavior rather than addressing other causal factors surrounding the drinking context. For example, strategies to deter drunk driving stress the abstinence from drinking before and during driving rather than other causal factors of accidents—e.g., unsafe cars and highways. As Gusfield (1976) points out, such an emphasis unnecessarily narrows the range of potential prevention strategies. Gusfield's observation is relevant to other areas of alcohol casualties. Numerous federal agencies that do not deal with alcohol per se exercise broad jurisdiction over casualties associated with alcohol abuse. Natural alliances among these agencies and alcohol prevention strategists are not being exploited.

This study explores these two areas in detail. For convenience, the agencies studied are divided into four categories of jurisdiction: land-based, transportation-based, safety-based, and economic-based agencies. Land-based agencies determine the availability of alcohol within their boundaries; safety agencies may reduce the risk of casualty in contexts in which alcohol may be a factor; and economic-based agencies have power to regulate the economic structure of the alcohol market. Transportation agencies have characteristics of both land-based and safety-based agencies; however, they are distinctive because of the unique nature of their powers.

The general characteristics of each category are first outlined and then followed by a description of the major agencies involved. The agencies' jurisdiction and powers are discussed in general, as is their authority over alcohol distribution and alcohol-related problems. Current alcohol-related procedures and regulations are also outlined.[3]

For each category (except that of economic-based agencies, which is only briefly described) at least one agency has been chosen for an in-depth analysis of its potential role in federal alcohol policy. An assessment of the likelihood of the agencies' responding favorably and changing their current policies is presented. Assuming that reform is judged possible, specific reforms are suggested, followed by the effects that can be expected.

Judging the advisability of any particular change in policy requires an assessment of the possible benefits and costs (which includes, in addition to financial expenditures, possible detriments to other positive goals). Resistance to change may be caused because of potential conflict with

[3] A list of the regulations covered in the discussion in each section is provided in the appendix. Exact citations are available on request to the authors.

other important societal values. However, agencies often resist public-health-oriented change, even when there are no costs or conflicting values, merely because it appears to fall outside their immediate mission. In such cases, reform may be both reasonable and important.

For the prevention strategist, the more persuasive the evidence that a particular reform will reduce alcohol problems, the more likely that possible costs of the reform can be outweighed. However, other types of reforms may provide benefits, short of proven problem reduction, that may be pursued because of the minimal costs involved. Symbolic reform includes policy initiatives that provide a uniform federal approach to alcohol, even though the reform will have little actual impact. These can have an important though unmeasurable significance in terms of heightening awareness of particular problems and legitimating other policy initiatives. Experimental programs to determine the advisability of a particular reform may be justified even if results are uncertain and provided costs can be accurately judged and are acceptable.

Statistical analyses and compilations of the role of alcohol in various casualties, which usually involve only minimal costs, are also important as a means to further understanding the scope and role of alcohol in societal problems. As a recent, extensive study has documented (Aarens et al. 1977), there is a vital need to compile accurate data in this area. Most agencies that are attempting to cope with alcohol-related societal problems currently do not consider alcohol as part of their jurisdiction or concern. Encouraging statistical compilations, then, could provide valuable information and contacts. Finally, in some cases, advocating policy changes or addressing particular issues in agencies not specifically involved with alcohol may be necessary, even when the evidence indicates that no policy change in the near future is likely. They may highlight previously ignored issues and problems and create debate and further study.

As this discussion indicates, determining what reforms are appropriate depends on a given agency's view of its own mission as well as its receptiveness to new directions. Key factors are the agency's statutory mandate and responsibility. For example, if an agency is mandated to collect statistics to determine the causal factors of particular accidents, there may be a strong basis for urging the agency to include alcohol as a potential variable. We have outlined the scope of statutory responsibility (and have also talked with staff to determine each agency's own view of this responsibility) in an attempt to determine the importance of a given set of alcohol-related problems to agency goals and responsibilities. This provides a sound basis for determining what potential strategies would be most appropriate.

As with any proposal for governmental change, the actual implementation of a given reform may involve complex negotiations and unforeseen obstacles. Practical guides may prove to be useful, in particular suggesting whom to approach inside and outside the agency to promote new directions in alcohol policy. We outline the realistic boundaries of potential new prevention strategies among these agencies. Providing tactical guides for implementation falls beyond the scope of this paper.

It must be stressed that the areas covered in this paper are at an experimental stage generally. The relationship of the structure of the alcohol market and alcohol-related problems is a recent subject of inquiry. Very few studies have been conducted, primarily because of a lack of interest on the part of the industry and the state regulators. Even less is known about the role of safety agencies in reducing alcohol-related problems. Our effort here is to describe the various agencies' potential role, providing both potential new strategies and potential new allies in the government's effort to reduce alcohol's role in societal problems.

FEDERAL LAND-BASED JURISDICTION OF ALCOHOL DISTRIBUTION

INTRODUCTION

Although state regulations dominate the retail market of alcoholic beverages, there is a vast expanse of land controlled by the federal government. Alcoholic beverage control within the boundaries of federal lands is placed with the federal agency in charge, and those agencies may choose to exercise their authority either exclusively or concurrently with the states (unless the federal legislation provides otherwise). Court decisions have explicitly held that the 21st Amendment does not necessarily give states jurisdiction over the distribution of alcoholic beverages on federal land.[4]

The major federal agencies with authority to regulate alcohol distribution are: the Department of Defense; the National Forest Service; the Bureau of Land Management; the National Park Service; the Army Corps of Engineers; and the Bureau of Indian Affairs. Taken together these agencies control an alcohol distribution network that exceeds, both in land area and in population affected, the jurisdiction of any single state ABC. Problems associated with alcohol constitute at least

[4] See note 2 above.

a minor concern of the agencies and all have adopted at least some regulatory provisions in an attempt to cope with them. By far the most important of these agencies and departments is the Department of Defense, which has, through the various armed services, adopted extensive regulations and experimental programs.

These agencies may have concerns with alcohol problems ancillary to their distribution jurisdiction, such as treatment or rehabilitation programs for employees or safety regulations to protect citizens, some of whom may be in danger because of drinking (e.g., regulatory access to fire danger areas to limit the danger of accidental forest fires). However, one of their primary duties is to administer jurisdiction over the everyday activities of a particular area, and this study examines their decisions concerning alcohol use and distribution as one aspect of this primary duty.

As discussed in the introduction to this section, the alcohol distribution network has been put under scrutiny as a possible tool for preventing alcohol-related problems. Federal policy has not previously been examined in a systematic way. The activities of the Defense Department, because of its scope and importance, are discussed last and in some detail. Following an analysis of the scope and exercise of each agency's authority is a discussion that focuses on the Navy and potential areas of reform that could benefit from a uniform federal policy to prevent alcohol-related problems.

NATIONAL PARK SERVICE

Jurisdiction over most National Park Service areas (as is jurisdiction over all areas managed by the Forest Service, the Army Corps of Engineers, and the Bureau of Land Management) is shared by the federal and state governments. Older, larger parks are under exclusive federal jurisdiction, although state law may apply in specific circumstances. Because of its flexibility, concurrent jurisdiction is generally favored by National Park Service officials.

The National Park Service maintains 320 areas, and in 1978 283 million people visited park lands. There are 400 park concessions, of which 151 sell alcoholic beverages, and sales from these reached $5.8 million in 1978, which was 3 percent of total concession sales. These figures do not include alcohol carried in by visitors—the National Park Service has no estimates concerning the extent of this practice.

The National Park Service regulates the sale and use of alcohol in all of its park areas through several regulations: (1) a prohibition on operating a vehicle or vessel while intoxicated or under the influence of alcohol; (2) a prohibition on being so intoxicated as to endanger oneself, others, property, or others' enjoyment of the park; (3) a prohibition on the sale of alcohol to or possession of alcohol by a person under 21 years of age, unless state law permits otherwise; (4) a requirement that a permit from the regional director be obtained before the sale of alcohol can be made from privately owned land within large parks; (5) a requirement that licensees conform to local and state law as if the land on which they operate were located outside the park area. The pricing of alcohol is subject to National Park Service approval and is judged primarily by comparison with similar facilities under similar conditions (e.g., length of season, accessibility, cost of labor, and type of patronage).

BUREAU OF LAND MANAGEMENT

The Bureau of Land Management manages nearly 600 million acres of federal land, which includes much desert and Alaskan wilderness as well as forest and water areas. Concessions operate in some areas and a fair percentage of these sell malt liquor. Exact figures are unavailable. The bureau exercises no control over these beverage sales. It issues no licenses and levies no franchise fees.

The bureau relies heavily on local law and law enforcement, and there are no provisions relating to alcoholic beverages in its regulations. However, it does regulate numerous other activities on bureau lands. These include prohibitions against the use of audio devices (any machine that makes noise), restrictions on vehicle operation, prohibitions against pets in certain areas, and other prohibitions deemed necessary to protect the interest of public health, safety, and comfort. Thus, it clearly has jurisdiction to regulate alcohol-related problems on these lands if it chooses to do so. However, its enforcement staff is very small in relation to its area of jurisdiction.

Few data are collected; thus, the level of alcoholic beverage consumption within its lands, the volume sold from its concessions, and the connection of alcohol use to accidents or vandalism are unknown.

ARMY CORPS OF ENGINEERS

The Army Corps of Engineers maintains concurrent jurisdiction over 457 water projects and 4,200 recreational areas. It leases about one-half of the recreational areas to state and local governments.

Prior to 1971, by departmental directive, the corps prohibited the sale and storage of all alcoholic beverages within its areas. Changes were made, however, in response to local, state, and congressional pressure. The corps now has a more flexible policy regarding alcoholic beverages.

The present directives, which are in the process of revision, prohibit the sale or storage of alcoholic beverages except in areas where it is a custom and where it is permitted by state law. The express purpose of this restriction is to preserve a family atmosphere on recreational lands. Where sales are permitted, concessions may sell beer and wine for off-premise consumption. On-premise sale of distilled liquor is permitted when it accompanies dinner at lodges and hotels. Corps officials stated that information concerning the number of areas that have chosen this option was not readily available.

The corps maintains statistics on fatalities and accidents in their areas, but irregularities in reporting methods and recordkeeping among the states and counties make the accurate maintenance of alcoholic beverage variables impossible. Federal law provides funding for enforcement in these localities, without which many areas would be unable to carry out such activities independently. However, cooperation in enforcement between the corps and local officals is strictly informal and loosely coordinated.

NATIONAL FOREST SERVICE

In 1978, 218 million "visitor days" were spent in the 104 national forests, where alcoholic beverages are served in approximately two-thirds of the 625 resort areas. No official figures are available on the number of outlets or on sales volume. The National Forest Service exercises some control over alcoholic beverages by issuing lease-permits to concession operators, which are subject to special conditions.

Forest service policies prohibit the off-premise sale of distilled liquor. On-premise sale of all liquor is permitted in facilities, provided they are part of a legitimate resort activity or service. Sale of malt beverages is left to prevailing local laws and customs. According to one official, the arrangements are intended to be as flexible as possible, and local and state governments carry out most law enforcement under the concurrent jurisdiction.

Despite the numerous research projects the National Forest Service maintains, we found no research being conducted to determine the level of alcohol use in forest areas and its connection with forest fires, unextinguished camp fires, or damage within forest areas.

BUREAU OF INDIAN AFFAIRS

The federal government, by constitutional provision, has exclusive jurisdiction over all native American reservations. This authority may be delegated to tribal councils, and often is. Because of the federal trust obligation to the tribes, the government may not delegate any of its powers to the states without tribal consent. The Bureau of Indian Affairs (BIA), a part of the Department of the Interior, was established to administer federal responsibilities to the tribes.

The federal government and the BIA, pursuant to this general grant of power, established exclusive jurisdiction over the sale and possession of alcohol on reservations as early as 1802.[5] In that year, the president was given the authority to ban sales of alcohol on reservations. From 1802 to 1953, alcohol regulations became increasingly restrictive, banning all possession and distribution and making violations of alcohol laws grounds for extended prison terms and denials of treaty annuities.

As has been documented elswhere (Mosher 1975), the history of increasingly stringent BIA alcohol control laws has been a history of symbolic measures that had little relationship to genuine concern with native American alcohol-related problems. During the 19th and much of the 20th century, violations of alcohol laws were extensive, fueling a vigorous bootlegging trade. The BIA did little to deter this illegal but profitable trafficking, although selective enforcement against tribal members who purchased from bootleggers could lead to serious consequences for the tribe and the individual.

In 1953, the BIA voluntarily lifted its ban on possession and distribution on those reservations on which tribal councils decided to assert their own authority. In effect, the new regulation turned over authority to regulate alcohol to the tribes. This move was not a signal of increased tribal autonomy, however. Rather, it was part of a move to detribalize the reservations, to promote integration of native Americans into the American mainstream, and to delegate authority to the states. Detribalization did in fact occur in several states, and the BIA's change of alcohol policy there signaled the beginning of state authority. Detribalization was fiercely opposed, however, and was eventually largely abandoned, although authority to control alcohol distribution has remained with the tribes.

[5] The statute provided (act of March 30, 1902, ch. 13 § 21, 2 Stat. 139):

> The President of the United States (is) authorized to take such measures, from time to time, as to him may appear expedient to prevent or restrain the vending or distributing of spirituous liquors among all or any of the . . . Indian tribes.

Tribal regulation of alcohol is extremely varied, depending on tribal norms concerning alcohol (May 1977). The variations are too extensive to be summarized here. It can be assumed that any attempt by the BIA to reassert its authority in this area will be violently opposed by the tribes as signaling unwarranted federal intervention into tribal affairs.

DEPARTMENT OF DEFENSE

The armed forces have an extensive network of alcohol beverage sales; it is sufficiently large to make them one of the most important retailers in the country. They also act as regulators and, because of the size of the sales network, can be compared to ABC boards in the various states. The federal government has exclusive jurisdiction over all alcohol sales within military reservations. Thus, state law does not apply. All three service branches have regulations to encourage local commanders to cooperate with local officials in matters pertaining to alcohol beverage sales, but the regulations state specifically that the armed services are not subject to local control.[6] Alcohol sales at military outlets are therefore exempt from state and local taxation.

The Department of Defense alcohol outlets serve some 8 million people—2 million active members, 3 million dependents, 1 million civilians, and 2 million national guard and reserve members (Killeen 1979). Over one-half of active members are under 25 and 40 percent are single—which means that the major customers of military sales are precisely those within one of the current NIAAA target groups (Killeen 1979). In sum, the armed forces' policies on alcohol sales have a major impact on the country, and they must be included when discussing policy issues concerning the regulation of alcohol availability.

All three services (the Army, Navy, and Air Force—the Marines are not included in this paper) operate both on-premise and off-premise outlets. The club (or open mess) system, a major component of the military's recreational services, provides the primary on-premise outlets. It sells all types of alcoholic beverages for consumption on the premises and sells beer for off-premise consumption. The Air Force has 320 open messes worldwide, which had a total volume of alcohol sales of $61.9 million in fiscal 1977; the Navy operates 311 clubs, which had a total volume of $49.5 million; and the Army has 654 on-premise outlets worldwide (304 within the United States), which had a volume totaling $68

[6] Because federal agencies (except the BIA) could voluntarily relinquish authority over alcohol distribution to the states, this qualification may have been deemed necessary to ensure that federal control over disputes is retained.

million. (The Army totals include sales of peanuts and popcorn; alcohol sales themselves are not broken down further.)

In addition, all services provide some additional on-premise outlets. Beer, and in some cases other alcoholic beverages, may usually be sold at bowling alleys, dining halls, and recreational centers during special activities. The Air Force permits beer vending machines in dormitories in some cases and permits the sale of 3.2 (low alcoholic content) beer in several facilities.

All three services operate package stores. The stores sell all types of alcoholic beverages and may or may not be found in conjunction with post-exchange stores. The network is extensive: the Army operates 200 stores, which had a total volume of alcohol sales of $119 million in fiscal 1977; the Air Force operates 163 stores, which had a volume of sales totaling $72.9 million; and the Navy operates 111 stores, which had sales amounting to $84.8 million. The total of $276.7 million in package store sales is about 75 percent of the total of $356.1 million in alcoholic beverage sales for the service in fiscal 1977—figures that are equivalent to or exceed many state sales figures. Since these figures derive from discount pricing policies (see below), the volume of beverages sold by the services is even higher than indicated by these figures. (The military statistics for quantities sold were not available.)

The package stores are managed by the clubs, and most profits from the stores are plowed into the club system. The club system, in addition to running bars, provides various forms of related recreation and entertainment. The Government Accounting Office (GAO) reports that the package stores had a net income of $51.4 million in fiscal year 1977, of which $34.8 million was distributed to the clubs (Comptroller General of the U.S. 1979a). Many of the clubs operate at a deficit, and the package store profits are used to make up the difference. According to a recent GAO report, this policy has led to problems of mismanagement in the clubs, as there is little incentive to cut costs (Comptroller General of the U.S. 1979a).

Restrictions on Availability

All three services enforce mandatory age limits on drinking (prohibiting the sale to, purchase by, or consumption of alcohol by those under a specified age). The Army and Air Force permit 18-year-olds to drink unless the state, country, or territory in which the club or store is located enforces a different age limit. (Army regulations permit local commanders to decide whether 18-year-olds may drink beer regardless of local law.) The Navy permits 18-year-olds to purchase beer (if not contrary to local law) but prohibits those under 21 to purchase other alcoholic beverages. The Air Force recently amended its regulations to permit 18-year-olds to purchase and consume beer at Air Force outlets regardless of local law.[7]

The services also attempt to restrict the resale of alcoholic beverages by the purchaser. As discussed below, military alcohol sales are generally much less expensive than civilian sales. Proper identification must be shown prior to any sales, and in the Air Force the purchaser must sign a sworn statement that the purchase is for personal use only. Local commanders are directed to oversee the per-capita consumption of authorized personnel (i.e., compare sales figures with total personnel permitted to purchase) in order to determine whether unauthorized sales and resales are occurring.

Other regulations of drinking behavior concern specific consumer activities. These include restrictions on where consumption or possession may occur (e.g., no possession in recreation centers and craft facilities unless in conjunction with special programs; prohibition of open containers in any automobile). Perhaps the most important of these restrictions is the Navy prohibition of any alcoholic beverage consumption aboard planes and ships.

Deglamorization Program

Until recently, the prohibitions described above were the major restrictions placed on alcohol availability. However, as the Defense Department has become more aware of the degree of alcohol-related problems found among its personnel, it has taken a serious look at availability. During the last 7 years, it has instituted "alcohol deglamorization" programs, which are found in all three services. Most of the program components rely on cooperation from the particular services and the local commanders. The department is now preparing a directive to require or recommend many additional reforms. (Unfortunately the contents of the directive were not available in time to include in this paper.)

The chief components of deglamorization are summarized as follows:

(1) *Restrictions on reduced-price drink periods (happy hours).* These restrictions include limits on the number and length of happy hours;

[7] It is remarkable that there has been so little contact between NIAAA and the military prevention programs, at least on this point. NIAAA, which has become very concerned with the age-group most prominent in the armed services, has considered opposing the lowering of the drinking age in various states. Yet the recent moves by the Air Force to lower beer availability to those 18 years of age regardless of state law was not even commented on by NIAAA.

requirements that snacks and food be served and that soft drink prices be reduced an equivalent amount; and prohibitions on excessive reductions of price.

(2) *Prohibition of certain serving practices.* These include prohibitions on: stacking drinks; two-for-one sales; free drinks; "last call" or countdown sales techniques; service of doubles without an appropriate increase in price; the inclusion of alcohol in the price of meals; and the service of intoxicated persons. On-premise outlets are encouraged not to begin service until noon and to do so only in conjunction with a noon meal. Nonalcoholic beverages must always be available.

(3) *Regulations of base operations.* When possible, family-oriented facilities must be provided that are not centered around the service of alcohol. Alternative recreational facilities should be available during the club's operating hours. The Air Force also provides that "special attention and creativity" should be given to developing and promoting nonalcoholic drinks and food and that amusement machines and "diversions" should be available.

(4) *Other deglamorization policies.* Deglamorization usually includes general policy statements concerning the expectation of moderate drinking; the expectation that personnel will refrain from drinking before or during working hours (with the exception of moderate drinking during a meal); and restrictions on club advertising. The armed forces also offer training to club personnel who actually serve patrons. Information concerning this training was not available in time for this paper.

The various deglamorization provisions vary from service to service and are mostly voluntary, although some are mandatory. Their implementation is left largely, if not entirely, to the discretion of the base commanders, who are often given power to make specific exceptions. Moreover, there has been considerable resistance to the policies, particularly from club personnel. The clubs do not want to lower their income, since they operate on nonappropriated funds and are expected to break even. Alcohol sales (both at the clubs and at the package stores) make up a large part of their profit margin. There has been no effort to ease this financial strain in conjunction with the deglamorization program.

As a result of these factors, deglamorization has not been instituted on a large scale. Because of the various exceptions and loopholes in the regulations, most base commanders have either ignored the policies or have given only marginal support to them. A GAO report on military alcohol abuse programs found only isolated instances of compliance (Comptroller General of the U.S. 1976a). The report states (p. 31):

> Many military installations we visited had not taken any action to deemphasize and discourage alcohol use. We found that (1) hard liquor was sold freely at noon in base clubs, (2) happy hours were widely advertised, (3) drinks were on sale at 25¢ apiece, (4) special low prices on "drink of the week" were provided, and (5) free bottles of champagne were given on individuals' birthdays. [The practice of providing free birthday drinks is specifically permitted in the Air Force and Army regulations.]
>
> We also found instances where special committees recommended discouraging alcohol consumption by reducing happy hours at base clubs or reducing the number of drinks for each individual; however, command personnel rejected these recommendations as too severe or unnecessary.

GAO personnel report that in general the status of deglamorization remains the same today. A survey conducted by GAO for another report supports this: 20 percent of military personnel were unaware of the deglamorization program; 60 percent did not change their drinking habits at all; 10 percent reported decreases in consumption; and 4 percent reported increases in consumption (Comptroller General of the U.S. 1979b).

The Defense Department, in conjunction with deglamorization policies, instituted guidelines for determining the necessity of new package stores. The criteria to be considered include: (1) the estimated number of customers; (2) the importance of estimated contributions of package store profits to providing, maintaining, and operating clubs and other recreational activities; (3) the availability of wholesome family social clubs to military personnel in the local civilian community; (4) geographical inconveniences; (5) limitations of nonmilitary sources; (6) potential disciplinary and control problems due to local law and regulation; (7) highway safety; (8) community response. The actual enforcement of these guidelines is difficult to determine. For our purposes it is interesting to note that both economic factors and factors relating to alcohol-related problems are included.

Pricing Policies

The Defense Department has promulgated a directive that requires that sales of alcoholic beverages in package stores cannot be less than 10 percent below the lowest prevailing rates of civilian outlets in the area. Exceptions may be granted, but only by approval of the secretary of the particular service after a substantial showing that special factors warranting an exception are present.

Determining actual prices is suprisingly difficult, given this seemingly elementary policy. First, the lowest available price in the area must be located. In license states, this may vary widely, particularly in urban areas. The Army regulations define a local area as any place within 25 miles or 30 minutes' drive from the military installation. Local and state taxes are not included in the price computation, which, in many states, means an additional 5 percent or more discount. GAO officials report that this tax exemption is used in monopoly states as a means to exclude state store overhead and profit margins (which are viewed as a type of tax). Defense Department personnel deny this. The Air Force and Army have recently deleted this minimum price requirement altogether with regard to beer.

DISCUSSION

Several observations can be made concerning federal land-based control of alcohol distribution: (1) a great volume of alcohol is sold under the auspices of the federal government in a variety of settings and types of establishments; (2) authority to regulate sales is widely dispersed and generally ignored, except in the Defense Department outlets, where sales are often encouraged; (3) the relationship of alcohol sales and use on federal lands to potential risks is generally ignored; and (4) there is virtually no contact between federal agencies with authority to regulate retail sales and those agencies charged with formulating federal policies toward alcohol use and abuse.

Other than Defense Department agencies and the BIA, there appears to be little expectation of major changes in current attitudes, at least in the absence of congressional action to force policy initiatives or to centralize alcohol distribution authority—equally unlikely prospects. Agencies such as the National Forest Service simply do not consider alcohol distribution and problems as part of their mandate. The tendency to turn over responsibility to private parties in accordance with rules of local jurisdictions is both convenient and understandable.

Despite these impediments, some minimal proposals could prove to be fruitful. Some agencies still maintain restrictions on availability (such as the National Forest Service's ban on off-premise distribution). Recent studies have documented the increasing availability of alcohol in an expanding number and types of settings (Douglass and Freedman 1977, Mosher 1979b,c), and other studies have suggested that there may be a link between increased availability and drinking problems (e.g., see Bruun et al. 1975). These studies suggest that the agencies should be encouraged to maintain current availability policies rather than to continue market expansion, at least until the relationship between availability and alcohol-related problems is better understood. Contacts with these agencies suggest that the only pressure they receive concerning alcohol distribution is for continued expansion. Providing some counterbalance to this trend, which would entail minimal costs, would appear appropriate, at least on a symbolic level.

The BIA presents a special case because of its unique position vis-à-vis the native American tribes. The competing interest in delegating authority to tribal councils (authority that is jealously guarded) appears to outweigh the potential benefits of reasserting BIA alcohol control policies, particularly given the BIA's dismal history of control prior to 1953. This does not rule out the possibility of providing advice to tribes that are instituting changes in their own control system or that need assistance in coping with alcohol-related problems. Programs along the lines suggested below regarding the military might be appropriate for funding.

The Defense Department appears to provide better opportunities for experimentation and change. Unlike the other agencies studied, the military has, in the last 9 years, commissioned outside studies and instituted major new programs to determine the extent of alcohol problems and to combat them among its service and civilian personnel (Cahalan et al. 1972, Cahalan and Cisin 1975, Comptroller General of the U.S. 1976a, Killeen 1979, Long et al. 1976, Manley et al. 1979, Polich and Orvis 1979, Schuckit 1977). Alcohol is now recognized as a serious problem that affects a large percentage of military personnel—an important development in Defense Department attitudes. According to Killeen (1979, p. 356) there were virtually no efforts to deal with alcohol abuse prior to 1965 and only small-scale programs between 1965 and 1971.

Increased interest in and concern about alcohol abuse has provided an impetus to examine the network of military alcohol distribution. As the deglamorization guidelines suggest, Pentagon officials recognize that practices of service tend to encourage problem drinking and to obstracize nondrinkers. Although the GAO and our own inquiries suggest that deglamorization directives have had only minimal impact at most local bases, the interest in controlling availability as a means of controlling alcohol problems is unusual (if not unique) among alcohol control agencies, including state ABCs.

Various local naval stations as well as officials at the Pentagon were contacted during the course of this study to evaluate the potential for policy reforms concerning alcohol marketing within the military. The Navy was chosen because, of all the services, it has shown the most concern and enthusiasm for instituting prevention programs. In addition, naval personnel have a reputation for heavy drinking.

Survey data show that the Navy does indeed have serious alcohol problems among its personnel. The Navy issued a report in 1975 concluding that 39 percent of enlisted personnel and 23 percent of officers had drinking problems described as "critical," "very serious," or "serious" (Cahalan and Cisin 1975; also reported in Comptroller General of the U.S. 1976b). The study also found that 27.4 percent of enlisted personnel and 22.4 percent of officers reported at least some lost work time or work inefficiency during the 6 months preceding the study due to drinking. The figures for enlisted personnel (who represent the heaviest drinkers generally) are substantially higher than comparable figures for civilians in the same age bracket. The GAO cited an additional study substantiating Cahalan and Cisin's findings. The Naval Air Training Command in 1974 issued a report that found that 32 percent of officers and 37 percent of enlisted personnel have had drinking problems. That report concluded that alcohol was the primary drug problem among naval personnel (Comptroller General of the U.S. 1976b).

The Navy fully recognizes and publicizes the alcohol problems among its personnel. A military-sponsored magazine recently published an article providing statistics on naval alcohol problems that were obtained from the Navy (Fowler 1979). According to the article, of the approximately 540,000 naval personnel, there are currently 200,000 problem drinkers and 80,000 alcoholics; excessive drinking costs more than $189 million in poor job performance. As a measure of particular alcohol-related problems, approximately 3,000 of the 12,000 drunk drivers arrested in San Diego annually are Navy and Marine Corps members. The article also notes that costs to the society generally are high in terms of both drunken driving accidents and discharged alcoholics.

The publicity concerning these alcohol-related problems can be viewed as part of the Navy's attempt to address directly the issues involved. Our contacts with naval officials found that the recently instituted Naval Alcohol Safety Action Program (NASAP) is enthusiastically endorsed at all levels. NASAP is a prevention-oriented effort that began in 1974 as an adjunct to other, more treatment-oriented programs. It is specifically designed to promote the early identification of alcohol problems and expanded education and publicity campaigns. NASAP officials claim that a 25-percent increase in work effectiveness will occur if NASAP programs are established on local bases. An installation has been established at San Diego with the express purpose of training counselors and advisors to work in NASAP programs around the world.

Without evaluating NASAP claims of success, one can still note that a major change in attitudes has occurred. High-level and middle-level officers, who until recently ignored alcohol abuse, now support the concept of NASAP. Moreover, participation in NASAP programs is becoming a more accepted alternative for personnel and is less stigmatized. Naval officials stress that its management is more centralized and prevention-oriented. Thus the NASAP's prevention focus is more easily accepted than in other services.

NASAP has not embraced deglamorization regulations concerning service practices or pricing reforms in off-premise outlets as an integral part of its program. The program serves an advisory role and will recommend deglamorization guidelines for administering clubs and stores when requests are made. Part of this reluctance comes from the resistance to deglamorization, which is common in all military services. Although local officials in general stated that they supported deglamorization, a Pentagon official gave a different view, which has been confirmed in contacts with other military branches. The enlisted personnel resent changes in club procedures as just another imposition from supervisors. Significant changes cause defections by regulars, particularly those with the worst problems, to off-base establishments. Finally, the financial arrangements, whereby alcohol sales help pay for other club activities, create a lack of interest at the local level.

Despite this lack of enthusiastic support, all persons contacted agreed that at least some bases have instituted some or all of the deglamorization policies. Bases with transient populations, particularly those that receive personnel from ships that have been at sea for extended periods, are less likely to comply because of opposition by patrons. Bases with stable populations, however, have a higher compliance rate.

We were unable to determine levels of compliance except through random contacts. Compliance is dependent on local commander initiative and there is no centralized method for reviewing local decisions. Three base representatives contacted in the San Francisco Bay area told us that the deglamorization policies were being followed. Independent observation evaluation has not been conducted to determine the accuracy of the representatives' claims.

It is our conclusion that certain policy initiatives within the Navy concerning services and selling practices would receive at least some support at all levels and could serve to build more extensive contacts and changes. The following initiatives are suggested:

(1) *A concerted effort to reform the financing of alcohol sales*. The GAO and Congress have both been active in attempting to divorce alcohol profits from club-related recreation and to raise off-premise

prices to reflect costs outside the base.[8] As the above discussion shows, there are essential elements in promoting deglamorization policies. The concept of the self-supporting club has emerged historically without any awareness of the problems it raises in promoting excessive alcohol use and abuse. It may have reflected a congressional sense that funds for purely recreational expenditures should not be made available. But promoting alcohol sales can be argued to be a false funding source—excessive alcohol use creates hidden costs for both the Navy and society, costs that have already been recognized by the Navy. Today, adequate recreational facilities associated with military clubs can be justified, and funding does not need to be dependent on alcohol sales. Divorcing alcohol sales from club-related recreation would diffuse opposition to deglamorization programs at the local level because a necessary service—club-related recreation—would not be put in jeopardy.

The profits from alcohol sales could go to the federal treasury, or they could be redirected to the NASAP program (a suggestion given by a NASAP official); independent funding could be provided for clubs and recreational activities. This would involve alcohol profits in a program that has expressed an interest in implementing deglamorization and other prevention programs, a program that justifies its existence in part by decreasing the cost of alcohol abuse and alcoholism. However, it runs the risk of creating new vested interests in alcohol profits.

Raising the package store price for alcohol also provides a promising strategy for aligning the military's alcohol sales authority with its interest in reducing alcohol problems. In addition to the GAO findings, anecdotal reports indicate that discounts may be much more substantial than those authorized by the Defense Department. Package store sales are subject to large discounts, which encourage unauthorized sales and resales. Providing the cheapest off-premise alcohol within a 50-mile radius is an obvious invitation for abuse and for possible drunken driving. Following this strategy does not mean that providing discount pricing as a benefit to enlisted personnel should be abandoned. The discounts

[8] According to a recent United Press International release (reported in the *San Francisco Chronicle*, March 3, 1980, p. 18), a congressional panel of the House Armed Services Investigators Subcommittee has strongly criticized current practices. Panel chairman Dan Daniel stated that the panel has issued an ultimatum to the military—that by the beginning of fiscal 1981 the package store profits should not be used to subsidize the military clubs and instead should be used for general morale, welfare, and recreational activity. The recommendations here differ somewhat from the panel's concern. The panel would use alcohol profits to support other recreational activities, such as gymnasiums, rather than to support clubs. Our suggestion would be to divorce alcohol profits from recreation completely and to put the funds into the general treasury or into alcohol-prevention programs.

could be transferred to essential items so that nondrinkers are not penalized. It should be noted that local alcohol retail businesses would probably welcome a change in this unfair government competition.

Pricing of on-premise alcohol is not subject to any regulation except the mainly voluntary restrictions found in the deglamorization program. A GAO study of military clubs shows that low prices for drinks and special discount policies are one of the prime incentives for military patronage of clubs, particularly among men (Comptroller General of the U.S. 1979a). Slightly lower prices in clubs may be justified to encourage patronage in order to monitor more carefully the drinking context and to deter drunken driving off base. In fact, having other nondrinking entertainment and services may provide the same benefits. However, these factors do not justify the use of sales techniques, such as free drinks, stacking drinks, etc., that encourage excessive drinking rather than patronage.

(2) *Establishing and evaluating various server reforms*. The deglamorization guidelines are really methods to influence drinking contexts, and very little is known about the relationship of drinking contexts to drinking problems. Thus, the Navy's efforts in this area are extremely important as sources for potential experiments. Implementation across the board is probably neither practical nor advisable. In some clubs, such as those catering to sailors who have been at sea for months, many of the reforms could be counterproductive. However, limited experimentation with careful evaluation (of both the type of reform and type of establishment) would be very valuable to the field. As results are understood, adjustments could be made and programs could be expanded.

In addition, recent programs to train servers to recognize problem drinkers and to institute various reforms to ensure that patrons do not become drunken drivers could be funded and evaluated (Mosher and Wallack 1979, Mosher 1979a). Many of the reforms in a recent California experiment, conducted by the California Alcoholic Beverage Control Department, are along the lines of the deglamorization guidelines (Mosher and Wallack 1979).

It should be emphasized that there are virtually no programs in the country in which governmental bodies are attempting to institute reforms in serving practices with a preventive orientation. The potential problems of dealing with military personnel, given the command structure's endorsement of prevention as a part of their military duties, are much smaller than the obstacles to instituting similar reforms in civilian life, where alcohol control authorities do not in general acknowledge prevention as within their sphere of concern. Evaluation of the Navy's efforts and encouragement for continuing and expanding efforts is therefore important.

Support would be enhanced by promoting these changes as part of the NASAP program. There was at least lukewarm support for such a tactic. Support could also be established by minimizing costs and opposition—beginning with procedures that do not directly deny drinks to nonintoxicated patrons and that do not directly threaten funding. Much of the effort could be geared toward attitude change, a slow process, but one that is already occurring in the NASAP program. Emphasis could be placed on the fact that some current practices (stacking drinks, drinking contests, etc.) actually encourage drinking, which certainly creates a reasonable suspicion that they contribute to drinking problems. Deglamorization creates a neutral attitude toward sales and will tend to reduce the costs of alcohol problems both for the Navy and for society as a whole.

FEDERAL SAFETY-BASED JURISDICTION OF ALCOHOL-RELATED PROBLEMS

INTRODUCTION

The History of Safety Regulation

One of the most expanding aspects of federal jurisdiction during this century concerns regulatory efforts to enhance the safety of our environment. Although the 19th century witnessed the growth of safety-related inventions (such as the safety pin and safety match), safety was considered an individual's responsibility. The concept of "contributory negligence" in legal cases reflected this view—one could not recover for injuries caused by another person's carelessness if it was shown that the person injured contributed in any way to his or her own injury.

Safety as a federal concern began in the first decade of this century. Initially, governmental action was prompted by the use of dangerous additives in foods. The Pure Food and Drug Act was passed in 1906 in an attempt to protect citizens from adulteration and fraud by food processors. The rationales for governmental intervention, which are still relied on in part today, were that the consumers are unable to protect themselves from deceptive industry practices, and the dangers were so great that federal action was warranted.

Other national initiatives to promote safety soon followed. In 1931, the National Safety Council was formed as a private agency. Its purposes included the promotion of safety throughout the country and the compilation of data concerning various societal risks. The Food and Drug Act was expanded several times; agricultural and public health chemicals came under government scrutiny in 1947. By the late 1960s there was a proliferation of safety agencies to perform the tasks taken on—the Occupational Health and Safety Administration (OSHA), the Consumer Product Safety Commission (CPSC), the Federal Drug Administration (FDA), the Environmental Protection Agency (EPA), the National Highway Transportation Safety Administration (NHTSA), the Nuclear Regulatory Commission (NRC), and the National Transportation Safety Board (NTSB). The Coast Guard and the Bureau of Mines, among other federal agencies, were given expanded safety responsibilities.

The federal concern for safety has made our society much safer than it was at the turn of the century in at least some regards. Working conditions in many industries have become much less hazardous; dangerous pesticides and canning chemicals have been banned; the number of fatal accidents (including automobile deaths) has been reduced by nearly one-half proportionate to the population. The life expectancy of Americans has risen dramatically partly because of government's emphasis on enhancing safety.

However, technological solutions have themselves created risks. Today's transportation network creates numerous dangers on the ground, in the water, and in the air. Our increasing dependence on chemicals and nuclear technology creates great risks that cannot be measured. Many of our modern risks tend to justify government action for the same reasons that the trend toward federal intervention began—consumers cannot effectively protect themselves against the dangers, and the risks may be potentially very great. The activities of the various safety agencies, then, have become a crucial part of the government's role in serving and protecting its citizenry.

The move toward increasing the government's regulatory safety powers has not been without debate, however. While reducing potential risks of harm may be a proper goal for government, providing overprotection has its own social costs. Individual freedoms may be jeopardized; creativity, both of individuals and of business, may be stifled; the viability of small businesses may be eroded by the prohibitive costs of safety requirements. Determining when and how to intervene to promote safety, then, is an important topic of current federal practice.

Lowrance, in his groundbreaking study, *Of Acceptable Risk* (1976)—one of the few efforts to broach this topic—discusses two discrete issues that have to be addressed before deciding to regulate a potentially risky activity. First, the risk must be measured. This is a largely factual inquiry, and Lowrance describes the various methods for measurement. It also includes determining the causal factors that contribute to particular injuries. Second, the acceptability of risks must be determined. This is a policy decision, and it requires the balancing of social values. One aspect of this policy analysis includes determining which causal factor(s) should be manipulated if government action is warranted.[9]

The choice of which factor to regulate and what level of safety is necessary are difficult, political decisions. In addition to protecting the safety interests of consumers, safety agencies must also balance competing industrial interests. Often, the industries' interests themselves will be in conflict, as the available strategies for reducing risks will have different financial impacts on different industries. Lobbying pressure may become intense, as those being regulated seek to influence and direct the actions of the regulators, often by obtaining friendly appointments to key administrative positions. Powerful industries may seek congressional action if they perceive that a particular safety agency has become overzealous. Thus, despite our increasing emphasis on safety, the political forces that affect the decision-making process create a danger that the decisions will reflect political rather than safety concerns. The experience of the Consumer Product Safety Commission, described below, illustrates the political problems that may affect the role of safety agencies.

Alcohol as a Factor in Safety Regulation

One fact that appears repeatedly in a variety of accidents of concern to safety agencies is the role of excessive drinking. It is well recognized that excessive drinking contributes to most common accidents—travel, employment, recreation, and home injuries (Aarens et al. 1977). Unfortunately, the extent of alcohol involvement is not well documented primarily because the agencies responsible for reducing risks in particular situations and compiling statistical data do not analyze alcohol

[9] William Haddon (1973a,b) has developed a set of 10 strategies for reducing "energy damage," a term he coined to include both accidental and intentional damage to persons and property. These strategies are based on an analysis of the "processes" that typify occurrences causing harm and identifying different possible points of intervention. Although it uses different terminology, Haddon's approach is parallel to the one discussed here—several different prevention-oriented interventions can be suggested by analyzing the various factors that contribute to an accidental occurrence and that cause unwanted harm.

casualty in any consistent or thorough way. For the most part, federal safety agencies do not consider alcohol as part of their jurisdiction.

Alcohol policy makers, in turn, have ignored safety agencies as potential allies in reducing alcohol-related problems (except, arguably, for automobiles) because only one strategy is used—changing drinking behavior. Most commonly, alcoholism is viewed as the problem to be addressed, and treatment and rehabilitation programs for alcoholics are relied upon.[10] When safety agencies and alcohol policy makers do coordinate activities, alcoholism programs are virtually the exclusive strategy employed. These programs have proven to be of only limited value in reducing risks, however, because alcoholism treatment is not a very specific policy instrument for changing drinking behavior in particular situations (e.g., before driving a car). Moreover, many drinking-related accidents have been found to be unrelated to alcoholism.

The failure to judge the role of alcohol in accidents and to study potential strategies to reduce its effect in risk situations has denied prevention strategists a potentially valuable tool. As Lowrance suggests (1976, p. 59), there may be other causal factors in a given situation that, with appropriate action, could be manipulated to reduce the risk of harm. By doing so, alcohol-related accidents could be reduced without any change in actual drinking behavior. This suggests that the activities of various federal safety agencies may be crucial in the effort to prevent alcohol problems. It also suggests that prevention strategies may involve attempts to change political priorities in regulating particular risky ac-

[10] See the section on federal transportation-based jurisdiction below regarding the Federal Railway Administration and the Federal Aviation Administration (FAA) for examples of this type of liaison. OSHA, while not studied in detail here, was found to have the same emphasis.

There are some safety agencies that, because of the extent of potential danger, cannot tolerate any drinking or alcoholism in certain situations, for example, the FAA (discussed later in this paper) and the Nuclear Regulatory Commission (NRC). The NRC permits dismissal of any personnel with security clearances when he or she is found to be "a user of alcohol habitually and to excess, and has been without adequate evidence of rehabilitation." Security guards must have a general health examination prior to employment and prior to renewal of licenses. Other employees do not receive any medical screening, and there are no regulations concerning drinking on the job.

The potential dangers associated with nuclear reactors would appear to justify more awareness of alcohol's potential harmful effects on employees. A newspaper report publicized an incident in Oregon in which security guards were found dealing in drugs on the job site. In partial response, the NRC announced it will require checks for alcohol addiction among all security guards beginning in March 1980.

As the NRC experience illustrates, safety agencies may have the power to regulate the availability of alcohol in particular situations. Transportation-based agencies with such dual powers are discussed in the section on federal transportation-based jurisdiction.

tivities. Aarens et al. (1977, pp. 600, 602) summarize the new direction being suggested:

Research can be used to find manipulations of the environment which will reduce alcohol-related casualties and crimes. In our view this is at once the most neglected and the most promising use for research. It requires painstaking and detailed studies of serious events with attention to the occurrence and sequencing of contributing factors and to potential strategies of prevention or intervention. It is not at all oriented around the quick and easy single figure of alcohol's involvement in the event. . . .

The common assumption in consumer product safety research is that, except for children and disabled persons, the consumer is in full command of his mental and physical faculties in using the products tested. No recognition is given to the fact that the products will at least occasionally be used by consumers in an inebriated condition. A serious aim of research into alcohol's role in casualties and crime should be to make the world safer for drunkenness. Whether this will result in more drunkenness is an interesting empirical question, but hardly a justification for tolerating continuing deaths and injuries.

We attempt in this paper to address this potential new strategy. This section analyzes the Consumer Product Safety Commission (an agency whose exclusive role is to promote safety) in detail, discussing (1) its structure, purpose, and history; (2) its view of the role of alcohol in its activities; (3) its attempts to reduce a particular alcohol-related problem—residential fires; and (4) possible policy initiatives that can be taken to further the goal of reducing alcohol-related problems. The section titled federal transportation-based jurisdiction discusses similar issues concerning transportation-related safety agencies (except those concerned with automobile traffic). The discussion of the Federal Railway Administration includes a section on occupational safety measures, a third type of safety agency responsibility. OSHA (the primary agency prompting occupational safety) and auto safety agencies are not discussed here in the interest of brevity.

THE CONSUMER PRODUCT SAFETY COMMISSION (CPSC)

Authority

The Consumer Product Safety Commission's purpose is to protect the public against unreasonable risk of injury associated with consumer products; to assist consumers in evaluating the comparative safety of consumer products; to develop uniform safety standards for consumer products and to minimize conflicting state and local regulations; and to promote research and investigation into the causes and prevention of product-related deaths, illnesses, and injuries. It is empowered to regulate manufacturers, distributors, retailers, private labelers, importers, and other distributors of consumer goods. It may set safety standards, ban hazardous products, require bookkeeping, examine records, call for reports, inspect business premises, impose labeling and warning requirements, and demand safety certification. Enforcement is affected by court injunctions, seizure of hazardous products, criminal sanctions, and civil penalties.

The CPSC was first established in 1973 by passage of the Consumer Product Safety Act. The legislation was originally proposed in 1970 by the National Commission on Product Safety, which concluded that "the exposures of consumers to unreasonable consumer products is excessive by any standard of measurement" (cited in Green and Moulton 1978, p. 698). It reported that the costs of accidents involving consumer products each year amounted to 20 million injuries (of which 110,000 caused permanent disabilities and 30,000 caused mortalities) and $5.5 billion in financial losses (Comptroller General of the U.S. 1977).

The Consumer Product Safety Act gave the commission additional powers by transferring to it authority over four other statutes—the Flammable Fabrics Act, the Federal Hazardous Substances Act, the Poison Prevention Packaging Act, and the Refrigerator Safety Act. The Consumer Product Safety Act exempts various products from its coverage—most notably, tobacco products, motor vehicles, pesticides, firearms, boats, food, and electronic and medical products—although some of them do fall under the commission's jurisdiction by means of one of the other acts (tobacco products are a notable example). Originally, in fact, the Consumer Product Safety Act provided that the transferred acts (which generally granted broader power to the commission than the Consumer Product Safety Act) had precedence if they could alleviate the risk of a particular injury.

Each act provides limits on the type of action the CPSC can take. Under the Consumer Product Safety Act, the commission has authority to set consumer product safety standards (following a complex procedure, discussed below) and to compel repair, replacement, or refund of purchase price. However, the major enforcement provision relies on notification by manufacturers and distributors, a requirement that is loosely construed. The other acts give the commission authority to require labeling or banning of hazardous substances, to develop special packaging standards for certain household substances so as to protect children, and to issue cease and desist orders against those transporting any product that creates an unreasonable risk of harm.

Organization and Procedures

The CPSC is an independent agency headed by five commissioners appointed by the president who serve staggered 7-year terms. The president also appoints the chairman with the consent of Congress, and the chairman retains all administrative and executive powers—to make appointments, allocate work among the staff, and allocate funding. The commissioner's authority is subject to the general policies of the commission, although the full commission must approve five major executive appointments. Until recently, these executives reported directly to the chairman once the appointments were approved.

There are three advisory councils: the Product Safety Advisory Council, the National Advisory Committee for the Flammable Fabric Act, and the Poison Prevention Packaging Technical Advisory Committee. Consultation with the second two councils is compulsory, while consultation with the first is advisory. The composition of the committees is equally representative of business and consumer interests, with some government representation. Under earlier chairmen, the advisory boards were seldom consulted despite their statutory mandate and exerted little influence over commission policy.

Under the Consumer Product Safety Act, the commission has the power to set mandatory consumer product safety standards or to enforce voluntary standards by following a complex "offeror" procedure. First, the commission must announce in the *Federal Register* the need for a new product standard together with all relevant information it has on the product. It must solicit standards from the public and entertain all responses that conform to general guidelines for responding. An existing voluntary standard may be submitted, provided it includes the procedures by which the standards were set and the organizations that participated in its development and approval. The commission may accept an existing standard, pay offerors to develop a new standard, or develop its own standards independently. Usually, outside consultants are hired when this third course is chosen.

This procedure is designed to encourage industry to establish its own safety standards and contrasts sharply with procedures followed under the other four acts. For example, under the Hazardous Substances Act, the formal rule-making procedure is started by publishing a proposed regulation in the *Federal Register*. After considering comments, the commission must act on its proposal. Objections and requests for public hearings must be made within 30 days of the commission's action.

Major conflicts often occur over whether performance or design standards should be adopted. Performance standards, favored by industry, provide the level of safety that is required without specifying particular design requirements. Thus, variations in design are permitted. Design standards, favored by consumer groups, set specific design specifications.

Judicial review of commission standards is much stricter than for other regulatory agencies. If a CPSC ruling is challenged, the court will affirm the commission's power only if it determines that the commission's findings are supported by substantial evidence. Other agencies need only show that their actions are neither "capricious" nor "arbitrary."

Problems Besetting the Consumer Product Safety Commission

The commission has been ineffectual from its inception for a variety of reasons. First, the five acts that comprise its jurisdiction are overlapping and contradictory. The CPSC might decide to proceed under one act only to find in court that it should have followed another act's procedures. Second, most actions fall under the Consumer Product Safety Act, which has extremely cumbersome procedures for setting standards, which cause protracted delays. According to the General Accounting Office, as of mid-1978, only three independent standards had been set—architectural glass, safety matchbook covers, and swimming pool slides—and the latter two were later struck down by courts (Comptroller General of the U.S. 1977). It took an average of 834 days to set these three standards. The delays are violations of the act, which permits only 330 days for completing standard setting. Third, as the two court reversals indicate, the strict standard for court review puts a severe burden on the commission to justify any action that it takes. This puts the commission on the defensive, as it is easily subject to reversals of its decisions.

Finally, presidential appointments have seriously hampered the commission's effectiveness. Richard Simpson, the first chairman (a Nixon appointee) opposed the concept of an independent product safety commission and testified against its creation in Congress. Simpson never appointed an executive director during his 2½ years and failed to set any objectives or goals. Because of the extraordinary powers of the chairman, his decisions made it impossible for the commission to fulfill its purposes. The appointment of John Byington as Simpson's successor (by President Ford) created new problems. According to U.S. Civil Service Commission (reported in U.S. Congress 1978), Byington appointed an inordinate number of high-level administrators; he was also charged with favoritism and misuse of consultants. In 1978, the Department of Justice disclosed that it had found evidence that Byington had accepted kickbacks, and he resigned soon after (reported in Green and Moulton 1978).

A GAO study in 1977 summed up the commission's problems. It found that the commission suffered from inefficient management, misplaced priorities, lack of enforcement, lack of adequate data collection, offeror process delays, and failure to promulgate standards (see Comptroller General of the U.S. 1978b).

Recent Reforms

According to the GAO (Comptroller General of the U.S. 1977, 1978b) and a recent article (Green and Moulton 1978), numerous reforms have been recently instituted to help solve many of the commission's problems. All five acts are now equally applicable to products falling under any of their jurisdiction, with minor qualifications. The offeror process has been modified to permit the commission discretion to develop its own standards without following offeror procedures, provided that public participation is ensured.

According to numerous groups contacted during the course of this study, these reforms have improved its performance, although it is still too early to evaluate the commission's effectiveness. Despite recommendations by the Office of Management and Budget (OMB) and some members of Congress to abolish the CPSC, it has been refunded until 1981.

The Commission and Alcohol-Related Problems

One of the purposes of the commission is to investigate the causes and prevention of product-related injuries. One would expect, given this statutory mandate, that the commission would take at least some interest in the role of alcohol consumption in product injuries. Anecdotal information, at least, indicates that the chances of injury are multiplied when alcohol has been consumed.

The CPSC, despite its mandate, has in fact ignored alcohol involvement in all but one of its investigations. When we contacted the agency and inquired about alcohol information, we were told that we had contacted the wrong government agency. When pressed, commission staff told us that accident reports could be examined individually, but there were no compilations of alcohol involvement except in the area of residential fires. Upon reviewing the accident reports, we discovered that alcohol information was not standardized, so that alcohol involvement was reported only when the reporter volunteered the information.

According to CPSC staff, our discoveries conformed to the commission's traditional mode of investigation. It focuses on technological and engineering inadequacies of the products involved without a view toward potential misuse. Its regulations and findings are based on injury statistics associated with particular products. The physical condition of consumers who are injured is generally ignored.

The GAO has recently criticized the CPSC for its inadequate data collection, particularly its failure to investigate and report the human factors involved in product injuries (Comptroller General of the U.S. 1977, p. 21):

> The Commission's collection and analysis of product-related injury data have been directed toward the mechanical factors associated with such injuries. . . . However, it has directed little attention toward determining how human factors—the way people use products—are involved in product-related injuries. Without adequate product exposure data . . . and data on how people actually use products, it is difficult for the Commission to identify adequately the unreasonable risk of injury.

The GAO concludes (p. 25):

> The Commission's product injury data base, including its surveillance and in-depth investigation activities, does not adequately support its hazard identification and hazard analysis needs. The Commission needs injury information that it can use to determine the causes of a product-related injury, the product's involvement in the incident, and the user's (people) involvement in the incident.

Recently, in a partial reversal of this investigative procedure, the CPSC has created an independent Human Factors Division. Originally housed in the Office of the Medical Director, this division of seven analysts is assigned to research the use of a particular product rather than the product itself. For example, it has conducted a study of skateboards and has recommended design modifications based on expected use and misuse by young children. Although the product can be considered safe if properly used, the human factor of potential misuse became the basis for a recommendation of reform. The staff reports that if alcohol were to be a consideration in CPSC actions, the Human Factors Division would be the place for involvement.

Residential Fires

One area in which the CPSC has taken action concerns a well-documented alcohol-related problem. Residential fires caused by cigarette or other tobacco product ignition are surprisingly common in the United

States. According the U.S. Fire Administration (Overbey 1979), there were 70,000 smoking-related fires in 1978 that caused 1,800 deaths, 4,000 injuries, and $180 million in economic losses. These fires constitute approximately 11 percent of all residential fires, 32 percent of residential fatalities, 21 percent of injuries, and 9 percent of dollar losses. A study in Maryland from 1972 to 1977 supports these figures (Berl and Halpin 1978). There were 295 fatal fires that caused 414 fatalities during the study period. Of the fires ignited by smoking materials, 135 (45.8 percent) resulted in 184 (44.4 percent) fatalities. Residential fires are the primary source of fire fatalities in the United States, and tobacco-related fires have contributed substantially to the U.S. per-capita rate of fire deaths, one of the highest in the world.

Surveys show a clear association between amount of drinking and amount of smoking (Cahalan et al. 1969, p. 148), and that association carries over to smoking-related fires. In an analysis by the CPSC (1978, pp. 27-29) of 102 fire fatalities in Maryland caused by upholstery furniture fires between 1971 and 1977, all but 6 were known to have been ignited by smoking materials (the ignition sources for the others were unknown). Autopsies of 48 of the 102 victims showed alcohol in the blood. Of these victims 32 were under age 15, none of whom showed evidence of alcohol. Thus, among adults, nearly 70 percent of the fatalities involved alcohol (see also U.S. Department of Health, Education, and Welfare 1972, pp. 25-48).

Berl and Halpin (1978), in a separate report of all Maryland fires between 1972 and 1977 (not limited to upholstery-related fires, which represent approximately 25 percent of all fire fatalities during this period), found a similar correlation of tobacco use, alcohol use, and age. As indicated earlier, smoking materials were the ignition source in 44.4 percent of all Maryland fire fatalities between 1972 and 1977. Thirty-five percent of the victims had a blood alcohol content (BAC) of 0.10 or greater; for the victims between ages 30 and 59, 67 percent had a BAC of 0.10 or greater and 41 percent had a BAC of 0.20 or greater.

Alcohol may be involved at various stages of a fire accident. It may create drowsiness, thus causing loss of awareness of a burning cigarette, or it may impair detection of a fire and possible escape. The statistics suggest that such ancillary effects are present. Although smoking-related fires cause only 11 percent of all residential fires, they cause 32 percent of all fatalities, an indication that these fires are likely to go undetected until escape is impossible.

The commission has identified residential fires as an unacceptable risk requiring government intervention. The statistics indicate that there are at least five possible factors to be considered: (1) the act of smoking;

(2) the act of drinking; (3) the structure of the residence itself; (4) the ignition source (tobacco products); and (5) the combustible furniture. In accordance with its emphasis on product design, the commission has not considered the first two factors for possible initiatives (although it has compiled statistics on their involvement). Building designs are also considered outside its jurisdiction. However, its handling of the last two factors provides insights into the commission's political weaknesses.

Cigarette papers in American cigarettes are treated with a citrate compound or phosphate salts to ensure longer burning—up to 45 minutes after a cigarette is put down on a flat surface. Nitrate compounds in the tobacco itself also heighten this effect. Research has shown that upholstered furniture and mattresses will not ignite if a cigarette goes out within 5 minutes. Thus many fire experts and most of the major fire protection lobbying groups are advocating a strategy that would require tobacco companies to develop a cigarette that extinguishes if not smoked within a short period of time.[11]

The industry opposes this idea on the basis that no method of creating such a cigarette would be acceptable to consumers. Critics charge that the industry is ignoring almost 30 separate patents, has refused to provide researchers with necessary information concerning cigarette ingredients, and has failed to conduct any research itself on the feasibility of self-extinguishing designs. Some critics also accuse the industry of opposing reforms to maximize profits—more cigarettes will be sold if they burn when not smoked (and are therefore not available for relighting).[12]

The CPSC, during its first years of existence, studied this strategy as well as the strategy of requiring fireproof upholstered furniture and mattresses. However, in 1976, it requested that Congress remove tobacco products from its jurisdiction. (The commission was being sued at the time by consumer groups to ban nonextinguishing cigarettes.) We were unable to determine the role of the tobacco industry in effecting this change, either at the CPSC or in Congress, but it can be assumed

[11] Groups supporting this strategy include: the American Burn Association; the Fire Marshalls Association of North America; the United States Fire Administration; the International Association of Fire Chiefs; the National Fire Protection Association; Action on Smoking and Health; and the International Association of Fire Fighters (see *Congressional Record*, October 16, 1979, p. E5061).

[12] See, e.g., comments by Gordon Damant, chief of California's Bureau of Home Furnishings, CBS Radio News Special (1979), O'Malley (1979).

[13] There was some conflict concerning the sequence of events prior to this legislative enactment. According to Alan Schoen, in the CPSC's General Council's office, authority over cigarette design was terminated because the commission was never intended to have such jurisdiction. A District of Columbia Federal District Court had ruled otherwise prior

that extensive lobbying was involved.[13] There was no input or protest made by any organization specifically concerned with alcohol-related problems.

The upholstery and bedding industries have tried repeatedly to reverse this decision and to fend off what they consider oppressive and ill-advised regulations of their products. A recent popular article details the difficulties confronting the CPSC in reducing fire danger through an upholstery strategy (O'Malley 1979). In fact, a CPSC staff member claims that Congress severely hampered its efforts because of these difficulties (O'Malley 1979, p. 60). Aside from the practical concerns, it is probable that new standards for home furnishings will have little immediate effect. Although statistics are not available, it is likely that smoking-related home fires mainly involve used furnishings and that persons most in danger do not have the financial resources or inclination to invest in new, safer products.

The drive for self-extinguishing cigarettes has continued, despite CPSC inaction. Representative Joe Moakley of Massachusetts and Senator Alan Cranston of California have recently introduced bills into Congress that would require that all cigarette manufacturers have their products extinguish within 5 minutes if not smoked. The Cranston bill would return the responsibility for developing necessary regulations to the CPSC. In Los Angeles, a lawsuit has been filed against a tobacco company for injuries caused by a cigarette-upholstery fire on the theory that the industry is acting negligently in its current practices. Recent publicity of the issue has helped heighten public awareness.[14] However, alcohol prevention strategists have not been involved in any of these recent events.

Discussion

This history illustrates several points: (1) alcohol-related problems may, in some cases, be best prevented by addressing causal factors other than drinking behavior; (2) the CPSC is politically weak and, as a result, its decisions may reflect political forces in addition to safety-based factors;

---

to the 1976 amendment. Schoen stated that the commission therefore supported the law. One review article stated that the CPSA "requested" the change. O'Malley (1979) indicated that Congress independently made the change due to tobacco industry pressure, although implying that at least some members of the commission supported the bill.

[14] The American Burn Council has established a nationwide campaign to publicize this issue. In addition, the media have been active. There has also been media coverage of Senator Cranston's bill (see, e.g., CBS Radio News Special 1979, Waas 1980, O'Malley 1979).

(3) alcohol policy makers have ignored the CPSC; (4) overburdensome safety regulations may be the result of the CPSC's abandoning a more promising strategy for political reasons.

It appears that the commission is currently undergoing administrative reforms and reevaluation, including reviews of data collection methods (Comptroller General of the U.S. 1978b). A heightened awareness of alcohol involvement in product-related accidents could be included in these reforms. Statistical analyses are needed, which will require changes in reporting forms and compilation designs. Although our contacts with the commission were not encouraging, its revamping may provide some avenues for discussion and accommodation.

The CPSC's lack of interest in alcohol rests primarily on a narrow view of its proper functions. Its statutory mandate clearly includes research and investigation into the cause and prevention of product-related injuries. CPSC has established a complex, nationwide network for collecting injury-related data; providing a small number of questions concerning alcohol involvement on existing forms could be a valuable source of information. Because the network now exists, additional expenses would be limited to revamping existing forms and processing the data that are collected. Thus, there appears to be a strong argument for reform.

As the GAO has indicated, the commission has also ignored the potential risks of negligent use. As a result, the commission may be ignoring important safety-related strategies that would be particularly relevant to alcohol-related injuries. For example, some products may be highly dangerous even if well designed because they are often used by intoxicated or negligent persons. Simple design changes might substantially reduce these risks, but until adequate studies are conducted, there is no way to assess such reforms.

The commission has already recognized the need for such a focus in regard to children (e.g., the child-proof medicine cap).[15] The formation of an independent Human Factors Division may signal the beginning of reform as to potentially negligent adult use. A relatively small amount of additional funding for this division could prove to be a valuable first step toward evaluating a new strategy for reducing alcohol-related problems. There does not appear to be any conflicting problem in doing so other than commission indifference and perhaps a lack of funds. Despite the lack of encouragement from our contacts wth the commission, the

[15] It should be noted that child protection initiatives may have an incidental effect of protecting intoxicated persons. For example, a depressed person considering a drug overdose might be unable to open a medicine bottle because of intoxication.

recent CPSC administrative actions indicate that attention and encouragement from the National Institute on Alcoholism and Alcohol Abuse could prove influential in changing CPSC priorities.

Finally, alcohol policy makers need to address the problem of alcohol-related residential fire injuries and deaths. As the statistics show, this is appropriate for prevention efforts. As has been discussed, the CPSC has failed to act on a promising strategy that is now being advocated by major fire protection groups, including the U.S. Fire Administration and the International Association of Fire Chiefs and Fire Fighters. Congressional and public support for that strategy appears to be growing. The public airing of this issue by other interested federal agencies, while not guaranteeing immediate results, may help to reverse previous decisions.

FEDERAL TRANSPORTATION-BASED JURISDICTION

INTRODUCTION

The federal government has become increasingly involved in the regulation of all forms of transportation. Because most forms of travel involve complex technologies and may cross state lines, federal presence is essential for creating uniform national standards. Federal jurisdiction may now include: employee practices in the transportation industry; equipment specifications and maintenance; passenger and operator practices; and safety requirements. Although the authority may be shared with particular states, federal transportation powers are extensive and may override state regulations, at least when interstate travel is involved.

Alcohol-related transportation problems are common to all parts of the industry and are recognized as both important and difficult to address. The nature of alcohol involvement may vary. For example, alcohol use by operators of transporting vehicles is a problem for all forms of transport. Alcohol use by trespassers and pedestrians poses a particular problem for the railway industry, and passenger drinking is a potential problem for airline companies. The federal jurisdiction for coping with these problems falls into the two categories already discussed. Regulations may be adopted affecting the availability of alcohol in particular situations, and safety and environmental measures may be taken to lessen the impact of existing drinking behavior.

This section first presents the activities of the National Transportation Safety Board, a safety agency with authority to study all forms of transportation safety measures. Many of its activities are relevant to discussions found in the preceding section. Detailed discussions of federal activities in the rail and air transport industries follow as well as a brief description of the Coast Guard's jurisdiction of waterway transport. In each case, federal powers over alcohol-related problems are addressed. (Automobile, bus, and truck transportation are omitted as they are outside the scope of this paper.) The section concludes with a discussion of possible regulatory actions, particularly in the railway industry. These policy initiatives illustrate both the scope and the type of reforms within existing federal authority that may reduce alcohol-related problems.

NATIONAL TRANSPORTATION SAFETY BOARD

Jurisdiction

The National Transportation Safety Board (NTSB) was created by the National Transportation Act of 1966, which also created the Department of Transportation (DOT). The board functioned as an autonomous agency under the DOT until 1974, when the Independent Safety Board Act established it as an independent agency that reports exclusively to Congress. Prior to 1966, the board was a division of the Civil Aeronautics Board and its activities were confined exclusively to investigations of air accidents. It has had jurisdiction over surface modes of transportation (including pipelines) since 1966, although its emphasis remains on air-related investigations.

The board's legislative mandate is to improve transportation safety generally. Its jurisdiction to do so covers the following areas, among others: investigating major transportation accidents; reporting on the facts, circumstances, and causes of such accidents; making specific recommendations to government agencies and industry for promoting transportation safety; conducting special safety studies and investigations; assessing accident investigatory methods and establishing requirements for reporting accidents to the board; and evaluating the transportation safety consciousness of other government agencies. The board has no regulatory power (other than the power to require certain statistical reporting methods). However, its recommendations, usually based on detailed investigations, are often adopted. One board employee stated that 85 percent of its recommendations are acted on and that the board vigorously follows up on its recommendations, publicizing the issues involved. If a party refuses to adopt a specific recommendation of the board, it must give a detailed explanation within 90 days.

Despite its wide range of responsibilities, the NTSB is one of the smallest federal agencies and it often relies on other public and private entities for investigatory assistance. The board consists of three members, appointed by the president, who serve 5-year terms. There are four offices: managing director, general counsel, public affairs, and administrative law judges. The Office of the Managing Director houses the Bureau of Accident Investigation, the Bureau of Technology, and the Bureau of Plans and Programs; these bureaus perform the basic work of the board.

Investigations of air accidents account for a substantial portion (60 percent) of the budget of these bureaus, although air fatalities account for only 3 percent of all transportation fatalities. All airline accidents and all general aviation accidents involving at least one fatality are investigated—approximately 12,000 accidents annually. The Bureau of Technology, through its Human Factors Division, assists in these investigations. The Marine Accident Division investigates only those major waterway accidents involving at least six deaths, the sinking of a major vessel, or damages exceeding $500,000—approximately 80 investigations per year. The Railroad Accident Division limits its investigations to those accidents involving passenger trains or property damage exceeding $150,000, approximately 30 per year. (The pipeline and highway investigatory divisions, which are both small, are not discussed in this paper.)

Investigations of Alcohol Involvement

The three investigation divisions discussed above have had at least some exposure to alcohol involvement in transportation accidents. Because of the emphasis on air accidents, the reports on aviation safety are more thorough and provide valuable information concerning alcohol use among general aviation pilots involved in accidents. However, as discussed below, the GAO found that its procedures were inadequate (Comptroller General of the U.S. 1978a). The Human Factors Division of the Bureau of Technology, in conjunction with the Air Investigation Division, made specific recommendations to the FAA concerning implied consent regulations for pilots (discussed below). A employee of the Human Factors Division reported that there is no systematic approach to studying alcohol problems or alcohol as a factor in accidents.

The awareness of alcohol involvement in rail and marine accidents is even more fragmented. Only three accidents involving alcohol have been investigated—two train accidents and one marine accident. The marine accident involved a collision between a ferryboat and a tanker on the Mississippi River in which 67 people died (National Transportation Safety Board 1979b). The captain of the ferryboat was intoxicated, and his intoxication, in conjunction with his lack of good judgment and failure to heed waterway rules, was found to be the primary cause of the accident. Yet none of the NTSB recommendations concerned alcohol, and there was no discussion of alcohol as a problem in water transportation. The train accidents involved intoxicated engineers, and significant recommendations were made to the Federal Railroad Administration and the railroad company (discussed below) to change regulations and rules concerning drinking behavior (National Transportation Safety Board 1974, 1980). However, the investigations have not prompted any further study of alcohol as a factor in rail accidents. The NTSB has also conducted a study of trespasser casualties on railroads and found intoxication to be a significant factor. Numerous recommendations were made concerning improving safety for trespassers.[16]

In its 1978 annual report to Congress, the NTSB devoted a special 3-paragraph section to alcohol and safety. After listing its own fragmented involvement, the board (1978, p. 105) recommended "the Department of Transportation should develop and implement a program designed to reduce dramatically the incidence of alcohol-related accidents." No other mention of alcohol involvement was made in the report; the 1977 report was completely silent on the subject.

In sum, the NTSB, although lacking regulatory authority, has extensive jurisdiction: to study the causes of transportation accidents; to provide recommendations; to assess the safety accomplishments of other agencies; and to urge congressional action. It has a record of steady success in promoting safety and has conducted numerous detailed investigations and studies. Its statistical data are invaluable to persons working in the safety field. However, although the board recognizes the importance of alcohol involvement, it has not developed a strategy for determining the role of alcohol in transportation accidents or for reducing its impact. Adequate statistics are not forthcoming despite the board's powers to compel accurate and extensive reporting from the various federal agencies; investigations and recommendations usually do not address alcohol impairment (with some exceptions in air transport); and no concerted effort has been made in any of the transportation areas to study the role of alcohol in accidents despite its explicit authority and responsibility to do so. Its only recommendations concerning further study—one of its primary roles—was for another agency to take action.

[16] The NTSB 1978 annual report (p. 63) summarized this study. It analyzes 2,679 railroad pedestrian accidents during a 20-month period and provides recommendations for improving railroad pedestrian safety. According to the NTSB, the Federal Railway Administration responded by acknowledging the need for railroad pedestrian safety and by initiating a program to select candidate areas for safety improvement and to evaluate the costs and benefits of providing restrictive fencing. NTSB personnel stated that intoxication was a common characteristic of these pedestrian injuries.

THE FEDERAL AVIATION ADMINISTRATION

Jurisdiction

The Federal Aviation Administration (FAA) has primary jurisdiction over the use of the airways within the United States. Although its jurisdiction may be shared in particular instances with various state agencies, its overall authority is more comprehensive. Air commerce is defined in the Federal Aviation Act to include: "interstate, overseas, or foreign air commerce or the transportation of mail by aircraft or any operation or navigation of aircraft within the limits of any Federal airway or any operation or navigation of aircraft which directly affects, or which may endanger safety in interstate, overseas or foreign air commerce."

This definition gives the FAA potential jurisdiction over virtually any flight within the United States. It has been found to include: the certification of pilots, both commercial and private; the behavior of crew members and passengers while in flight; and the design of airplanes and airports. The use of alcoholic beverages may become involved in any of these aspects of FAA jurisdiction.

Alcohol Problems in Air Transportation

The dangers of using alcoholic beverages while operating an aircraft are well documented (Comptroller General of the U.S. 1978a, Aarens et al. 1977, NTSB 1979a). The physiological effects of alcohol are magnified when flying—the effects are twice as great at 10,000 feet and three times as great at 15,000 feet as they are at sea level. Since the operation of aircraft is more demanding and complex than the operation of an automobile, it is reasonable to assume that even stricter rules than those applying to driving under the influence of alcohol are needed for regulating pilot behavior.

A recent GAO report (Comptroller General of the U.S. 1978a) substantiates this assumption. The report cites a National Transportation Safety Board study, which found that 430 fatal accidents, or 10 percent of all fatal accidents during the 11-year period of 1965-1975, involved alcohol impairment of pilot judgment and efficiency. The GAO report argued that this figure was underestimated because of an unrealistically high BAC requirement before alcohol impairment was assumed. It cites tests conducted by the FAA's Civil Aeromedical Institute indicating that 818, or approximately 20 percent of fatal aviation accidents during that period, involved alcohol impairment. This figure may still be underestimated because of FAA's failure to obtain adequate BAC testing. The

GAO also found that, in one sample of 163 accidents involving pilot impairment, 63 percent involved alcohol. (For a summary of various studies that support these findings see Aarens et al. 1977, pp. 76-80).

Aviation accidents involving alcohol impairment are concentrated almost exclusively in private or general aviation flights. Commercial airlines have an excellent record of maintaining sobriety among their pilots before and during flights.

Drinking by passengers on commercial airlines is a second alcohol-related problem area in which the FAA plays a major role. Although figures are unavailable, it can be assumed that there is a high volume of consumption.[17] Drinking has become one of the main activities of passengers flying commercially, and air travel has increased dramatically in recent decades. The alcohol beverage industry, recognizing this potential market, has many of its advertisements directed at this drinking context.

The FAA does not act as the actual dispenser or retailer on airplanes; however, it does have regulatory authority similar to state ABC boards; the airline companies have the role of retailer. Until 1972, the airlines imposed a voluntary two-drink maximum; since then no maximum has been enforced. Free drinks and free refills are often used to promote carrier popularity and first-class patronage. The FAA and its predecessor, the Civil Aeronautics Board (CAB), have taken no action concerning drink limits or the dispensing of free drinks. An association of flight attendants has made two requests for a one-drink maximum per 1½ hours of flight, to which the FAA has not responded.[18]

In addition to increasing the health dangers that are associated with increased consumption generally, drinking during commercial flights may contribute to auto accidents by passengers who drive from airports after arrival. Although there has been no study of this potential danger, a recent program of the California ABC strongly suggests that driving from an airport while intoxicated is a common occurrence.[19]

Recent publicity concerning assaults on commercial flights points to another danger of in-flight alcohol service. According to the Association

[17] The FAA does not collect data concerning how much alcohol is sold during commercial flights. The information is apparently only available from airline companies on an individual basis. (This conclusion is based on conversations with FAA personnel and with William Jankman of the American Airlines Transportation Association.)

[18] This information was obtained from Del Fina, Association of Flight Attendants.

[19] See Mosher and Wallack (1979) for a discussion of this project. The project, in part, recorded information from drunk driver arrest reports in Los Angeles County, which included a question concerning where the driver had been drinking. The Los Angeles International Airport was among the top five locations recorded.

of Flight Attendants, there were 60 assaults on its members in 1978, up from a figure of 25 in 1975.[20] About half of the assaults involved drunken passengers. Attendants complain that airline companies do not screen passengers to keep intoxicated persons off planes and have failed to support attendants who refuse to serve intoxicated passengers. In some instances, they charge, the airlines have instead taken disciplinary action against the attendants. There is apparently a tension between company policies, which see attendants as primarily food and drink servers, and union policies, which see its members' role as one of ensuring passenger safety. The FAA has shown little interest in the assault problem.

FAA Regulations and Response

The FAA has promulgated the following regulations in an attempt to cope with alcohol-related problems:

(1) no commercial airline may serve any alcoholic beverages to any person in the aircraft who appears intoxicated, who is carrying a deadly weapon, or who is in criminal custody;
(2) no commercial airline may permit a person who appears to be intoxicated to board an aircraft;
(3) no passenger on a commercial airline may carry on board any alcoholic beverage;
(4) no person may act as pilot or crew member (commercial or private) within 8 hours after the consumption of any alcohol or while under the influence of alcohol (the "8-hour rule");
(5) no private aircraft may carry any person who is "obviously under the influence of intoxicating liquors."

In addition, the FAA determines the qualifications for receiving a pilot's license. A medical certificate is necessary, and the regulations provide that an applicant will be disqualified if he or she has an established medical history of alcoholism. Alcoholism is defined as a "condition in which a person's intake of alcohol is great enough to damage his physical health or personal or social functioning, or when alcohol has become a prerequisite to his normal functioning."

The FAA's Office of Aviation Medicine, in a memo dated November

[20] These statistics were obtained from Del Fina of the Association of Flight Attendants. An FAA spokesman reported 75 violations of laws prohibiting interference with crew members in 1978 (Frank 1979). The Association of Flight Attendants has hired a law firm to prosecute private suits, on behalf of attendants who have been assaulted, against the passengers involved.

10, 1976, described the procedures for enforcing the alcoholism regulation (FAA 1976). The memo first notes the difficulty in defining "alcoholic" and differentiating "alcoholism," "incipient alcoholism," and "alcohol abuse." However, it points to certain objective criteria that should be seriously considered. These include: (1) a history of more than one arrest for driving an automobile while intoxicated; (2) the need for medical detoxification; (3) the development of neuropathy; (4) alcoholic psychoses, including delirium tremens; or (5) seizures in withdrawal. Recently, the FAA, in conjunction with NIAAA, has conducted a reinstatement program for those who successfully complete treatment.

Finally, the FAA has conducted voluntary education programs concerning alcohol and flying (Comptroller General of the U.S. 1978a, pp. 20-23). These include distribution of literature concerning the dangers of alcohol to pilots and presentations during the general aviation accident prevention programs. The contents of these programs are left largely up to the individual coordinators, and the inclusion of alcohol varies from region to region. Alcohol is also a suggested topic for the biennial flight review program (a program to test pilots' awareness of current general operating and flight rules, and includes a flight test). Again, the inclusion of alcohol information varies from region to region.

According to the GAO, these regulations are inadequate in numerous respects. It argues that the 8-hour rule is inadequate because a pilot can still be under the influence after abstaining for that time. In fact, many commercial airlines require a 24-hour abstinence period, and the FAA's own handbook for pilots suggests a 12-hour period. The regulations also offer no guidance in defining "under the influence." The FAA has failed to set mandatory minimum alcohol levels and implied consent regulations, despite specific NTSB recommendations to do so. The GAO argues that such provisions would help the FAA identify alcohol's role in accidents, improve enforcement, reduce investigatory time, and serve as a deterrent to pilots considering drinking before or during flight.

The GAO also criticized the review procedures for issuing medical certificates. Its investigation showed that many of those pilots involved in alcohol-related flying accidents had received state convictions for alcohol-related traffic violations (9 of 35 in 1975). Furthermore, the GAO estimated that as many as 12,000 licensed pilots have had their driver's licenses revoked for driving while intoxicated without FAA knowledge. Applicants for FAA medical certificates who have past alcohol-related convictions were found to routinely omit their criminal records, and the FAA has taken no precautions to check the applications, although the means to do so are available.

The GAO also roundly criticized the FAA educational programs.

According to GAO, alcohol is deemphasized by many program coordinators, and there was no effort on the FAA's part to ensure that educational materials were universally distributed.

According to agency personnel, the FAA, in response to the GAO recommendations and criticisms, have taken the following actions: (1) updated seminars (every 5 years) for medical examiners, which include material on alcohol-related problems; (2) searched for laboratory reports that might help in the detection of alcoholism and chronic alcohol-related problems; (3) undertaken a legal review of regulations. These actions are minimal at best and do not address most of the GAO concerns.

FEDERAL RAILROAD ADMINISTRATION

Recent headlines have accentuated the need for increased railway safety. In November 1979 a Canadian city of over 200,000 persons had to be evacuated due to a derailment of a train carrying chlorine gas. During the same month, similar derailment accidents (in sparsely populated areas) were reported in Florida (involving propane), Michigan (involving ammoniated hydrogen fluoride), and Indiana (involving a variety of toxic gases).[21] As these accidents indicate, the United States is increasingly dependent on trains for transporting dangerous chemicals and nuclear waste. Statistics show an alarming increase in the number of accidents in which trains carrying hazardous materials release dangerous chemicals into the atmosphere, forcing evacuation in surrounding areas.[22] A single accident can have profound effects, as serious as the

[21] These incidents were reported in four separate news articles in the *San Francisco Chronicle*: November 6, 1979, p. 6; November 12, 1979, p. 2; November 13, 1979, p. 2 (two articles).

[22] The following table shows the steady increases in accidents and consequences resulting from the operation of trains carrying hazardous materials (from FRA 1976, 1977, 1978):

	1975	1976	1977
Total accidents (trains carrying hazardous materials)	690	798	864
Total railroad cars damaged	847	903	1,072
Total cars releasing hazardous materials	131	166	173
Total persons evacuated	3,495	19,369	14,105

This represents a 25.2-percent increase in the number of accidents; a 26.6-percent increase in number of railway cars damaged; and a 32.1-percent increase in the number of cars releasing hazardous materials into the environment in a 2-year period. The wide fluctuation in evacuation statistics shows the dependency of such figures on the place of accident. The large increases suggest that the risk of a major accident in a populated area will increase with the number of accidents.

Three Mile Island incident. Even discounting the potential for a catastrophic accident, rail travel has proven to create serious dangers to employees, passengers, automobile travelers, and pedestrians. In 1977, the most recent statistics available, there were 76,574 separate train accidents in which 1,530 people were killed and 67,867 were injured. The NTSB states that the rate of accidents, including those involving hazardous materials, is steadily increasing (NTSB 1979a, p. 26, NTSB 1978, p. 32).

Jurisdiction

The Federal Railroad Administration (FRA), a part of the Department of Transportation, administers and enforces federal laws and regulations designed to promote safety. It has financial and safety jurisdiction over AMTRAK and safety jurisdiction over USRA (Con Rail) and other lines. It does not regulate regional or urban rail systems, except for interstate common carriers who carry commuter traffic. Its jurisdiction over safety includes accident investigations, regulation of equipment design and use, and regulation of railroad procedures, including hours of service, employee practices, and record inspection. The FRA is also charged with developing program and policy positions for the Department of Transportation, which includes authority to design appropriate studies, analyses, and evaluations.

Despite this wide range of authority and the potential dangers to society, the FRA has taken virtually no regulatory action concerning alcohol-related problems on railroads. Its primary strategy has been to delegate all responsibility to union and management groups, attempting to provide support and encouragement to railroad employee assistance programs. These programs determine whether an employee is in need of treatment for alcohol abuse or alcoholism and, if so, either provide necessary treatment or refer the employee to an appropriate agency. Overregulation is cited as the prime justification for the FRA's hands-off policy. As a result, there are no federal standards for employees' drinking practices on the job or before coming to work and no procedures for handling offenders.

Industry Rules

The railroads themselves do have rules pertaining to employee drinking. Rule G of the General Rules in the Consolidated Code of Operating Rules of the Association of American Railroads (cited in Mannello and Seaman 1979, p. ix) provides that operating personnel shall not possess or use intoxicants or be intoxicated while on duty or on company property, and shall not use intoxicants while on call. (Most railroad operating personnel are on call 8 hours after any duty period. Some management personnel are on 24-hour call, which means, technically, that any drinking among this group constitutes a rule violation.) Although Rule G does not apply to all railroads (including AMTRAK), most or all railroads have similar or identical company policies, at least implicitly, and in most cases, the company rules include exempt and nonoperating personnel (Mannello and Seaman 1979). The potential consequences for most railroad workers are very severe—dismissal can result after only one offense, provided a lengthy and complicated appeals procedure is followed.

Alcohol-Related Problems Within the Railroad Industry—The University Research Corporation Study

Until recently, the FRA had failed to conduct any studies concerning the extent and potential costs of alcohol-related problems. In 1979, a report by Mannello and Seaman of the University Research Corporation (URC), which was commissioned by the FRA and was the result of 3 years of study, provided the first extensive research in this area (referenced as Mannello and Seaman 1979; hereafter referred to as URC 1979). The drinking practices and problems among employees and their responses to voluntary treatment programs in seven major railroads were studied in detail, relying primarily on questionnaire and interview data. The sample railroads account for nearly half of all railroad workers and range from very large to relatively small companies. Although the primary purpose of the study was to evaluate employee assistance programs, many of its findings are relevant here.

The findings are startling. The report states that 19 percent of those studied, or an estimated 44,000 of 234,000 of the workers for the 7 railroads (23 percent or an estimated 16,000 of 72,000 of all operating personnel), are problem drinkers. In some companies, the rates of operating personnel, the most important employees in terms of potential safety hazards, reach over 34 percent. Railroad workers are twice as likely to drink on "binges" as men nationally. An estimated 28,000 workers (12 percent) were drunk on the job at least once in 1978. On-the-job drinking was found to be extensive. Twenty percent, or an estimated 46,000 workers, came to work hung over at least once in 1978, and an estimated 7,000 were too hung over to perform their duties. Another 30,000 employees were drunk while on call. The national average for adult men being intoxicated on the job at least once a year is only 1.5 percent, indicating that drinking among railway employees is extremely high compared with other industries. In total, the URC report estimated that there were 174,000 violations of the drinking rules in 1978.

Despite the high incidence of drinking on the job, the number of rule violations reported to the companies was remarkably low. There were only 900 disciplinary notices served on employees, and only 383 of these led to official proceedings resulting in dismissal (0.4 percent of all on-duty drinking violations). Interviews with supervisors indicated that both benign neglect and extreme selectiveness were involved. Supervisors are reluctant to report violations because of the severity of the potential punishment and the pervasiveness of the drinking practices. Fifty-five percent of all supervisors stated that they never report a first offender. They also indicated that the personality of the offender could affect the decision. Thus, disciplinary action concerning drinking may well reflect an employee's work difficulties unrelated to drinking problems.

Given the extent of drinking and lack of rule enforcement, it is not surprising that the URC study found extensive alcohol-related problems among the seven railroads. It found that injuries and property damage attributable to alcohol were extensive, although the report stressed the lack of adequate data. The financial cost of loss of productivity due to drinking on or before the job was also estimated to be quite high. Questionnaire results show that workers attribute approximately 5 percent of all injuries to alcohol involvement. The URC study found that this figure was a substantially underreported one. Moreover, it does not include trespasser injuries, which, according to the NTSB, may involve intoxicated persons.

The URC study had to rely on worker questionnaire reports because neither the FRA nor the railroad companies have any procedures for determining the role of alcohol in railway accidents. The URC study found that railroad medical and safety officers make no investigation to determine causes of accidents. Reports are instead generated from the field, where reporting of alcohol involvement is discouraged. Safety officers' estimates of alcohol involvement ranged from 1 percent to 50 percent. One officer stated that accident reports sent to the FRA failed to mention drinking even when it was involved.

There are many possible explanations for the industry's failure to report alcohol involvement in accidents. According to the URC report, the railroad companies are concerned that such reports could lead to civil liability to those injured (customers, trespassers, or employees). Evidence that a company permitted or failed to take reasonable steps to prevent heavy drinking by train operators involved in accidents virtually ensures a finding of company negligence in civil lawsuits.[23] Alcohol involvement reports would also result in poor public relations and potential endangering of federal funds.

FRA statistical analyses substantiate the URC findings and show steadily increasing safety problems (FRA 1977, 1978). In 1976, there were 1,684 fatalities and 65,387 injuries in 73,510 separate occurrences connected with railroad operation (excluding rapid transit). Although fatalities have dropped during the last 10 years, injuries have become much more frequent. The rate of accidents per 1 million train miles has nearly doubled since 1967, from 7.72 to more than 14 in 1978 (NTSB 1978, p. 26).

FRA divides all untoward occurrences into "accidents" and "incidents." Accidents involve those occurrences that result in damage of $1,750 (as of 1976)[24] or more; in 1976 they accounted for approximately 14 percent of all occurrences, 7 percent of all fatalities, and 1 percent of all injuries. Incidents accounted for all remaining occurrences and, because they account for most injuries and fatalities, are obviously important for determining human safety standards. Yet analyses of "human factors" and alcohol involvement are limited to accident statistics only. These show that 2,360 of the 10,248 accidents (23 percent) in 1976 were caused by human factors but only 2 of these (a derailment and a collision) were attributed to "impairment of efficiency and judgment due to drugs and alcohol." Neither of these reported alcohol-related accidents caused any injury or fatality. For 1977, the FRA reported that there were *no* accidents attributed to drug and alcohol impairment. As for "incidents," there are no human factors statistics in the reports.

The FRA is obviously attributing alcohol-related accidents to other factors. There is no compilation of blood alcohol levels in accidents and apparently no requirement that potential drinking involvement be reported. Yet all parties, including the industry itself, admit that the statistics ignore a basic problem. For example, according to J. C. Kenefick (1977, p. 4), President of Union Pacific Railroad, the railroad industry probably has an alcoholism rate among its employees of over 10 percent and such employees have an accident rate three to four times higher than other employees.

[23] Legal negligence doctrines do not necessitate such a result; the primary determination is whether the operator acted reasonably or unreasonably under the particular circumstances. Evidence of intoxication, however, is often crucial to jury deliberations. It should also be noted that the finding of negligence on the part of an operator does not necessarily result in a finding of liability, since a number of defenses may be applicable (Dooley and Mosher 1978).

[24] This figure was raised to $2,300 in 1977 and $2,900 in 1980.

Environmental Factors Contributing to Alcohol-Related Problems[25]

Employment practices in the railroad industry contribute substantially to potential alcohol problems among operating employees. First, most personnel are on call 8 hours after every work shift. Workers who leave work and drink before sleeping (a norm among a large percentage of American workers) will very likely be drowsy and unfit for work 8 hours later. The on-call procedure virtually ensures that drinkers, even those obeying the rules, will be a potential hazard if called to work 8 hours after going off duty. This practice contrasts with airline procedures for commercial pilots, which emphasize extended dry periods before going on duty and longer off-duty periods.

Second, sick leave practices probably tend to accentuate drinking problems. Unlike most industries, railroads usually do not provide any sick leave to their employees. Thus, employees with alcohol-related maladies (drunkenness, hangovers, etc.) face a choice of no pay or appearing for work. Finally, train operators are often stationed at away-from-home terminals. These layovers, according to the URC report, have traditionally been occasions for drinking; almost half the employees surveyed felt that at least moderate drinking should not be prohibited. Thus, away-from-home layovers provide a drinking context between shifts and increase the likelihood of intoxication on the job.

These employee practices do not necessarily encourage drinking itself; rather they increase the probability that drinking employees will be on the job while under the influence of alcohol. For example, providing a longer rest period before being on call would give drinking employees a chance to sleep off a night's drinking. Sick leave benefits would not reduce an employee's likelihood of drinking (even on the sick day itself) but they would establish an industry policy that incapacitated workers would not lose wages for failing to appear for work. Thus, the emphasis here is on structuring working conditions so that an employee's drinking is less likely to interfere with his or her job performance.

UNITED STATES COAST GUARD

The Coast Guard's jurisdiction includes all navigable waters in the United States and federal impoundments (lakes, rivers, etc., on federal lands). To an extent, its jurisdiction overlaps that of the Corps of Engineers, the National Park Service, and the National Forest Service. Its

[25] Much of the analysis for this topic is based on conversations with Robin Room, Director, Social Research Group, School of Public Health, University of California, Berkeley.

authority over alcohol use includes sales and consumption on board vessels and the operation of clubs on its bases. The Coast Guard's clubs were not investigated for our study. It does have a deglamorization program; however, the study of the operations of the Department of Defense, which are much larger, is adequate for exploring such programs for the purposes of this paper.[26]

Boating accidents and drownings are a major source of injury and death in the United States. There were almost 8,000 drowning deaths in 1975, most of which concerned persons under 25 years of age (Aarens et al. 1977, pp. 80-86). In 1978, there were 1,321 fatalities and at least 1,761 injuries (many were not reported) in the United States caused by noncommercial recreational boating accidents and 4,268 commercial casualties (U.S. Coast Guard 1979a,b).

Although the data are sparse, available information on boating accidents and drownings suggest that alcohol is a major factor. In one study of statistics collected by the coroner's office in several large cities, over 40 percent of the drowning victims had measurable BAC levels. In a survey of boating-related drownings in Maryland, 79 percent had positive BAC levels and the average level was 0.077. A Coast Guard official stated that a study in North Carolina (a state that maintains an excellent reporting system) showed that almost half of all recreational craft fatalities had significant BAC levels. Other studies cited in Aarens et al. (1977, pp. 80-86) support these results.

There are physiological factors concerning alcohol use that are relevant here. Alcohol produces a "pseudo-warmth" effect, which may encourage swimmers to remain in cold water too long, causing overexposure and subsequent drowning. Boating accidents are frequently caused by poor judgment and lack of coordination, effects that are associated with alcohol use. Alcohol may adversely affect the swallowing and breathing reflexes, decreasing a drowning person's chances of survival. The effects of alcohol may be magnified in boating accidents because of fatigue and the stress of exposure to wind, sun, water, and vibration. A Coast Guard official mentioned one study that found that fatigue and alcohol together produce a "quantum jump" in debilitation.

The Coast Guard, through its Office of Boating Safety, regulates the operation of recreational crafts. There are no regulations that are specific regarding alcohol. The Federal Boat Safety Act provides civil penalties for negligence and criminal penalties for gross negligence in the oper-

[26] These efforts are directed at Coast Guard employees themselves. There is an alcohol and drug program within the Personnel Office, which, among other activities, conducts educational sessions for supervisory personnel.

ation of boats. These would encompass accidents arising from injurious operation by intoxicated persons but are not specifically designed as measures for reducing the number of persons who are intoxicated while boating. The Coast Guard has taken a policy position that each state is responsible for promulgating laws governing craft operation and that each state has the primary duty of enforcing such laws. In fact, it interprets the Federal Boat Safety Act as mandating this position. However, one Coast Guard official stated that he knew of few states that regulate the use of alcohol on recreational boats.

An examination of the Federal Boat Safety Act does not support this interpretation; it gives the Coast Guard a clear mandate to promote safety and the power to promulgate regulations to that end; there are no provisions prohibiting it from enacting alcohol-specific regulations. Its lack of interest may be justified in many areas by its lack of resources. However, the Coast Guard maintains large stations on many major waterways, and its extensive involvement in recreational boat travel would justify more attention to alcohol use.

The Coast Guard also has authority to regulate various aspects of commercial resale operations, including vessel design; license qualifications for various personnel; license review when infractions are reported; and operation practices. Yet Coast Guard regulations governing commercial vessels lack any alcohol-specific provisions. Licensing requirements for pilots, chief engineers, and mates contain no provisions concerning chronic alcohol problems, and there are no specific prohibitions against working under the influence of alcohol. Licensing provisions require good character, habits, etc. A search is made for criminal records, particularly for violent crimes and narcotics convictions, but prior history of alcoholism or heavy drinking is not probed for. Overall, it appears that alcohol problems are ignored in determining the fitness of those in charge of safe and efficient ship operation.

There are also no specific regulations prohibiting drinking or being under the influence of alcohol while on duty. If the master or chief engineer discovers someone drunk on duty it is within his discretion to "log" him. If logged, the case is reviewed by the Coast Guard. Charges may be brought and a hearing held to consider license revocation. An official of the investigation division stated that such charges are not uncommon; however, the infraction is not technically related to intoxication but rather is classed as nonperformance of an assigned duty. Thus, collecting data on its frequency is impossible.

Ship masters generally allow crew members to bring alcohol aboard for use during a voyage, as many galleys do not provide alcoholic beverages of any kind. There is no control or review of this practice, nor is there any regulation of serving practices on boats that serve alcohol on board.

Despite this lack of regulation, the Coast Guard Office of Accident Investigation stated that alcohol is frequently reported as a factor in boating accidents. However, an analysis of its statistics on commercial vessel accidents (U.S. Coast Guard 1979a,b) shows that alcohol involvement is not systematically reported and is probably scattered among several categories of potential causes. Furthermore, the figures are considered unreliable because reporting and examination practices vary between localities—many fail to perform autopsies or ascertain blood alcohol content except in unusual circumstances or in the case of suspected alcohol use.

Educational materials, which the Coast Guard publishes and distributes to various groups—the Coast Guard Auxiliary, the Boy Scouts, Red Cross, etc.—include some warnings of the effects of alcohol use on boating safety. They include such suggestions as not to raise "cocktail flags" until boating is over. However, there are no efforts to have these materials distributed generally to boat owners and operators. The Coast Guard has also been involved in setting requirements for boat design.

The Coast Guard in 1975 commissioned a study of alcohol and boating safety that pointed to the need for development of standards for alcohol use during boating operation (Stiehl 1975). It gave several recommendations for conducting necessary research for this purpose, although the Coast Guard has apparently taken no steps to accomplish this.

DISCUSSION

There appears to be a heavy emphasis on safety and a general recognition of alcohol involvement in transportation accidents among most of the federal agencies we studied. It is therefore surprising that there has been so little effort to evaluate the scope of the problem or to plan and implement programs to contend with it. Part of this failure may be due to unnecessarily limited views of the agencies' authority and duty. The FAA has neglected to institute restrictions on availability for airline passengers, apparently because it does not feel the issue warrants attention. The FRA believes that the availability of alcohol on the job is purely a matter for the industry, a position that contrasts sharply with FAA regulations. The main strategies involve encouraging and implementing rehabilitation and treatment programs, which are the general federal approaches to alcohol-related problems.

As previously discussed, this view, while a necessary component of an overall strategy, is limited and ignores important policy possibilities. Alcohol policy makers need to tap the resources of transportation industries and stimulate new directions for reducing alcohol-related problems. While a full analysis of the possible initiatives is beyond the scope of this study, some specific first steps can be outlined. No recommendations concerning Coast Guard activities are given, although reforms in statistical methods are obviously relevant.

National Transportation Safety Board

The NTSB has full authority and responsibility to conduct studies of alcohol involvement in transportation accidents and to require adequate statistics from the regulatory agencies involved. It has numerous ongoing studies and has already taken a position that there is a need for a detailed study of alcohol involvement in transportation accidents. Thus, the board may be an excellent agency to approach for just such a study. We were unable to determine the likelihood of success of such a request, but board personnel appeared receptive. Budgetary problems appear to be a hindrance to ready acceptance, but board publications make clear that an alcohol study would be an appropriate endeavor.

In conjunction with such a study, a thorough revamping of the methods of reporting alcohol involvement would be beneficial. This is particularly true with regard to railroad accidents, as the FRA appears very reluctant to take any action on its own. The NTSB has full authority to require adequate reporting, and NIAAA could provide both encouragement and advice. It should be noted that the NTSB, while lacking regulatory authority (except that it can require particular reporting techniques), is a respected agency capable of generating public and congressional support for its recommendations.

Federal Aviation Administration

Many of the FAA reforms recommended by the GAO report appear to have merit. Without addressing these broader issues, two specific issues can still be considered. First, a possible drink maximum on commercial flights—either an absolute maximum or a sliding maximum depending on hours in flight—could provide an excellent means for addressing alcohol-related problems on both a symbolic and a practical level. The use of free drinks for promotional purposes should also be examined. There can be few policy objectives served by continuing the FAA's hands-off approach to the availability of alcohol other than providing a small additional income to the airline industry. Counterbalancing this concern are the possible health and safety problems associated with passenger drinking. Furthermore, the current FAA policy in effect encourages excessive drinking, and reform could have significant symbolic value for highlighting NIAAA's concerns. The FAA's policies concerning tobacco products demonstrate the feasibility and potential value of a maximum-drink initiative. Although our contacts with the FAA were not encouraging, efforts to seek reform could still be attempted. Potential supporters include flight attendant associations and some members of Congress.[27]

Second, the lack of implied consent provisions for private pilots similar to those used for drunk drivers should be addressed. The NTSB and GAO have both urged FAA action, and, in fact, the FAA has agreed to act. Many states have their own provisions, but they are ineffective for air traffic without a national policy. This is an important issue for alcohol policy makers, at least on the level of providing more accurate data on alcohol involvement.

Federal Railroad Administration

Recommendations for FRA action are more problematical. Despite the evidence of alcohol impairment among train operators and the dangers it creates to the American public, the FRA, which is charged with the duty to promote safety within the railroad industry, has provided virtually no leadership to reduce alcohol-related accidents and injuries. A 1974 NTSB report is illustrative. The report (National Transportation Safety Board 1974) concerns a 1973 collision of two trains, which caused over $1 million in damage and two fatalities. It concluded (p. 16) that the accident was caused by "the failure of the crew [of one of the trains] to stop their train, which was being operated at excessive speed by an engineer under the influence of alcohol [0.16 percent blood alcohol level]." The NTSB made numerous recommendations, which included: prohibiting use of alcohol for a specified period prior to reporting for duty; establishing procedures for conductors to examine crew members coming on duty; determining what environmental conditions contribute to crew members falling asleep at the controls; developing equipment modifications so that a train will stop if the engineer falls asleep; and

[27] There is some support for FAA action concerning availability of alcohol on commercial flights. *The Globe* (January 1979, page 22) reports that Congressman Charles E. Bennet has introduced a bill to end free drinks on the airlines and that Congressman Harold E. Johnson, Chairman of the House Committee on Public Works and Transportation, has ordered a study into the matter. The *Christian Science Monitor* carried an editorial entitled "Keep Liquor Out of Planes" in a September 1978 issue.

providing better training to new employees (pp. 16-17).

The FRA acted on none of these recommendations, finding them either unfeasible or not "cost-efficient." Its response to the alcohol-involvement problem has been to encourage railroad companies to institute voluntary treatment programs. In April 1975, Asaph H. Hall, administrator of FRA, stated (1975, pp. 4-5):

> Last year the National Transportation Safety Board recommended to FRA . . . that safety regulations covering use of alcohol and drugs be issued. We looked into the matter and came up with a draft of a proposed rulemaking. . . . The draft proposal was put before the new Railroad Operating Rules Advisory Committee for comment. The response from both labor and management was nearly unanimous—federal regulations are not necessary; management and labor can handle the situation satisfactorily among themselves.
>
> My own review of the situation bears out this statement. It *does* appear that the rail industry is moving positively to solve the problem without adding further outside government pressure. The last thing we want to do is to rush into federal regulation when it is not required.

In 1979, after much industry activity to implement and expand employee assistance programs, a serious accident involving two deaths and over $1 million in damages occurred. The NTSB determined that the engineer was intoxicated at the time of the accident and that his intoxication was a "significant causal factor" of the accident (National Transportation Safety Board 1980, p. 10). The NTSB repeated its 1973 recommendations and noted that while it "approves and supports" the rehabilitation programs, they do "not prevent employees from working while impaired by alcohol" (National Transportation Safety Board 1980, p. 11). Significantly, the only reports of alcohol-related accidents involving railroads that we could obtain came from the NTSB rather than the FRA.

Unfortunately, the FRA's lack of leadership does not appear to be limited to this one area. The NTSB reports (1979a, p. 95):

> During 1977 and 1978, it became apparent that the Federal rail safety programs and regulations had not achieved the degree of safety required by the public and Congress. The Congressional Office of Technology Assessment in its major evaluation of railroad safety issued May 1978, indicated a number of findings: Train derailments and total accidents are continually increasing in spite of Federal regulations and enforcement efforts; accident data have not been adequately used in the determination of priority objectives to address improvements in rail safety; and alternative approaches to the regulatory process have not been considered.

Most disturbing is the FRA's failure to implement adequate safety measures concerning the transport of hazardous materials. The NTSB, in its 1978 annual report, states (p. 57):

> FRA activities have not resulted in significant improvements in the safety of rail transportation of hazardous materials. The Board found that . . . organizational structure and program arrangements and procedures have impeded safety effectiveness.

Recent accidents indicate that the hazards are at least as great today in spite of repeated NTSB pressure for FRA action.

Despite the FRA's reluctance to take safety initiatives, the widespread use of alcohol among railroad employees, the present rate of deaths and injuries associated with rail travel, and the profound dangers posed by rail transport of hazardous materials all point to the need for policy reform. The reforms discussed below need exploration by those concerned with reducing alcohol-related problems.

Statistical Analyses The FRA, as indicated earlier, has made no effort to ascertain the role of alcohol in railroad accidents. In fact, its methods make it virtually impossible to determine the role of human error generally. Human factors statistics do not even include train "incidents," which account for most injuries and fatalities. Instead, they analyze "accidents" only, and the distinguishing factor between accidents and incidents is a monetary figure. This illustrates the FRA's failure to emphasize human safety concerns. Obtaining accurate statistics is essential and is relevant to other recommendations. They may also lead to additional policy initiatives. Both the NTSB and the FRA could provide accurate and simple methods for reporting accidents so that alcohol information is recorded.

As indicated earlier, there are difficulties beyond statistical methodology for obtaining accurate alcohol information. The railroad industry has primary responsibility for reporting accidents, and they are reluctant to include alcohol as a causal factor for a variety of reasons. Thus, there may be a need to include at least some outside, governmental supervision of reporting techniques to ensure that statistics are accurate and complete. Sampling techniques of different classes of accidents could be used to avoid the necessity of maintaining a large network of accident investigators. The FAA and the NTSB experiences in reporting alcohol involvement in aviation accidents could provide some guidance in this area.

Regulation of Employee Drinking There are numerous problems with current alcohol rules affecting employee drinking practices. First, industry rules prohibit employees from being intoxicated on duty, but they do not provide a "dry" period prior to reporting for duty. Thus, even a rule-abiding engineer could be intoxicated on the job. The FAA's 8-hour rule for private pilots (and the airline industry's 24-hour rule for its crew members) is illustrative of an alternative for dealing with this problem.

Second, current on-the-job rules deter detection and provide unwarranted punishments for violations. The URC report shows that employees are likely to cover up a coworker's drunkenness for fear of causing a dismissal even if the failure to report endangers lives. Supervisors are unlikely to report violations unless the violator is generally unproductive or disliked. These practices cause discrimination, impair referrals to treatment programs, and increase the likelihood of drunken workers on the job.

A number of alternatives could help alleviate this situation. For example, the NTSB has recommended that conductors be required to check all train employees to ensure they are physically capable of performing their duties. In addition, the URC has recommended that punishment for violations (which, technically, calls for possible dismissal) should be made more lenient. Care should be taken, however, not to lessen punishment without taking other forceful action to reduce availability of alcohol before and on the job, including a before-work "dry" period and regular supervisory checks. To reduce possible punishments alone could take on a symbolic meaning—that on-the-job drinking is an accepted practice.

Continuation of current rules, which are widely violated and whose violations are seldom reported, does not serve treatment programs and offers few benefits other than convenience and the appearance of concern. Given the industry's reluctance to change its policies, the FRA's lack of initiative appears negligent at best.

Employee Practices As discussed earlier, the lack of sick leave, the "on-call" status for those who have been off duty 8 hours, and the extensive use of away-from-home layovers all contribute to alcohol problems among railroad employees. Reforms in these work conditions, which represent a classic example of reforming environmental factors rather than drinking behavior per se (as discussed earlier), could provide new strategies for reducing alcohol-related problems. It should be noted that other modern industries have either abandoned or limited most of these practices. Although it might be difficult or impossible to eliminate away-from-home layovers, it might still be possible to emphasize scheduling so that crews are given round-trip daily assignments. The primary arguments against this are the convenience and short-term efficiency of the existing schedules—weak rationales, considering the potential safety dangers.

Reforms in Safety Equipment The NTSB has recommended that "dead-man" controls be required on trains. This device, which was once a common part of railroad equipment, automatically stops a train when not depressed or held. Thus, an engineer is required to actively hold the control, and if he falls asleep, becomes drowsy or loses motor control (common attributes of heavy drinking), the train stops. The FRA, in response to our inquiries, stated that no regulation will be forthcoming because such controls are "not cost-effective." One serious accident involving hazardous chemicals would undoubtedly change the equation being used.

We were unable, within the scope of this study, to determine the feasibility of such a device. Certainly, further study is indicated. At a minimum, the potential use of "dead-man" controls on trains that carry hazardous materials should be investigated.

The FRA has repeatedly failed to act on these and similar recommendations. NIAAA initiatives are therefore likely to find a reluctant, if not hostile, agency. Nevertheless, providing support to the NTSB, which is currently conducting various investigations, may well be beneficial. Given the potential dangers and the extent of alcohol use and abuse, it is imperative that alcohol policy makers enter this arena, at least to open debate and discussion.

FEDERAL ECONOMIC-BASED JURISDICTION

INTRODUCTION

Previous sections of this paper have emphasized the participation of various federal agencies in regulating the consequences of alcohol consumption and the availability of alcohol at the retail level. However, the discussion has tended to ignore federal economic powers over the availability of alcohol and the alcohol production industry.

The Bureau of Alcohol, Tobacco, and Firearms is the primary agency exercising economic controls, through excise and import taxes on alcohol beverages themselves as well as other economic levers. Indeed, the federal taxing power has been a focus of recent proposals in the field of alcohol problems prevention. Several researchers have argued that raising government taxes on alcohol would reduce alcohol consumption and alcohol-related problems (e.g. Bruun et al. 1975, Popham et al. 1975, 1976).

As indicated earlier, BATF regulations are beyond the scope of this study. However, various other agencies exercise economic controls on alcohol. For example, military pricing policies (which have been discussed previously) are a form of economic control. The Federal Trade Commission (FTC), through its antitrust powers, has recently acted to prohibit various illegal business practices by alcohol producers.[28] Because they involve producers' attempts to influence retail practices, the FTC powers could have an effect on the availability of alcohol generally.

The relationship of governmental economic policies (other than tax policy) and alcohol problems has been largely ignored in current literature, although a recent study of California practices provides some insights (Bunce 1979, Morgan 1980). For our purposes, two agencies, the Internal Revenue Service and the Small Business Administration, are briefly examined to highlight the extent and variety of governmental economic regulations and to suggest that they may be an important component of an overall policy to prevent alcohol problems. Our purpose is to suggest new areas of potential research and debate rather than to make concrete proposals for reform.

INTERNAL REVENUE SERVICE[29]

Although the Internal Revenue Service (IRS) has only indirect control over drinking behavior, it still wields considerable influence. Tax law may have very significant indirect effects on all aspects of American life, such as decisions to get married or divorced, basic decisions concerning how to conduct a business, and the price of a concert ticket.

At least three aspects of the IRS tax code have an economic impact on drinking practices and marketing techniques.

(1) *Business gifts*. The Tax Code permits deductions for gifts related to one's business not to exceed $25. Gifts up to $100 given to employees for years of service or safety achievements may be deducted. These

[28] See In the Matter of Heublein, Inc., Allied Grape Growers, United Vintners, Inc., Heublein Allied Vintners, Inc., *Federal Trade Commission Initial Decision*, July 2, 1979, paras. 258, 643 (Docket No. 8904); Bunce (1979, pp. 55-56). BATF and the Security Exchange Commission also have authority over illegal business practices by alcohol producers. See, e.g., BATF, Offer in Compromise by Anheuser-Busch, Inc., March 30, 1978, Announcement 78-48.

[29] For a thorough discussion and analysis of the "business" alcohol market and federal tax policy, see Mosher (1980).

provisions implicitly include gifts of alcoholic beverages. Alcoholic retailers and manufacturers may also give away tax-free samples in order to promote their products.

(2) *Advertising deductions*. Alcohol producers, like other businesses, are permitted to deduct all advertising expenses as a cost of doing business. Because of the size of producers' advertising budgets, these deductions amount to a very large tax break, although the exact amount is unavailable.[30] In effect, they provide governmental encouragement of alcohol advertising, which appears to conflict at least symbolically with NIAAA efforts to promote a media campaign on the dangers of alcohol.

(3) *Business entertainment and meals*. The Tax Code provides that expenses for food and beverage furnished to any individual under circumstances that are "a type generally considered to be conducive to a business discussion" may be classed as a business expense and thus deducted from gross income for tax purposes. Alcohol served as a form of entertainment may also be deducted if a "business" connection can be established. Although such expenses must be "ordinary and necessary" in the conduct of the business, the code specifically states that expenses to promote good will (which can include buying alcoholic beverages) are permissible. Employers may also claim business deductions for food and beverages served on business premises primarily for the benefit of their employees.

The IRS regulations, designed to implement the Tax Code, make the business drinking deductions even broader. It states that there is no requirement that business actually be discussed for the deduction to apply so long as the surroundings are such that they provide an atmosphere in which there are no substantial distractions to discussions generally (i.e., no floor shows, etc.). Further, the deduction may be claimed for beverage expenses apart from meals, for example, at cocktail lounges or hotel bars. The regulations provide no limits on the number of drinks that may be deducted in one setting or in one day. Rather, claims for deductions are challenged only if there is inadequate documentation of cost or business purpose.

[30] According to a U.S. Senate hearing, the alcoholic beverage industry spent over $200 million in advertising in 1974 (U.S. Congress 1976). Advertising expenditures have increased dramatically since that time. For example, Anheuser-Busch's advertising budget increased from $49 million in 1976 to $115 million in 1978 (O'Hanlon 1979). O'Hanlon states that the trend toward increased advertising expenditures is continuing to accelerate. The distilled spirits industry, which is prohibited by trade agreement from advertising on broadcast media by trade agreement, spent $81 million in magazine advertising alone in 1976, 5 percent of all magazine revenues for that year (*Advertising Age*, January 31, 1977).

These business drinking policies can condone serious abuses. For example, potential sellers of goods may try to ply buyers with alcohol in order to loosen their judgment and increase sales. This practice, although unethical and perhaps illegal, appears to be a tax deductible expense. In fact, it appears that alcohol consumption at business conventions or other business meetings or trips can generally be claimed as deductions provided that a business associate is included in the drinking episode.

The IRS does not compile the amounts claimed by taxpayers for particular types of tax deductions. The Bureau of Economic Analysis (U.S. Department of Commerce) estimates that U.S. corporations and federal, state, and local governments spent an aggregate of $10 billion on alcohol in 1979 (Mosher 1980). The alcohol industry estimates that at least two-thirds of this amount ($6.7 billion in 1979) can be attributed to corporate purchases (*Spirits Magazine* 1960). Virtually all of this expense can be assumed to have been taken as a business tax deduction, saving corporations approximately $2.7 billion on their tax returns. This amount must be supplemented by alcohol deductions taken by individual taxpayers, for which there are no available estimates. Tax savings for business use of alcohol probably total between $4 and $7 billion annually (Mosher 1980).

Most corporate purchases result in drinking without cost, since the corporation, a fictitious entity, pays the expense and provides the alcohol to employees, associates, guests, and customers, usually without reimbursement. This is particularly important in light of current policy proposals to raise the price of alcohol as a means to reduce alcohol-related problems (e.g., Bruun et al. 1975; Popham et al. 1975, 1976).

Business lunch deductions became a political issue early in the Carter Administration as a symbol of unfair "tax loopholes." Mistermed the "three-martini lunch" (an inaccurate characterization because any number of drinks may be deducted), the issue was presented as a tax loophole rather than a public health issue and the proposal was eventually dropped for lack of support. The administration failed to separate the issue of alcohol deductions from meal deductions generally.

Tax policies reflect conscious decisions concerning governmental priorities, and changes are often made in tax deductions in order to encourage or discourage particular business and consumer practices. Tax deductions are considered "indirect government subsidization of activities" by tax analysts (Mosher 1980, p. 3) and provide an important means for establishing government priorities. For example, consumers can no longer deduct gasoline taxes on their returns, reflecting the government's increasing concern for reducing, or at least not encouraging, gas consumption. Child care expenses, once not deductible, are now deductible, but only if both parents are gainfully employed or seeking employment. A parent's decision to engage in volunteer work will not qualify for obtaining the deduction. This could be viewed as a partial accommodation to the women's movement.

Public health policies are also reflected in Tax Code provisions and IRS regulatory decisions. For example, part but not all of a family's medical budget is deductible. Certain types of treatment, deemed unscientific, are excluded. Donations to certain public health organizations may be deducted. Businesses that provide various health benefits or services to employees may deduct these expenses.

Thus, there is a strong precedent for translating governmental objectives, including public health priorities, into tax deduction policy. A public health perspective provides a strong rationale for examining the alcohol deductions. Present policies act to encourage providing drinks to business associates in a variety of settings, place no limit on the amount of drinking that is appropriate, are an indirect discount on the price of alcohol, and promote extensive advertising of alcohol products. They establish as a government policy, in conflict with policies advanced by NIAAA, that the service of alcohol is an "ordinary and necessary" part of various business dealings. Perhaps most ironically, a tax deduction can be claimed for providing a gift of alcohol to an employee in recognition of his or her safety achievements.

A number of possible reforms could be considered in an attempt to realign tax policy with NIAAA priorities. For example, limits could be placed on the number of drinks permitted on one business occasion or the places that may be frequented. More sweeping reforms are also available, such as declaring that the use of alcohol is not an ordinary and necessary business expense in any circumstance.

Such proposals are highly political and may generate strong opposition from restaurant and business groups. Although the IRS could, through regulation, implement at least minimal reforms, congressional direction would probably be needed. However, if one views alcohol deductions as a public health issue and analyzes their implications for NIAAA initiatives (at least on a symbolic level), airing the issue might provide a valuable educational tool.

SMALL BUSINESS ADMINISTRATION

The Small Business Administration (SBA) is fundamentally concerned with aiding, counseling, and assisting small businesses. It does so primarily by offering loans and by ensuring that small businesses receive a fair proportion of government purchases and contracts.

SBA loans are divided into two categories—direct loans to small businesses (Article 7a of the SBA Act) and loans to small business investment corporations (SBIC), which, in turn, provide loans to small business concerns. The direct loans are usually in the form of bank guarantees—the borrower obtains the loan from a private bank, and the SBA acts as guarantor. Until approximately 1968, SBA regulations prohibited the granting of any loan (including SBIC loans) to any business that received more than 50 percent of its gross receipts from alcoholic beverages. The regulation was repealed by order of the SBA administrator by publication in the *Federal Register*. SBA personnel today state that the change was due to the administrator's decision that alcoholic beverage outlets were no longer considered suspect businesses and were entitled to the benefits received by other small businesses. Lobbying efforts were not a factor in the decision, according to current personnel.

The regulatory revision took place during the ghetto riots of the late 1960s. Many ghetto businesses were burned and looted, and the SBA received many applications for assistance. Coincidentally, liquor stores and bars made up a large part of the businesses affected by the civil disturbances. Liquor establishments are one of the few small businesses that have proven to be successful with minority proprietorship. Certainly, civil rights concerns in the late 1960s had at least some influence in the decision to make loans available to liquor stores and bars. In addition, the timing coincided with nationwide policies that dramatically increased the availability of alcoholic beverages.

Current figures on direct loans granted by SBA indicate that alcoholic beverage businesses obtain a significant portion of SBA financing, particularly among minority applicants. For fiscal 1978, 636 loans amounting to $48 million were granted to bars and liquor stores, and 40 loans ($7.5 million) were granted to alcoholic beverage wholesalers. Of these, 262 loans (162 of which were liquor stores) were granted to minority applicants. This means that 2.1 percent of all SBA loans and 4.3 percent of all minority loans involved alcoholic beverage distribution. The percentage of alcoholic beverage concerns that have minority ownership is particularly striking: 38.8 percent of alcoholic beverage loans (and 50.9 percent of liquor store loans) went to minority applicants, compared with 18.9 percent of nonalcoholic beverage loans. At the end of fiscal 1978, SBA had outstanding direct loans to 1,764 alcoholic beverage businesses, 1,060 of which were liquor stores.

SBA officials state that there are no longer any special considerations for or restrictions on bars and liquor store applications except for those that might affect their potential for financial success. Applications are considered on an individual merit basis, and there are no records kept to determine the default rate of loans by the type of business.

As these statistics show, SBA has a significant involvement in the establishment of new alcoholic beverage outlets throughout the country. Yet its staff has no concern with alcohol-related problems and no contact with the policies of NIAAA. SBA policies provide a case study of conflicting federal policies (in this case, alcohol beverage control versus promotion of minority business) in which alcohol policy makers have had no input.

SBA could provide a source of information for evaluating the retail trade in alcoholic beverages through the study of loan repayment statistics. Many proposals for reform in the retail trade could be made conditions for SBA financing. For example, mandatory training of on-premise sales personnel or requirements that establishments maintain certain hours, be located off main highways, or provide for alternative modes of transportation could be required for a loan to be granted. These restrictions could be made on an experimental basis to determine whether they have a beneficial effect on alcohol-related problems. Such reforms could provide important new research data at minimal costs.

CONCLUSION

Perhaps the most important finding of this study is that significant portions of federal authority to regulate the availability of alcohol and to respond to alcohol-related problems rest with federal agencies not usually associated with alcohol policy. This has resulted in little or no coordination in the development of the federal response to alcohol, particularly in the area of prevention. In some cases, contradictory policies are being made and enforced; in others, promising new prevention strategies (and potential new allies) are not being pursued. This paper has outlined the scope of federal authority, the important actors, and the potential impact of new strategies of prevention policy.

Determining which actors should be approached and what strategies are appropriate is dependent on several factors. First, the current policies of particular agencies can be examined to determine whether alcohol is already considered within an agency's realm of responsibility. The Department of Defense, for example, is already operating an extensive alcoholic beverage market as well as large prevention and treatment programs. Alcohol is perceived both within and outside the services as a source of serious problems that the military must contend with in order to fulfill its military responsibilities effectively.

In other cases, the relevance of alcohol to the agency's mission may exist, but the agency does not yet acknowledge the relationship of alcohol issues to its decisions. For example, the IRS does not consider itself involved with alcohol policy, yet its actions have an impact on the perception of alcohol in the population, at least on a symbolic level.

If a particular agency is already addressing alcohol issues, political support and recommendations for improvement or further study can be offered. For example, the military may provide an excellent opportunity to learn more about the relationship of alcohol availability and alcohol-related problems, and experimental programs may be appropriate. In some cases, however, a first step may be necessary—convincing an agency that its actions do in fact have an impact on federal alcohol policies.

In many cases described in this paper, agencies with the capacity to address alcohol-related issues do not view alcohol as part of their mandate. This may be justified. For example, alcohol issues are insignificant to the main goals of the Bureau of Land Management and the National Park Service. Despite their supervision of a modest alcohol retail trade and their concern for at least some alcohol-related problems (e.g., drunk driving), they have only minimal enforcement and supervisory capacity and have important, unrelated responsibilities.

A particular agency's view of its own responsibilities may conflict, however, with a significant part of its statutory duties. For example, the Consumer Product Safety Commission is mandated to provide useful and extensive statistical data on the cases of accidents involving consumer products. Its accident forms and collection procedures virtually ensure that this responsibility cannot be met, particularly as to alcohol involvement. In fact, the commission in the past has considered alcohol as outside its scope of responsibility altogether, despite this statutory directive. The Federal Railroad Administration's duty to ensure safe rail transportation appears to be significantly hampered by its failure to determine the extent of alcohol involvement in accidents, particularly given the recent study of drinking among railroad employees. Agencies with overlapping responsibilities may fail to act, relying on others to do so: for example, the National Transportation Safety Board and the Department of Transportation, in regard to the general study of alcohol involvement in transportation accidents.

When a particular agency is not meeting statutory responsibilities, strategies that attempt to initiate policy changes and new programs may be appropriate. The tactics relied on depend on a given agency's current financial and political situation and its likely response. Alcohol policy makers in some instances may have to take as a first step simply opening an issue for debate, hoping that alliances can be formed and attitudes changed. More concrete actions may be appropriate if an agency is likely to be responsive.

Coordinating federal alcohol policy may in some cases require direct contacts with particular agencies or interested third parties. In others, however, there may be issues common to more than one agency. For example, a common thread through most of this paper concerns the need for better statistical analysis from a variety of federal agencies. Here, an interagency approach may prove to be useful. NIAAA is currently mandated by law to coordinate the Interagency Committee on Federal Activities for Alcohol Abuse and Alcoholism, which includes a prevention subcommittee. However, the full committee as presently constituted does not include many of the agencies discussed in this paper. Its emphasis has been on coordinating treatment efforts rather than on prevention; and it has not attracted attendance by high-level officials. This committee may have potential usefulness. However, if it is relied on as a means to address some of the strategies discussed in this report, NIAAA will need to place renewed emphasis on improving its effectiveness.

Agency responsibilities, according to the agency's own perception and according to its statutory mandate, will be important factors in evaluating new strategies associated with nonalcohol-specific agencies. If NIAAA decides that coordinating alcohol prevention policies among agencies with significant alcohol-related responsibilities is part of its mission, it will need to devote long-term staff and financial support to the effort. Important new contacts will be necessary, either directly with particular agencies, or through a committee approach. The process of implementation requires both negotiation and careful analysis of costs and possible opposition. We hope that our preliminary contacts and analysis provide a sound basis for further research and study.

REFERENCES

Aarens, M., Cameron, T., Roizen, J., Roizen, R., Room, R., Schnebeck D., Wingard, D. (1977) *Alcohol, Casualties and Crime*. Berkeley, Calif.: Social Research Group.

Berl, W., and Halpin, B. (1978) *Human Fatalities from Unwanted Fires*. Laurel, Md.: John Hopkins University Applied Physics Laboratory.

Bruun, K., Edwards, G., Lumio, M., Mäkelä, K., Pan, L., Popham, R., Room, R., Schmidt, W., Skog, O., Sulkunen, P., Österberg, E. (1975) *Alcohol Control Policies in Public Health Perspective*. Helsinki, Finland: The Finnish Foundation for Alcohol Studies.

Bunce, R. (1979) *The Political Economy of California's Wine Industry*. Berkeley, Calif.: Social Research Group.

Cahalan, D. Cisin, I., and Crossley, H. (1969) *American Drinking Practices: A National*

Survey of Drinking Behavior and Attitudes. New Brunswick, N.J.: Rutgers Center of Alcohol Studies.

Cahalan, D., Cisin, I., Gardner, G., Smith, G. (1972) *Drinking Practices and Problems in the U.S. Army, 1972*. Information Concepts, Inc., Report No. 73-6. Washington, D.C.: Information Concepts, Inc.

Cahalan, D., and Cisin, I. (1975) *Final Report on a Service Survey of Attitudes and Behavior of Naval Personnel Concerning Alcohol and Problem Drinking*. Washington, D.C.: Bureau of Social Science Research.

CSB Radio (1979) Special: Up in Smoke—Cigarettes and Safety. May 26, 1979, 8:30-8:54 p.m., E.D.T. New York City, N.Y. CBS News.

Comptroller General of the U.S. (1976a) *Report to Congress: Alcohol Abuse Is More Prevalent in the Military Than Drug Abuse*. MWD-76-99, April 8, 1976. Washington, D.C.: General Accounting Office.

Comptroller General of the U.S. (1976b) *Report to Congress: The Consumer Product Safety Commission Should Act More Promptly to Protect the Public from Hazardous Products*. HRD-78-122, June 1, 1976. Washington, D.C.: General Accounting Office.

Comptroller General of the U.S. (1977) *Report to Congress: The Consumer Product Safety Commission Needs to Insure Safety Standards Faster*. HRD-78-3, December 12, 1977. Washington, D.C.: General Accounting Office.

Comptroller General of the U.S. (1978a) *Report to Congress: Stronger Federal Aviation Administration Requirements Needed to Identify and Reduce Alcohol Use Among Civilian Pilots*. CED-78-58, March 20, 1978. Washington, D.C.: General Accounting Office.

Comptroller General of the U.S. (1978b) *Report to Congress: The Consumer Product Safety Commission Has No Assurance That Product Defects Are Being Reported and Corrected*. HRD-78-48. Washington, D.C.: General Accounting Office.

Comptroller General of the U.S. (1979a) *Report to Congress: Changes Needed in Operating Military Clubs and Alcohol Package Stores, Volume I and II*. FPCD-79-7, January 15, 1979 and April 23, 1979. Washington, D.C.: General Accounting Office.

Comptroller General of the U.S. (1979b) *GAO Staff Report: The Tax Status of Federal Resale Activities: Issues and Alternatives*. FPCD-79-19, April 19, 1979. Washington, D.C.: General Accounting Office.

Consumer Product Safety Commission (1978) *HIA Hazard Analysis Report, Upholstered Furniture Flammability 1978*. Washington, D.C.: Consumer Product Safety Commission.

Dooley, D., and Mosher, J. (1978) Alcohol and legal negligence. *Contemporary Drug Problems* 7(2):145-179.

Douglass R., and Freedman, J. (1977) *Alcohol Related Casualties and Beverage Market Response to Beverage Alcohol Availability Policies in Michigan: Final Report*. Ann Arbor, Mich.: Highway Safety Research Institute, University of Michigan.

Federal Aviation Administration (1976) Alcoholism and Airline Flight Crewmembers. Office of Aviation Medicine, Memorandum, November 10, 1976. Washington, D.C.: Federal Aviation Administration.

Federal Railroad Administration (1976) *Accident/Incident Bulletin,* Calendar Year 1976 No. 144. Washington, D.C.: U.S. Department of Transportation.

Federal Railroad Administration (1977) *Accident/Incident Bulletin,* Calendar Year 1976 No. 145. Washington, D.C.: U.S. Department of Transportation.

Federal Railroad Administration (1978) *Accident/Incident Bulletin,* Calendar Year 1976 No. 146. Washington, D.C.: U.S. Department of Transportation.

Feinburg, J. (1978) Consumer product safety: The current record of administrative interpretation. *Federal Bar Journal* 37(2):344-356.

Fowler, F. (1979) A toast to NASAP. *Driver: The Traffic Safety Magazine for the Military Driver* 11(1):1-13.

Frank, G. (1979) Passengers 75, Flight Attendants 0. *San Francisco Chronicle*, September 28, p. 29.

Green, M., and Moulton, D. (1978) The case of the battered agency. *The Nation* June 10:698-700.

Gusfield, J. (1976) The prevention of drinking problems. Pp. 267-293 in Filstead, Rossi, Keller, eds., *The Prevention of Drinking Problems: New Thinking and New Directions*. Cambridge, Mass.: Ballinger Publishing Company.

Haddon, W. (1973a) Energy damage and the ten countermeasure strategies. *Journal of Trauma* 13(4):321-331.

Haddon, W. (1973b) *Exploring the Options in Research Directions: Toward the Reduction of Injury*. U.S. Department of Health, Education and Welfare Publication 73-124. Washington, D.C.: National Institute of Health.

Hall, A. H. (1975) Remarks: The FRA position and proposal. Pp. 3-7 in New York State School of Industrial and Labor Relations, *Proceedings of 1975 Conference on the Detection, Prevention and Rehabilitation of the Problem Drinking Employee in the Railroad Industry*. No. PB-248906. Washington, D.C.: U.S. Department of Commerce, National Technical Information Service.

Kenefick, J. C. (1977) The great train robbery. Pp. 4-6 in Federal Railroad Administration, *Proceedings of the 1976 Conference: Employee Assistance Programs—An Alternative to Tragedy*. Report No. FRA-OPPD-77-1. Washington, D.C.: Federal Railroad Administration.

Killeen, J. (1979) U.S. military alcohol abuse prevention and rehabilitation programs. Pp. 355-381 in H. Blane and M. Chafetz, eds., *Youth, Alcohol and Social Policy*. New York: Plenum Press.

Long, J., Hewitt, L., and Blane, H. (1976) Alcohol abuse in the armed services: A review. I. Policies and programs. *Military Medicine* 141:844-850.

Lowrance, W. (1976) *Of Acceptable Risk*. Los Altos, Calif.: William Kaufmann, Inc.

Manley, T., McNichols, C., and Stahl, M. (1979) *Alcoholism and Alcohol Related Problems Among USAF Civilian Employees*. U.S. Air Force Institute of Technology Technical Report 79-4, August, 1979. Wright-Patterson Air Force Base, Ohio: U.S. Air Force.

Mannello, T. A., and Seaman, F. J. (1979) *Prevalence, Costs, and Handling of Drinking Problems on Seven Railroads: Final Report*. Washington, D.C.: University Research Corporation.

May, P. (1977) Alcohol beverage control: A survey of tribal alcohol statutes. *American Indian Law Review* 5:217-288.

Medicine in the Public Interest, Inc. (1976) *The Effects of Alcoholic Beverage Control Laws*. Washington, D.C.: Medicine in the Public Interest, Inc.

Morgan, P. (1980) The evolution of California alcohol policy: Alcohol problem management in the post-war period. *Contemporary Drug Problems* 9(1):107-140.

Mosher, J. (1975) Liquor Legislation and Native Americans: History and Perspective. Unpublished paper. Social Research Group, Berkeley, Calif.

Mosher, J. (1978) Alcoholic Beverage Controls in the Prevention of Alcohol Problems: A Concept Paper on Demonstration Programs. Prepared for the Division of Prevention, National Institute on Alcohol Abuse and Alcoholism. Social Research Group, Berkeley, Calif.

Mosher, J. (1979a) Dram shop liability and the prevention of alcohol-related problems. *Journal of Studies on Alcohol* 40(9):773-798.

Mosher, J. (1979b) Alcohol Beverage Control System in California. Unpublished paper. Social Research Group, Berkeley, Calif.

Mosher, J. (1979c) Retail Distribution of Alcoholic Beverages in California. Unpublished paper. Social Research Group, Berkeley, Calif.

Mosher, J., and Wallack, L. (1979) *The DUI Project: A Description of an Experimental Program to Address Drinking-Driving Problems.* Sacramento, Calif.: California Department of Alcoholic Beverage Control.

Mosher, J. (1980) Alcoholic Beverages on Tax Deductible Business Expenses: An Issue of Public Health Policy and Prevention Strategy. Unpublished paper. Social Research Group, Berkeley, Calif.

National Transportation Safety Board (1974) *Railroad Accident Report: Rear-End Collision of Two Southern Pacific Transportation Company Freight Trains, Indio, California, June 25, 1973.* Technical Report No. NTSB-RAR-74-1, adopted March 20, 1974. Washington, D.C.: National Transportation Safety Board.

National Transportation Safety Board (1978) *Annual Report to Congress 1977.* Washington, D.C.: National Transportation Safety Board.

National Transportation Safety Board (1979a) *Annual Report to Congress 1978.* Washington, D.C.: National Transportation Safety Board.

National Transportation Safety Board (1979b) *Marine Accident Report: Ferry M/V George Prince Collision with the Tanker S.S. Frosta (Norwegian) on the Mississippi River Luling/Destrehan, Louisiana October 20, 1976.* Technical Report No. NTSB-MAR-79-4, adopted March 22, 1979. Washington, D.C.: National Transportation Safety Board.

National Transportation Safety Board (1980) *Railroad Accident Report: Rear-End Collision of Two Southern Pacific Transportation Company Freight Trains, 02-Holat-21 and 01-BSM FK-20, Thousand Palms, California, July 24, 1979.* Technical Report No. NTSB-RAR-74-1, adopted March 20, 1974. Washington, D.C.: National Transportation Safety Board.

O'Hanlon, T. (1979) August Busch brews up a new spirit in St. Louis. *Fortune Magazine*, January 15:98.

O'Malley, B. (1979) Cigarettes and sofas: How the tobacco lobby keeps the home fires burning. *Mother Jones*, July:56-62.

Overbey, J. W. (1979) Letter of the Director, Analysis and Management Studies, United States Fire Administration to A. Jarvis, Legislative Assistant to Representative Moakley, September 5, 1979.

Polich, J., and Orvis, B. (1979) *Alcohol Problems: Patterns and Prevalence in the U.S. Air Force.* R2308-AF, June 1979. Santa Monica, Calif.: Rand Corporation.

Popham, R., Schmidt, W., and deLint, J. (1975) The prevention of alcoholism: Epidemiological studies of the effect of government control measures. *British Journal of Addiction* 70:125-144.

Popham, R., Schmidt, W., and deLint, J. (1976) The effects of legal restraint on drinking. Pp. 597-625 in D. Kissin and H. Begleiter, eds., *The Biology of Alcoholism, Volume 4: Social Aspects of Alcoholism.* New York: Plenum Press.

Room, R., and Mosher, J. (1979) A role for regulatory agencies in the prevention of alcohol problems. *Alcohol and Health Research World* 4(2):11-18.

Schuckit, M. (1977) Alcohol problems in the United States armed forces. *Military Chaplains Review* (Winter):9-19.

Spirits Magazine (1960) Is liquor's billion dollar business being recognized? *Spirits*, October.

Stiehl, C. (1975) *Alcohol and Pleasure Boat Operators: Final Report.* No. CG-D-143-75. Washington, D.C.: U.S. Department of Transportation.

U.S. Coast Guard (1979a) *Boating Statistics, 1978*. COMDTINST M16754.1. Washington, D.C.: U.S. Department of Transportation.

U.S. Coast Guard (1979b) *Statistics of Casualties, 1978*. Washington, D.C.: U.S. Coast Guard.

U.S. Congress (1976) *Media Images of Alcohol: The Effects of Advertising and Other Media on Alcohol Abuse*. Hearings before the Subcommittee on Alcoholism and Narcotics of the Committee on Labor and Public Welfare, United States Senate 94th Congress, March 8 and 11. Washington, D.C.: U.S. Government Printing Office.

U.S. Congress (1978) *Consumer Product Safety Commission*. Hearing before the Subcommittee on Civil Service of the Committee on Post Office and Civil Service, March 9. House Document 95-68. 95th Congress, Second Session. Washington, D.C.: U.S. Government Printing Office.

U.S. Department of Health, Education and Welfare (1972) *Fourth Annual Report to the President and the Congress on the Studies of Death, Injuries, and Economic Losses Resulting from Accidental Burning of Products, Fabrics, or Related Materials*. Washington, D.C.: U.S. Department of Health, Education and Welfare.

Waas, M. (1980) Life savers: Self-extinguishing cigarettes.*Washington Post Sunday Outlook*, June 1:2.

APPENDIX

RELEVANT REGULATIONS AND LAWS BY AGENCY

National Park Service
36 Code of Federal Regulation ("CFR") §§2.16, 5.2
Public Law 89-249, October 9, 1965; 79 Stat. 969-971

Bureau of Land Management
43 CFR §8363

Army Corps of Engineers
Army Corps of Engineers Regulation 1130-2-400, May 28, 1971

Forest Service
36 CFR §§3.14, 4.6

Bureau of Indian Affiars
Act of August 15, 1953, ch. 505 §2; 67 Stat. 586; 25 CFR §§11.1, 11.55

Department of Defense
DOD Directive 1330.15, April 11, 1972
DOD Directive 1010.2, March 1, 1972 (as amended)
DOD Instruction 1010.3, May 22, 1974

Army Regulation A210-65, December 1, 1978
Air Force Regulation 215-1, July 31, 1974 (as amended)
Navy Manual for Messes Ashore NAVPERS 15951
U.S. Navy Regulations 1973, art. 1150
SECNAV Instruction 1700.11B, March 29, 1973

Consumer Product Safety Commission
Consumer Product Safety Act, Public Law 92-573, October 27, 1972; 86 Stat. 1207

Nuclear Regulatory Commission
10 CFR §§55.11, 55.33

National Transportation Safety Board
49 U.S.C.A. §§1901 et seq.

Federal Aviation Administration
14 CFR §§67.1-67.31, 91.11, 121, 575

Federal Railroad Administration
Rail Safety Act of 1970 (Public Law 91-458, May 10, 1970)

Coast Guard
Public Law 92-75, August 10, 1971, 85 Stat. 214-226
46 CFR §§10.1 et seq.

Internal Revenue Service
26 U.S.C.A. §§162, 274
26 CFR §1.274.2

Small Business Administration
none

Biographical Sketches of Panel Members and Staff

MARK H. MOORE is Guggenheim professor of criminal justice policy and management at the John F. Kennedy School of Government, Harvard University. Previously he was special assistant to the administrator and chief planning officer of the Drug Enforcement Administration, U.S Department of Justice. He was also a consultant for the National Institute on Drug Abuse and the National Institute of Justice. His research interests include crime, criminal justice policy, and management. Recently he has focused on the regulation of "dangerous and abusable commodities," notably drugs, alcohol, and firearms. He has a BA from Yale University and MPP and PhD degrees in public policy from Harvard University.

GAIL BURTON ALLEN is a psychiatrist and director of the Comprehensive Alcoholism Treatment Program at St. Luke's-Roosevelt Hospital Center in New York. She has a long-standing interest in community health care issues, prevention, and service delivery. She is an assistant clinical professor of psychiatry at Columbia University and has served on numerous advisory groups, including the Council on Mental Health Delivery Services of the American Psychiatric Association. She received a BA from Wellesley College and an MD degree from New York University.

DAN E. BEAUCHAMP is associate professor of health administration at the University of North Carolina. His principal research interests are alcohol policy and the ethics of public health. His areas of research have included the evolution of the alcoholism movement and the idea of alcoholism in the United States, the ethical foundations of public health, and the issue of paternalism. He received a BA from the University of Texas and MA and PhD degrees in political science from Johns Hopkins University.

PHILIP J. COOK is associate professor of public policy studies and economics at Duke University. His principal research interest is the regulation of unhealthy, unsafe, and criminal behavior. He has written extensively on the value of life in public policy decisions, the preventive effects of punishment, and the role of weapons in violent crime. He received a BA from the University of Michigan and a PhD in economics from the University of California, Berkeley.

JOHN KAPLAN is Jackson Eli Reynolds professor of law at Stanford University Law School, where he teaches criminal law, evidence, and criminology. Previously he was professor of law at Northwestern University. His research interests are law and criminology and marijuana and heroin policy. He is a former member of the National Advisory Council on Alcohol Abuse and Alcoholism and has authored numerous articles on drug control and other subjects related to law. He has BA and LLB degrees from Harvard University.

NATHAN MACCOBY is the Janet M. Peck professor emeritus of international communication and director of the Institute for Communication Research at Stanford University. For the past 10 years he has been especially interested in communication and health and is the codirector for community studies of the Stanford Heart Disease Prevention Program. His principal interests have been in communication and learning of information, attitudes, and behavior, either through mass media or through other forms of communication. He has served on the editorial boards of *Public Opinion Quarterly* and the *Journal of Communication*. He is a fellow of the American Psychological Association, a member and past president of the International Communication Association, and a member of the Academy of Behavioral Medicine. He has a BA from Reed College, an MA from the University of Washington, and a PhD from the University of Michigan, all in psychology.

DAVID F. MUSTO is professor of psychiatry (in the Child Study Center) and of the history of medicine at Yale University. He also heads the section on history and social policy of the Bush Center in Child Development and Social Policy. His principal interests are the history of mental health issues, including drug abuse and alcoholism and the history of the American family. He was a member of President Carter's Strategy Council on Drug Abuse Policy from 1977 to 1981 and is a member of the council of the Smithsonian Institution. He received BA and MD degrees from the University of Washington and an MA degree from Yale University.

ROBIN ROOM is senior scientist and director of the National Alcohol Research Center at the Institute for Epidemiology and Behavioral Medicine, Institutes of Medical Sciences, and lecturer at the School of Public Health, University of California, Berkeley. A sociologist who has worked in alcohol and drug studies of the general population since 1963, he serves as a member of the World Health Organization Expert Advisory Panel on Drug Dependence and Alcohol Problems, as chair of the federal Alcohol Abuse Prevention Review Committee, and as chair of the epidemiology section of the International Council on Alcohol and Addictions. He received an AB from Princeton University and MA and PhD degrees from the University of California, Berkeley.

THOMAS C. SCHELLING is Lucius N. Littauer professor of political economy and chair of the program in public administration of the John F. Kennedy School of Government, Harvard University. Professor of economics at Harvard since 1958, he has been a faculty member of the Center for International Affairs and chair of the Kennedy School's public policy program. Prior to 1958, Schelling was an economist with the U.S. government in foreign aid programming and a professor of economics at Yale University. He has been a consultant to the U.S. Department of State, the U.S. Department of Defense, and the Arms Control and Disarmament Agency and a lecturer at the Foreign Service Institute and the several war colleges. He was a member of the Nuclear Energy Policy Study—the Ford/Mitre study. He was elected to the Institute of Medicine in 1980. He received the Frank E. Seidman distinguished award in political economy in 1977. He has a BA from the University of California, Berkeley, and a PhD in economics from Harvard University.

WOLFGANG SCHMIDT is director of the social policy research department of the Addiction Research Foundation of Ontario. He has served on the International Committee on Alcohol, Drugs and Traffic Safety of the World Health Organization and national committees for research on alcohol problems. He has had a career-long interest in the prevention of alcohol-related damage and its environmental determinants. His current areas of research include the role of alcohol and cigarettes in the mortality of heavy drinkers, a biomedical definition of safe alcohol consumption, and the economic determinants of variation in alcohol-related problem rates. He received a JD from the University of Graz and an MA from the University of Toronto.

NORMAN SCOTCH is director of the Boston University School of Public Health and chairman of the Department of Socio-Medical Sciences and Community Medicine in the Boston University School of Medicine. Trained in anthropology, sociology, and epidemiology, he has conducted research and written on stress and disease, alcohol use, and genetic counseling. In the alcohol field he has served the National Institute of Alcohol Abuse and Alcoholism (NIAAA) as a member of the research initial review group for four years, the last year as chairman. He spent two years on the prevention initial review group. He was also a member of the National Advisory Council of NIAAA for four years. He has BA and MA degrees from Boston University, a PhD in anthropology from Northwestern University, and an SM in hygiene and epidemiology from Harvard School of Public Health.

DONALD J. TREIMAN is professor of sociology at the University of California, Los Angeles. Currently on extended leave at the National Research Council/National Academy of Sciences, he is engaged in policy-related research on occupational classification and on the social utility of basic research. His research interests center on the comparative study of social stratification and social mobility and extend to broader issues in the comparative analysis of social structure. He has also served as a methodological consultant on a study of teenage drug use. He has a BA from Reed College and MA and PhD degrees from the University of Chicago, all in sociology.

JACQUELINE P. WISEMAN is professor of sociology at the University of California, San Diego. Her major research interests are deviant behavior, with an emphasis on alcoholism, and the social psychology of family relationships. She won the C. Wright Mills award for her study of the contrasting perspectives on treatment modalities available to skid row alcoholics. She is currently completing a cross-cultural study of the interactive dynamics of spouses of alcoholics, which was begun when she was a research fellow at the Finnish Foundation for Alcohol Studies. She has served as president of the Society for the Study of Social Problems and is currently on the governing council of the American Sociological Association. She received BA and MA degrees from the University of Denver and a PhD from the University of California, Berkeley.

DEAN R. GERSTEIN, who served as study director for the panel, is senior research associate with the Committee on Substance Abuse and Habitual Behavior. He formerly held research and teaching appointments in sociology at the University of California, Los Angeles, and in psychiatry at the University of California, San Diego. His principal interest is general theory in the social, psychological, and behavioral sciences. His publications include studies of heroin, polydrug, and tobacco use as well as analyses of the development of sociological theory. He received a BA from Reed College and MA and PhD degrees, in sociology, from Harvard University.